GeoEnergy Science and Engineering

GeoEnergy Science and Engineering

Editors

Reza Rezaee
Yujie Yuan

MDPI

Basel • Beijing • Wuhan • Barcelona • Belgrade • Novi Sad • Cluj • Manchester

Editors

Reza Rezaee
Energy Engineering
Curtin University
Bentley
Australia

Yujie Yuan
School of Engineering
Edith Cowan University
Perth
Australia

Editorial Office
MDPI
St. Alban-Anlage 66
4052 Basel, Switzerland

This is a reprint of articles from the Special Issue published online in the open access journal *Eng-Advances in Engineering* (ISSN 2673-4117) (available at: www.mdpi.com/journal/eng/special_issues/3945Y6QLJ8).

For citation purposes, cite each article independently as indicated on the article page online and as indicated below:

Lastname, A.A.; Lastname, B.B. Article Title. *Journal Name* **Year**, *Volume Number*, Page Range.

ISBN 978-3-7258-1010-9 (Hbk)
ISBN 978-3-7258-1009-3 (PDF)
doi.org/10.3390/books978-3-7258-1009-3

Contents

About the Editors

Reza Rezaee

Reza Rezaee has a PhD in Reservoir Characterization. His research focuses on integrated solutions for reservoir characterization, formation evaluation and petrophysics, as well as the application of artificial intelligence. Currently, he is focused on unconventional gas and natural hydrogen studies. He has over 28 years of experience in academia, being responsible for both teaching and research. During his career, he has been engaged in several research projects supported by major national and international companies, together with his supervisory work at various universities, acomplishing a wide range of achievements. During his research career, he has led several major research projects funded by various oil and gas companies such as WAPET, WMC, MESA, Santos, NIOC, PEDEC, DMP, Buru, Woodside, Carnarvon, Norwest, Devon Energy, SWU/SKL and ANLEC. He has received a total of more than USD 2.5M in funds through his collaborative research projects. He has supervised over 80 M.Sc. and PhD students during his university career. Moreover, he has published more than 310 peer-reviewed journal and conference papers and is the author of six books on petroleum geology, logging and log interpretation, and gas shale reservoirs. As a founder of the "Unconventional Gas Research Group" of Australia, he has established a unique and highly sophisticated research lab at Curtin University. This lab was established to research the petrophysical evaluation of tight gas sands and shale gas formations. He was announced as the top Australian researcher in the field of Petroleum Engineering by The Australian Research 2023 magazine.

Yujie Yuan

Yujie Yuan holds a PhD degree in Petroleum Engineering from Curtin University, Australia, and a B.Sc. degree in Petroleum Engineering from the China University of Petroleum (East China). She is currently a lecturer (Teaching and Research Scholar) at the School of Engineering at Edith Cowan University (ECU), Australia.

Preface

GeoEnergy science and engineering is an ever-evolving field at the forefront of sustainable energy development. It encompasses a diverse range of traditional methods and cutting-edge techniques aimed at harnessing the Earth's resources efficiently and responsibly. As we navigate toward a low-carbon future, the importance of GeoEnergy solutions cannot be overstated.

This Special Issue collects articles that investigate various aspects of low-carbon and sustainable exploration, production and storage of GeoEnergy resources. From innovative machine learning approaches to geomechanical feasibility analysis, each contribution offers valuable insights and practical solutions to address the complex challenges in GeoEnergy development.

The articles in this issue cover a wide spectrum of topics, including water saturation prediction, exploration of new hydrocarbon sources, reservoir adaptability evaluation, optimization of production well design and the application of machine learning algorithms to CO_2 miscible injection. Additionally, the Special Issue explores the potential of carbon capture utilization and storage in extractive industries, the geomechanical feasibility analysis of salt cavern gas storage, and the emerging role of discrete fracture network (DFN) modeling in stochastic modeling and geomechanical design.

Through these contributions, we hope to foster a deeper understanding of GeoEnergy science and engineering, and lead sustainable energy development practices.

Reza Rezaee and Yujie Yuan
Editors

eng

Article

Water Saturation Prediction in the Middle Bakken Formation Using Machine Learning

Ilyas Mellal [1], Abdeljalil Latrach [1], Vamegh Rasouli [1], Omar Bakelli [2], Abdesselem Dehdouh [1] and Habib Ouadi [2,*]

[1] Department of Energy & Petroleum Engineering, University of Wyoming, Laramie, WY 82071, USA; imellal@uwyo.edu (I.M.); alatrach@uwyo.edu (A.L.); vrasouli@uwyo.edu (V.R.); adehdouh@uwyo.edu (A.D.)
[2] Department Petroleum Engineering, University of North Dakota, Grand Forks, ND 58202, USA; omar.bakelli@und.edu
* Correspondence: habib.ouadi@und.edu; Tel.: +1-701-215-8911

Abstract: Tight reservoirs around the world contain a significant volume of hydrocarbons; however, the heterogeneity of these reservoirs limits the recovery of the original oil in place to less than 20%. Accurate characterization is therefore needed to understand variations in reservoir properties and their effects on production. Water saturation (Sw) has always been challenging to estimate in ultra-tight reservoirs such as the Bakken Formation due to the inaccuracy of resistivity-based methods. While machine learning (ML) has proven to be a powerful tool for predicting rock properties in many tight formations, few studies have been conducted in reservoirs of similar complexity to the Bakken Formation, which is an ultra-tight, multimineral, low-resistivity reservoir. This study presents a workflow for Sw prediction using well logs, core data, and ML algorithms. Logs and core data were gathered from 29 wells drilled in the Bakken Formation. Due to the inaccuracy and lack of robustness of the tried and tested regression models (e.g., linear regression, random forest regression) in predicting Sw as a continuous variable, the problem was reformulated as a classification task. Instead of exact values, the Sw predictions were made in intervals of 10% increments representing 10 classes from 0% to 100%. Gradient boosting and random forest classifiers scored the best classification accuracy, and these two models were used to construct a voting classifier that achieved the best accuracy of 85.53%. The ML model achieved much better accuracy than conventional resistivity-based methods. By conducting this study, we aim to develop a new workflow to improve the prediction of Sw in reservoirs where conventional methods have poor performance.

Keywords: petrophysical analysis; Bakken Formation; reserve estimation; tight reservoirs

Citation: Mellal, I.; Latrach, A.; Rasouli, V.; Bakelli, O.; Dehdouh, A.; Ouadi, H. Water Saturation Prediction in the Middle Bakken Formation Using Machine Learning. *Eng* **2023**, *4*, 1951–1964. https://doi.org/10.3390/eng4030110

Academic Editor: Alireza Nouri

Received: 20 May 2023
Revised: 3 July 2023
Accepted: 7 July 2023
Published: 11 July 2023

1. Introduction

Tight reservoirs hold significant oil and gas reserves; however, their complex rock and fluid properties present production challenges [1]. Despite the application of multistage hydraulic fracturing, less than 20% of the original oil in place can be technically and economically producible [2]. Reservoir simulation results tend to overestimate production from unconventional reservoirs due to uncertainties in characterizing rock and fluid properties variation. Therefore, accurate characterization of these reservoirs is crucial for production forecasting, gas storage, and enhanced oil recovery [3]. One of the significant challenges in estimating Sw in tight reservoirs is the reliance on resistivity-based methods that depend on formation-dependent variables known as Archie parameters [4]. These parameters can be difficult to estimate due to the high variation in the cementing minerals and volumes and to the presence of conductive minerals and high-salinity formation water [5,6]. To address this challenge, ML models have been widely applied in various conventional and complex sandstone and carbonate reservoirs to predict petrophysical properties when conventional methods fail to accurately estimate these properties, as studied by [7,8]. Some

ML algorithms have shown promise in accurately predicting Sw from well logs in tight reservoirs. However, the accuracy of the results is highly dependent on the quality of the well logs used and the complexity of the studied reservoir [9].

Due to the potential to improve the estimation of Sw in complex reservoirs using data-driven models, the application of ML has been the subject of several extensive studies over the past two decades. One of the very first studies on fluid saturation prediction using ML algorithms was conducted by Amiri et al. [10], who compared the Sw results of conventional models, such as the Indonesian and improved Indonesian models, with the Sw predicted from the imperialist competitive algorithm (ICA) artificial neural network (ANN) and backpropagation (BP) ANN models in the tight gas sand of the Mesaverde group located in various basins such as the upper Great River Basin, Piceance Basin, Uinta Basin, and Washakie Basin. Both ANN models scored a correlation coefficient (R2) of more than 0.92 in their study, which was much higher than the R2 of the conventional methods. Boualam [9], on the other hand, conducted a detailed petrophysical analysis of the Three Forks Formation, Williston Basin, which is a thin-bedded carbonate formation, and found that the estimation of Sw using resistivity-based methods was challenging. The author applied two ML modes (support vector regression (SVR) and ANN). Both models achieved correlation coefficients of approximately 0.78, demonstrating higher accuracy compared to the Sw estimates obtained from resistivity-based methods. Miah et al. [11] addressed the problem of predicting fluid saturations in low-resistivity, shaly sand formations using two ANN models, achieving a correlation coefficient of 0.9968. Hodavimoghaddam et al. [12] tested a total of four ML algorithms, XGBoost, LightGBM, CatBoost, and AdaBoost, without relying on the resistivity log. The study was conducted in a tight sandstone reservoir in Russia using triple combo logs from 11 wells. Most of the models provided a very good prediction of Sw.

Despite the promising results found in these studies, most were conducted in either conventional reservoirs with relatively simple lithologies or tight reservoirs with relatively simple mineralogy, pore size distribution, and constant volume of cementing minerals. Additionally, these studies relied on data from a limited number of wells to develop their predictive models, sometimes as low as data from two wells [13], which is not a good starting point for developing a data-driven model capable of generalization and performing predictions on new unseen data from new wells that can potentially have different petrophysical properties.

This study aimed to accurately predict Sw in the MBM using seven classification ML algorithms. It was motivated by poor correlation coefficient results from regression ML algorithms and resistivity-based methods tested on the same sample set. We collected 378 diverse samples from 29 wells in the Bakken Formation across six counties in North Dakota, USA. The dataset included gamma ray, deep resistivity, neutron porosity, bulk density, and Dean–Stark Sw measurements. The aim was to develop a robust ML model capable of capturing maximum variance and mapping the input variables to Sw accurately. This study discusses the results of the implemented ML models and their potential for predicting other petrophysical properties in highly heterogeneous reservoirs and outlines future work to enhance accuracy and generate a Sw map of the MBM. These findings benefit petrophysicists and petroleum engineers, improving the prediction of petrophysical properties and enabling accurate reservoir characterization and simulation in unconventional reservoirs.

2. Geological Settings

The Middle Bakken Member (MBM), as a member of the Bakken Petroleum System, is the main producing unit in the Williston Basin and the second oil-producing formation in the United States [14,15]. The formation is located between the Upper and Lower Bakken shale members, with a thickness ranging from 20 to 50 feet in most areas [16]. The geology of the MBM is complex and heterogeneous, with variability in mineralogy, pore types and distribution, and permeability along the formation [17]. The thin layers, bioturbations, and the variation in lithofacies (which can be up to eight lithofacies) of the MBM result in low reservoir quality despite the high volume of hydrocarbons. This presents a significant challenge for extracting oil and gas [18]. Besides the tightness and the geologic complexity of the formation, the MBM has a low resistivity reading despite the presence of a high volume of oil. Unfortunately, well logs cannot depict the variation observed in Figure 1 due to the low resolution of the logging tools and the high heterogeneity of the petrophysical and geological properties, which vary from one lithofacies to another [18,19]. Mineralogy also plays a substantial role in the characterization challenges [20–22] as the MBM has up to eight minerals, which consist of quartz, calcite, dolomite, illite, kaolinite, K-Feldspar, Muscovite, and plagioclase, where illite, kaolinite, dolomite, and calcite play the role of the cementing materials. This makes the characterization of the mineralogy along with the Sw challenging and the use of conventional methods inaccurate due to the linear models used for mineralogy estimation and the dependence of the Archie equation on the lithology variation [21]. Figure 1 shows the geological feature variation of the five lithofacies of the studied well in the MBM [23].

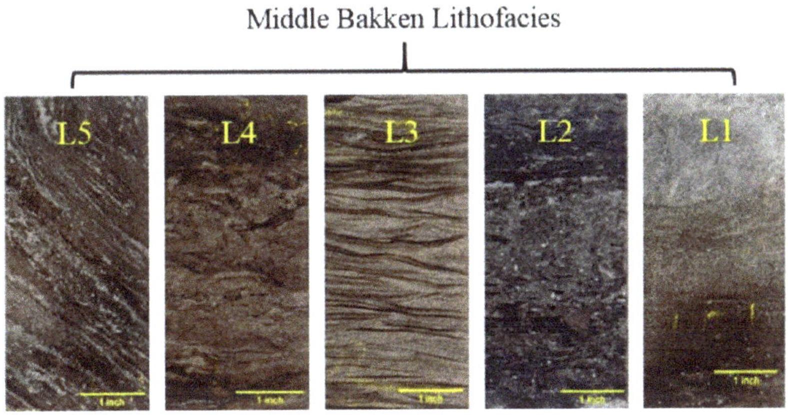

Figure 1. Pictures of cores taken from different Middle Bakken facies. L1, L2, L3, L4, and L5 represent the MBM lithofacies [23].

3. Materials and Methods

In this study, a total of 378 data points were collected from 29 wells drilled in the MBM. The datasets used to predict Sw consist of conventional logs (gamma ray, depth resistivity, bulk density, and neutron porosity), and Sw was calculated using the modified Simandoux equation, while the Dean–Stark Sw was used as the dependent variable in the dataset. These data were split into 80% training and 20% testing sets and used to train and evaluate the accuracy of different ML models. The best-performing model was then compared to conventional resistivity-based Sw estimation methods.

3.1. Petrophysical Data Processing

To accurately predict Sw using ML models, we selected well log values at the same depth as core data, as core plugs provide point values while well logs are continuous. Before conducting this study, petrophysical pre-processing of the dataset were performed. This involves three key steps, environmental correction, depth shifting, and log normalization. The objective of this phase was to generate a comprehensive and coherent set of continuous log and core data curves.

Firstly, environmental corrections were performed to correct resistivity, bulk density, and neutron logs for factors such as borehole properties, mud characteristics, temperature, pressure, and salinity. Secondly, block-shifting was applied to preserve core data integrity. Finally, log normalization ensured consistent analysis results across the selected wells, avoiding significant deviations in calculations such as clay volume from gamma ray logs. This comprehensive processing approach lays the foundation for an accurate prediction of Sw in the MBM.

3.2. *Machine Learning Model Description*

Seven classification ML models were selected, tested, and compared. The initial aim of this study was to formulate a regression problem and attempt to predict the exact values of Sw. However, after trying several models, refining them, and optimizing the parameters, the models gave poor results, with the best performer, the random forest regressor, only giving an R2 score of 0.53. Instead of a regression problem, the continuous saturation data were converted into 10 classes of 10% saturation intervals, and classification models were used instead. This approach takes into account the uncertainty of the Dean–Stark Sw measurement, accounting for potential over/underestimation caused by salt precipitation or incomplete solvent drying [24]. It is, therefore, a tradeoff between accuracy and precision; using classes with 10% saturation intervals indeed yielded good prediction accuracy, but we traded off the precision of these predictions.

3.2.1. Logistic Regression

Logistic regression is a supervised classification model that uses a linear combination of the input variables followed by a logistic function (i.e., sigmoid). Logistic regression is a binary classification model, with the dependent variable being binary (i.e., either 0 or 1). Nevertheless, it can be extended to K cases to support multiclass classification [25].

3.2.2. Support Vector Classifier

The support vector classifier (SVC), or support vector machine (SVM) classifier, is another supervised learning classification algorithm. SVC works by finding an optimal hyperplane in a hyperdimensional space that separates the different classes of our classification problem. This hyperplane is found by maximizing the distance between this hyperplane and its closest point from each class [26].

3.2.3. Random Forest Classifier

The random forest classifier is an ensemble model that uses predictions from multiple trained decision trees to develop more robust and accurate predictions. The decision trees that make up the random forest classifier are each trained on a random subset of the data, then used to predict the classes of new data. The aggregate of the trees' predictions is the outcome of the random forest [27,28].

3.2.4. AdaBoost

AdaBoost, or adaptive boosting, is another type of ensemble model that iteratively combines weak learners to achieve a stronger learner. After each iteration, wrongly predicted samples are weighted stronger than the rest (i.e., it modifies the sample distribution) to emphasize them more. Although AdaBoost also uses decision trees (like random forest), the different decision trees have different weights on the final prediction, with the weights assigned based on their performance, and they are not treated equally like in random forest [29,30].

3.2.5. Gradient Boosting Classifier

Gradient boosting is yet another type of ensemble method. It is different from AdaBoost in the way it adds and trains the weak learners; it does not change the distribution of the sample and instead trains the weak learner on the residual error of the previous learner. Gradient boosting also employs a gradient-based optimization scheme for a differentiable loss function [31].

3.2.6. Artificial Neural Networks

Artificial neural networks are types of models inspired by the structure of biological neurons. They consist of several layers: (1) input, (2) hidden, and (3) output layers. Each layer contains a specific number of nodes or neurons. Data are fed into the model through the input layer and move forward in the network, where each layer performs matrix multiplication followed by a nonlinear activation function and passes the output to the next layer until the output layer is reached, as shown by [32]. The matrices used for computation contain the parameters of the network or weights. The whole process of moving forward in the network from input to output is termed forward propagation. During the learning phase, the weights are initialized randomly, and the network output is calculated. A loss function is used to measure the prediction error; then, the backpropagation algorithm is used to perform auto differentiation on the loss function with respect to the model's weights. Gradient descent is used to optimize these weights to reduce the model's error. This process goes on until a minimal loss is achieved [33,34].

For this study, we used a neural network with 8 inputs (i.e., well logs and calculated inputs), a hidden layer with 32 neurons, and an output layer with 1 output and Softmax activation function (i.e., multiclass classification).

3.2.7. Voting Classifier

The voting classifier is an ensemble ML model that takes the predictions made by several separately trained estimators (e.g., SVC, random forests) and aggregates these predictions using two types of voting systems:

- Hard voting: the final prediction is based on the majority vote of the estimators.
- Soft voting: the final prediction is the average of the class probabilities predicted by each model.

The voting classifier does not perform classification but draws on the combined predictions of several models to obtain a more accurate estimate [35].

3.3. Data Scaling

Data scaling is an important step prior to feeding data into ML models [36]. Scaling the input variables affects the model performance and interpretability in several ways:

- Improved model performance by reducing the effect of variables' differences in scale.
- Faster model convergence, especially for neural networks with gradient descent optimization.
- Better interpretability by making it easier to compare the different coefficients head-to-head rather than being scaled.

Data were scaled and standardized by removing the mean from each variable and scaling it to unit variance. To scale a variable x, we use

$$z = \frac{(x - \mu)}{s} \quad (1)$$

where μ is the mean of the samples and s is the standard deviation. The scaler used for the training set should be the same scaler used for testing because the test set will necessarily have different means and standard deviations.

The flowchart in Figure 2 summarizes the data and methods used to estimate Sw in the MBM.

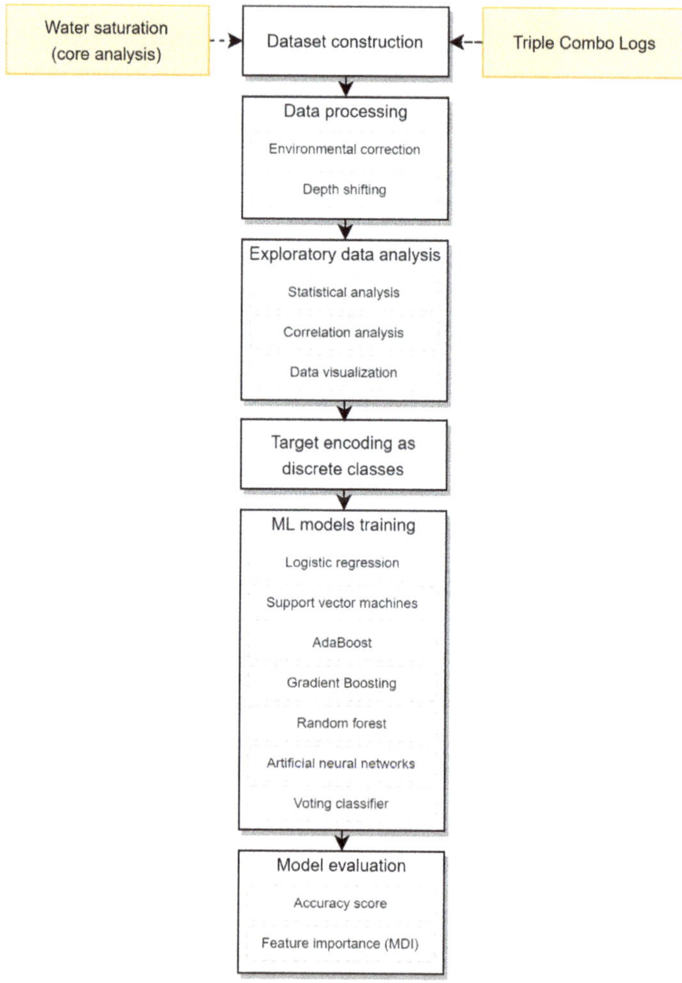

Figure 2. Flowchart for water saturation prediction from conventional well logs using ML models.

4. Results and Discussion

In this section, we present and discuss the results of seven ML models used to predict Sw in the MBM. ML models were applied due to the limitations of resistivity-based methods in such complex and heterogeneous reservoirs. Classification ML models were used over regression models due to their low performance and the inherent uncertainty of Sw measured in the tight core samples of the MBM. The measurements could be either over- or underestimated.

The selected ML models were processed, trained, and tested with the objectives to:

- Investigate the potential linear relationship between well logs, Sw calculated using the modified Simandoux method, and Dean–Stark Sw.
- Assess the performance of the Sw prediction models and compare their accuracy with that of conventional methods.
- Determine the well logs that have the highest feature importance among the applied ML models.

In the following subsections, we conducted a petrophysical analysis, followed by an analysis of the data used for prediction, to finally apply and test the performance of the ML models. The results were compared with regression ML models and resistivity-based methods.

4.1. Petrophysical Analysis

Before training the selected ML models, we conducted a petrophysical analysis to calculate Sw using the Archie, Simandoux, modified Simandoux, Indonesian, Waxman–Smits, and dual-water methods (Figure 3). We evaluated the accuracy of the models using Sw measured with the Dean–Stark method. Before estimating Sw using resistivity-based methods, we calculated the clay volume (Vsh) and calibrated the results with the Vsh measured from the X-ray diffraction analysis. Effective porosity was also calculated and calibrated with porosity measured from cores, and Archie parameters (a, n, and m) were defined. From the petrophysical analysis and among all the models used to calculate Sw, the modified Simandoux method provided the most accurate estimation. The results of the model were then used as input to train and test the ML models. Figure 3 shows the results of Sw calculated using the resistivity-based methods and calibrated with core data.

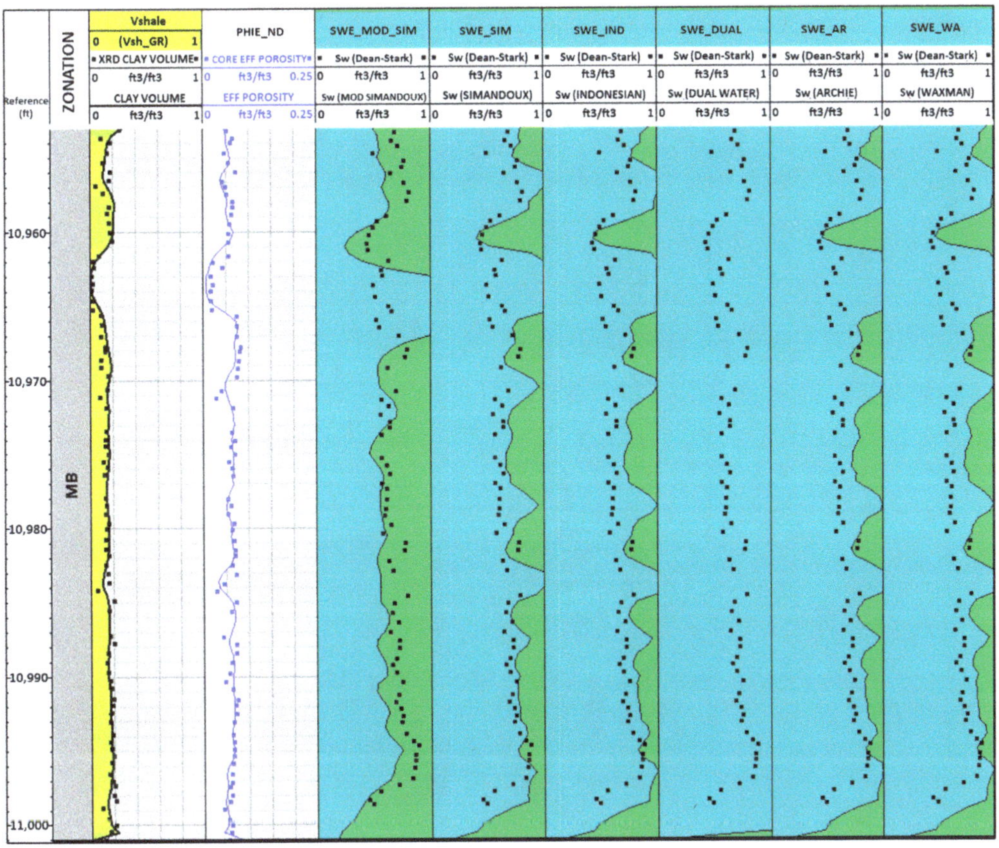

Figure 3. Petrophysical analysis of the Middle Bakken Formation. Tracks from left to right: track 1: reference depth, track 2: Thickness of the MBM, track 3: clay volume calibrated with XRD measured clay volume, track 4: effective porosity calibrated with core measured porosity, tracks 5 to 10: Sw calibrated with core measured Sw, oil saturation (green color), water saturation (blue color).

4.2. Exploratory Data Analysis

To investigate the relative importance of the input variables with the output variable, exploratory data analysis of the constructed dataset was conducted. This analysis enabled us to understand the dataset's patterns, linearity, and potential relationships. By examining measures of mean, standard deviation, and the Pearson correlation coefficient, we were able to identify outliers, discern trends, and establish a solid foundation for subsequent modeling and analysis techniques. Table 1 shows the different statistics of the predictive variables (i.e., well logs) and predicted variables (i.e., water saturation).

Table 1. Statistical summary of the predictive and predicted variables.

	GR	R	RHO	NPOR	PE	Vsh	Por	$Sw_{Simandoux}$	Sw_{core}
mean	80.79	11.02	2.62	0.0865	3.59	0.1101	0.0818	0.3016	0.3836
std	16.56	18.84	0.03	0.0263	0.39	0.0393	0.0221	0.1248	0.1466
min	26.86	2.11	2.51	0.0108	2.61	0.0133	0.0278	0.0200	0.0540
25%	73.25	4.59	2.60	0.750	3.33	0.0890	0.0675	0.2022	0.2593
50%	80.89	6.44	2.62	0.0851	3.55	0.1063	0.0800	0.3151	0.3695
75%	90.88	11.0	2.64	0.0993	3.83	0.1315	0.0958	0.3883	0.4690
max	124.47	254.43	2.69	0.1550	5.66	0.2435	0.1518	0.6682	0.7490

Figure 4 shows the distributions of the various variables and their distributions' kernel density estimates. Most of the variables show a normal distribution, with some exceptions. The photoelectric factor, modified Simandaux saturation, and core saturation exhibit a bimodal distribution, while resistivity demonstrates significant positive skewness with extreme values. These extreme values, located near the upper and lower Bakken members within the Middle Bakken, should not be considered outliers.

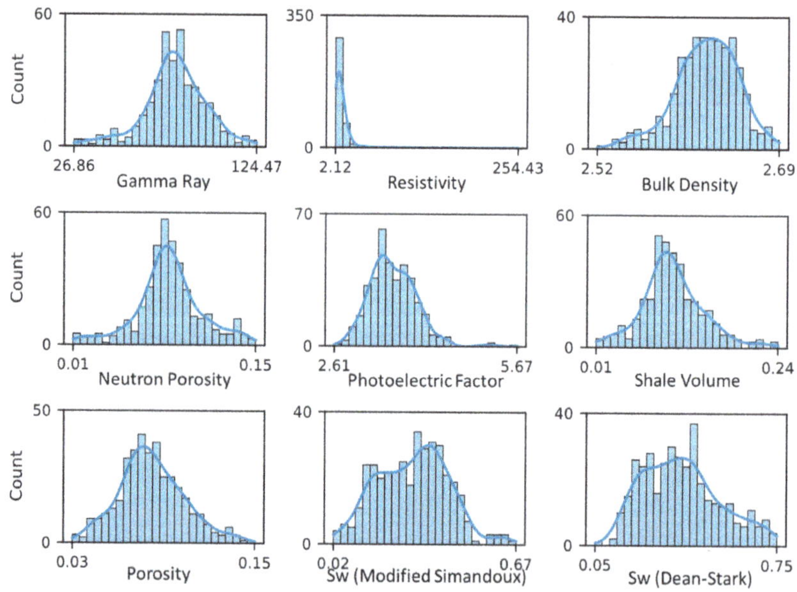

Figure 4. Histograms and skewed normal distribution curves of the input and output variables used to predict Sw.

Figure 5 displays the correlation matrix, providing insights into the relationships between the predictive variables (well logs) and the target variable. Notably, none of the predictive variables exhibit a strong linear correlation with the target variable. However, the Sw calculated from the modified Simandoux emerges as the variable with the highest linear correlation coefficient of 0.44, followed by bulk density at 0.36. While a high correlation coefficient indicates a potentially robust and accurate ML model, the lack thereof only implies the absence of a direct linear relationship between the predictive and predicted variables. Nonetheless, there still can be nonlinear relationships that can be found and exploited by ML models.

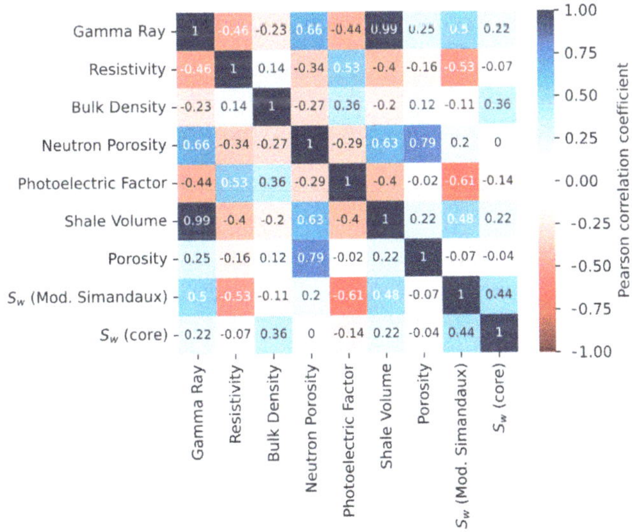

Figure 5. Correlation matrix of the input and output variables used for predicting Sw.

4.3. Machine Learning Model Performance

Figure 6 shows the accuracy scores of the various models on the test set. Among them, gradient boosting and random forest achieved the highest classification accuracy, 82.89% and 77.62%, respectively. The model parameters are summarized in Table 2. Using a soft voting scheme, these two models were subsequently combined into a voting classifier. This approach sums the predicted probabilities for each class and selects the class with the highest probability.

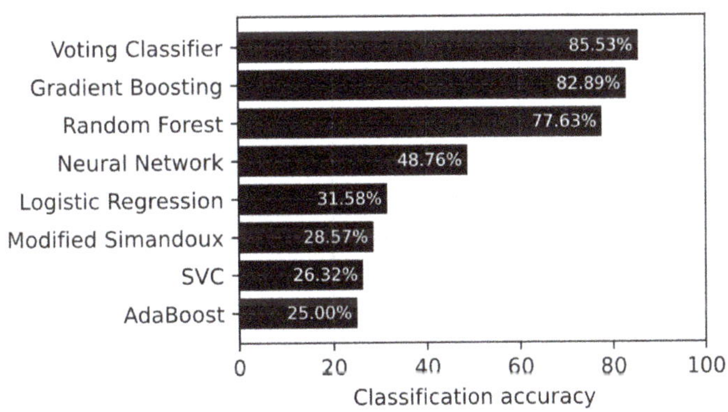

Figure 6. Classification accuracy of the applied ML models for the Sw prediction.

Table 2. Modeling parameters of gradient boosting and random forest classifiers.

	Loss	Learning Rate	Num. of Estimators	Criterion	Max Depth	Minimum Sample Split
Gradient boosting classifier	log_loss	0.1	100	friedman_mse	3	2
Random forest classifier	N/A	N/A	100	Gini	None	2

The voting classifier yielded the best accuracy among all models, reaching a score of 85.53%. However, the ANN achieved a modest accuracy of 48.76% despite attempting various network architectures and hyperparameter tuning. This lower performance can be attributed to the relatively small dataset, which consisted of only 378 samples. The limited data may have hindered the network's training process and its ability to generalize to unseen data. Data collection was limited by the publicly available core analysis reports; if a large enough dataset is to be collected in future work, the artificial neural network should be able to have a better generalization, consequently achieving higher accuracy. Additionally, the modified Simandoux Sw, calculated for the entire dataset rather than just the testing subset, exhibited a low accuracy of only 28.57%.

4.4. Model Evaluation

The confusion matrices shown in Figure 7 present the results obtained from two methods for predicting Sw: the voting classifier and the modified Simandoux equation. The voting classifier exhibits high prediction accuracy, proving its efficiency in estimating Sw in the selected wells drilled in the MBM. The equation mostly overestimates the Sw, yielding significantly higher values than the Dean–Stark Sw.

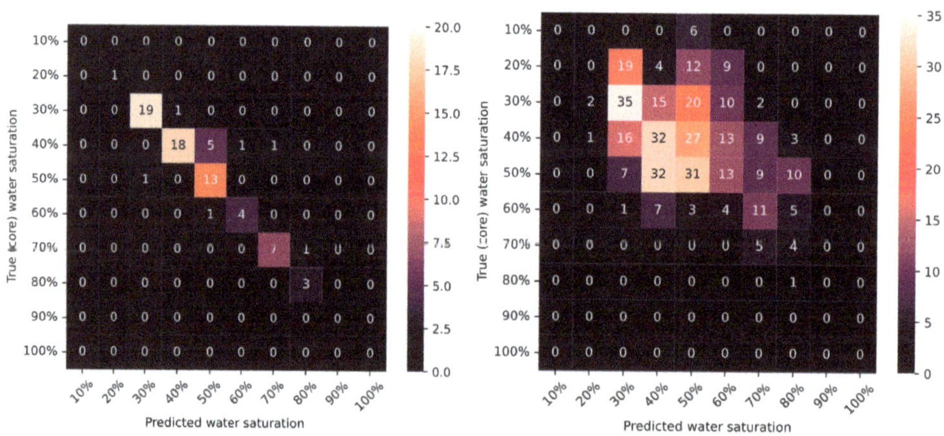

Figure 7. Confusion matrices of Sw predicted using voting classifier (**left**) and Sw calculated using modified Simandoux (**right**).

Despite its inaccurate estimation, modified Simandoux equation Sw calculations had the highest feature importance (measured by the mean decrease in impurity) for the ML model, followed by the resistivity log (Figure 8). The remaining logs had very similar feature importance scores. This indicates that the modified Simandoux calculations set the starting point for the ML models, which then used the well logs to refine the estimations and provide much more accurate predictions.

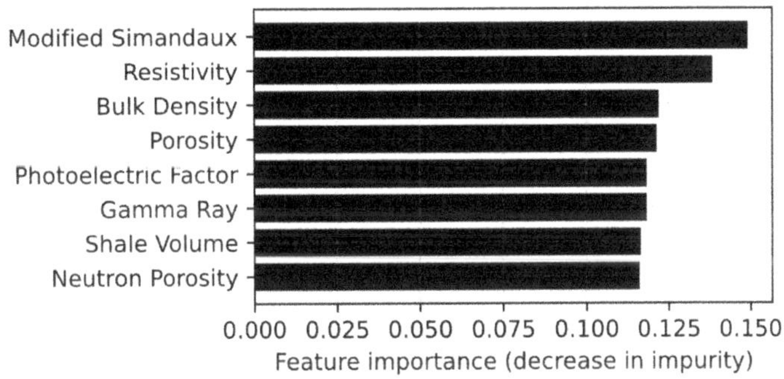

Figure 8. Feature importance of the different parameters used for Sw prediction.

Table 3 lists Sw prediction scores of various ML models from recently published studies using conventional well logs as input variables. The input variables used by these studies are identified with the R2 values achieved between predicted and measured Sw values. Note that all ML methods display a narrow range of high R2 values (ranging from 0.78 to 0.99; Table 2), but the SVM method achieved a lower R2 value (R2 = 0.78).

Table 3. Results from published studies that have used conventional well logs to predict Sw from ML models. The comparison was made with R2 values achieved by the models with the test subsets.

Author	Samples and Wells Number	ML Model	Formation	Results
Ibrahim et al. [36]	782 samples, 2 wells	ANN, ANFIS	Tight gas sandstone	$R^2 = 0.93$
Hadavimoghaddam et al. [12]	11 wells	XGBoost	Sandstone	$R^2 = 0.999$
Miah et al. [10]	182 samples	ANN and SVM	N/A (Bengal Basin)	$R^2 = 0.999$
Khan et al. [37]	150 samples	ANN and ANFIS	N/A (South Asian field)	$R^2 = 0.94$
Hamada et al. [38]	269 samples	ANN	Shaly sandstone	MSE = 0.012
Gholanlo et al. [39]	564 samples, 1 well	ANN	Carbonate	$R^2 = 0.87$
Boualam et al. [9]	2509 samples	SVM and ANN	Tight carbonate	$R^2 = 0.78$
This study	378 samples, 29 wells	Voting classifier	Ultra tight and multimineral formation	Accuracy = 85.53%

In this study, the classification approach outperformed both the initially proposed regression approach and resistivity-based methods, resulting in higher accuracy. The limitations faced by the initially proposed methods in accurately predicting Sw can be attributed to the following challenges:

- High heterogeneity of the MBM, including the presence of extremely thin laminations and significant variation in the volume of cement minerals.
- The low resolution of logging tools cannot accurately represent the high variation of physical properties (bulk density, neutron porosity, photoelectric factor, and resistivity) of such formations, which are used as inputs for Sw prediction.
- The uncertainty associated with laboratory measurement of Sw in tight cores using the Dean–Stark method, which undermines the accuracy of Sw prediction using ML regression algorithms, even when models show a high correlation coefficient.

Considering these challenges, the application of ML classification models has proved to provide the most accurate estimation of Sw. The proposed workflow introduces a methodology aimed at improving the estimation of Sw in unconventional reservoirs. Notably, this approach takes into consideration the inherent heterogeneity of the formation and the limitations posed by logging measurements and the Dean–Stark method.

The application of classification models yields prediction results as classes representing ranges of Sw, rather than continuous values. The robustness of the applied models relies on using data from wells distributed across the MBM to capture the reservoir heterogeneity. To the best of our knowledge, this is the first attempt to predict Sw using classification ML models in reservoirs where regression ML models and resistivity-based methods fail to accurately estimate Sw.

To improve this workflow, it is essential to quantify the effect of the uncertainty associated with logging and laboratory core measurements on the predicted Sw. This will lead to more accurate classes for a representative Sw classification. Overcoming these limitations will yield a more precise and dependable Sw prediction in heterogeneous reservoirs.

In conclusion, this study offers an effective solution for estimating Sw in reservoirs where resistivity-based methods and ML regression models are underperforming. It also establishes a basis for future research, enabling validation and improvement of the proposed workflow. Additionally, the suggested workflow can be extended to predict other rock properties, such as absolute permeability in the Bakken, by integrating it with the approach developed by Aimen et al., 2022 [19]. Future research should replicate this workflow in diverse fields, including the Bakken Formation and other complex reservoirs. Moreover, exploring alternative classification ML models could further refine the prediction process and enhance the overall outcomes.

5. Conclusions

In this study, we assessed the performance of seven classification ML models in predicting the Sw of the MBM as a tight, low-resistivity, and multimineral reservoir. Well logs, Sw calculated using the modified Simandoux method, and Dean–Stark Sw data were used to train and test the models. The performance results were then compared with the accuracy of the regression ML models and resistivity-based methods. The results of the ML models led to the following conclusions:

- The voting classifier model, based on gradient boosting and random forest, displays the highest accuracy of Sw in the MBM.
- The Sw calculated using the modified Simandoux method tends to be overestimated in the MBM. However, using it as input to train and test the classification ML models improved result accuracy.
- Petrophysical data processing, which consists of depth shifting, environmental correction, and log normalization, is crucial for accurate prediction of Sw.
- The voting classifier model, based on gradient boosting and random forest, can accurately match the Dean–Stark Sw within a specific range. Therefore, it can be a viable alternative to expensive laboratory tests.
- We propose applying classification ML models to predict other rock properties, such as permeability and shale volume. This suggestion stems from the recognition that these properties share similar limitations as water saturation.

Author Contributions: Conceptualization, A.L. and I.M.; methodology, I.M.; software, I.M.; validation, V.R. and H.O.; investigation, I.M. and A.L.; writing—original draft preparation, I.M.; writing—review and editing, A.D. and O.B.; supervision, V.R. All authors have read and agreed to the published version of the manuscript.

Funding: This research was funded by the LeNorman Family Excellence Fund.

Institutional Review Board Statement: Not applicable.

Informed Consent Statement: Not applicable.

Data Availability Statement: Data available upon request.

Acknowledgments: The first author would like to acknowledge the scholarship provided by the LeNorman Family Excellence Fund.

Conflicts of Interest: The authors declare no conflict of interest.

References

1. Sorenson, J.; Hawthorne, S.; Jin, L.; Bosshart, N.; Torres, J.; Azzolina, N.; Smith, S.; Jacobson, L.; Doll, T.; Gorecki, C.; et al. *Bakken CO_2 Storage and Enhanced Recovery Program—Phase II Final Report*; U.S. Department of Energy: Pittsburgh, PA, USA, 2018.
2. Shawaf, A.; Rasouli, V.; Dehdouh, A. The Impact of Formation Anisotropy and Stresses on Fractural Geometry—A Case Study in Jafurah's Tuwaiq Mountain Formation (TMF), Saudi Arabia. *Processes* **2023**, *11*, 1545. [CrossRef]
3. Kurtoglu, B.; Sorensen, J.A.; Braunberger, J. Geologic Characterization of a Bakken Reservoir for Potential CO_2 EOR. In Proceedings of the SPE/AAPG/SEG Unconventional Resources Technology Conference, Denver, CO, USA, 12–14 August 2013.
4. Malki, M.L.; Rasouli, V.; Saberi, M.; Mellal, I.; Ozotta, O.; Sennaoui, B.; Chellal, H. Effect of Mineralogy, Pore Geometry, and Fluid Type on the Elastic Properties of the Bakken Formation. In Proceedings of the 56th U.S. Rock Mechanics/Geomechanics Symposium, Santa Fe, NM, USA, 26–29 June 2022. [CrossRef]
5. Malki, M.L.; Rasouli, V.; Saberi, M.R.; Sennaoui, B.; Ozotta, O.; Chellal, H. Effect of CO_2 on Mineralogy, Fluid, and Elastic Properties in Middle Bakken Formation using Rock Physics Modeling. In Proceedings of the 56th U.S. Rock Mechanics/Geomechanics Symposium, Santa Fe, NM, USA, 26–29 June 2022. [CrossRef]
6. Malki, M.L.; Rasouli, V.; Mehena, M.; Mellal, I.; Saberi, M.R.; Sennaoui, B.; Chellal, H. The Impact of Thermal Maturity on the Organic-Rich Shales Properties: A Case Study in Bakken. In Proceedings of the SPE/AAPG/SEG Unconventional Resources Technology Conference, Denver, CO, USA, 13–15 June 2023. [CrossRef]
7. Laalam, A.; Boualam, A.; Ouadi, H.; Djezzar, S.; Tomomewo, O.; Mellal, I.; Bakelli, O.; Merzoug, A.; Chemmakh, A.; Latreche, A.; et al. Application of Machine Learning for Mineralogy Prediction from Well Logs in the Bakken Petroleum System. In Proceedings of the SPE Annual Technical Conference and Exhibition, Houston, TX, USA, 3–5 October 2022. [CrossRef]
8. Ouadi, H.; Mellal, I.; Chemmakh, A.; Djezzar, S.; Boualam, A.; Merzoug, A.; Laalam, A.; Mouedden, N.; Khetib, Y.; Rasouli, V.; et al. New Approach for Stress-Dependent Permeability and Porosity Response in the Bakken Formation. In Proceedings of the SPE Annual Technical Conference and Exhibition, Houston, TX, USA, 3–5 October 2022. [CrossRef]
9. Boualam, A. Impact of Stress on the Characterization of the Flow Units in the Complex Three Forks Reservoir, Williston Basin. Ph.D. Thesis, University of North Dakota, Grand Forks, ND, USA, 2019. Available online: https://commons.und.edu/theses (accessed on 30 April 2023).
10. Amiri, M.; Ghiasi-Freez, J.; Golkar, B.; Hatampourd, A. Improving Water Saturation Estimation in a Tight Shaly Sandstone Reservoir Using Artificial Neural Network Optimized by Imperialist Competitive Algorithm—A Case Study. *J. Pet. Sci. Eng.* **2015**, *127*, 347–358. [CrossRef]
11. Miah, M.I.; Zendehboudi, S.; Ahmed, S. Log Data-Driven Model and Feature Ranking for Water Saturation Prediction Using Machine Learning Approach. *J. Pet. Sci. Eng.* **2020**, *194*, 107291. [CrossRef]
12. Hadavimoghaddam, F.; Ostadhassan, M.; Sadri, M.A.; Bondarenko, T.; Chebyshev, I.; Semnani, A. Prediction of Water Saturation from Well Log Data by Machine Learning Algorithms: Boosting and Super Learner. *J. Mar. Sci. Eng.* **2021**, *9*, 666. [CrossRef]
13. Ibrahim, F.; Elkatatny, S.; Al Ramadan, M. Prediction of Water Saturation in Tight Gas Sandstone Formation Using Artificial Intelligence. *ACS Omega* **2022**, *7*, 215–222. [CrossRef] [PubMed]
14. Merzoug, A.; Ellafi, A. Optimization of Child Well Hydraulic Fracturing Design: A Bakken Case Study. In Proceedings of the SPE Oklahoma City Oil and Gas Symposium, Oklahoma City, OK, USA, 17–19 April 2023. [CrossRef]
15. Sorensen, J.; Jacobson, L.; Pekot, L.; Torres, J.; Jin, L.; Hamling, J.; Doll, T.; Zandy, A.; Smith, S.; Wilson, J.; et al. *Bakken CO_2 Storage and Enhanced Recovery Program—Phase I Final Report*; Energy & Environmental Research Center: Grand Forks, ND, USA, 2014.
16. Hester, T.; Schmoker, J. *Selected Physical Properties of the Bakken Formation, North Dakota and Montana Part of the Williston Basin*; U.S. Geological Survey: Montana, ND, USA, 1985.
17. Merzoug, A.; Chellal, H.A.K.; Brinkerhoff, R.; Rasouli, V.; Olaoye, O. Parent-Child Well Interaction in Multi-Stage Hydraulic Fracturing: A Bakken Case Study. In Proceedings of the 56th U.S. Rock Mechanics/Geomechanics Symposium, Santa Fe, NM, USA, 26–29 June 2022. [CrossRef]
18. Mellal, I.; Malki, M.; Latrach, A.; Ameur-Zaimech, O.; Bakelli, O. Multiscale Formation Evaluation and Rock Types Identification in The Middle Bakken Formation. In Proceedings of the SPWLA 64th Annual Logging Symposium, Lake Conroe, TX, USA, 10–14 June 2023. [CrossRef]
19. Laalam, A.; Ouadi, H.; Merzoug, A.; Chemmakh, A.; Boualam, A.; Djezzar, S.; Mellal, I.; Djoudi, M. Statistical Analysis of the Petrophysical Properties of the Bakken Petroleum System. In Proceedings of the SPE/AAPG/SEG Unconventional Resources Technology Conference, Houston, TX, USA, 20–22 June 2022; pp. 1–21. [CrossRef]
20. Chellal, H.; Merzoug, A.; Rasouli, V.; Brinkerhoff, R. Effect of Rock Elastic Anisotropy on Hydraulic Fracture Containment in the Bakken Formation. In Proceedings of the 56th U.S. Rock Mechanics/Geomechanics Symposium, Santa Fe, NM, USA, 26–29 June 2022.
21. Mellal, I.; Rasouli, V.; Dehdouh, A.; Letrache, A.; Abdelhamid, C.; Malki, M.L.; Bakelli, O. Formation Evaluation Challenges of Tight and Shale Reservoirs. A Case Study of the Bakken Petroleum System. In Proceedings of the 57th U.S. Rock Mechanics/Geomechanics Symposium, Atlanta, GA, USA, 25–28 June 2023. [CrossRef]
22. Shawaf, A.; Rasouli, V.; Dehdouh, A. Applications of Differential Effective Medium (DEM)-Driven Correlations to Estimate Elastic Properties of Jafurah Tuwaiq Mountain Formation (TMF). *Processes* **2023**, *11*, 1643. [CrossRef]

23. Sorensen, A.; Kurz, B.A.; Hawthorne, S.B.; Jin, L.; Smith, S.A.; Azenkeng, A. Laboratory Characterization and Modeling to Examine CO_2 Storage and Enhanced Oil Recovery in an Unconventional Tight Oil Formation. In *Energy Procedia*; Elsevier: Amsterdam, The Netherlands, 2017; pp. 5460–5478. [CrossRef]
24. Kazak, E.S.; Kazak, A.V. A Novel Laboratory Method for Reliable Water Content Determination of Shale Reservoir Rocks. *J. Pet. Sci. Eng.* **2019**, *183*, 106301. [CrossRef]
25. Rymarczyk, T.; Kozłowski, E.; Kłosowski, G.; Niderla, K. Logistic Regression for Machine Learning in Process Tomography. *Sensors* **2019**, *19*, 3400. [CrossRef] [PubMed]
26. Pisner, A.; Schnyer, D.M. Support Vector Machine. In *Machine Learning: Methods and Applications to Brain Disorders*; Elsevier: Amsterdam, The Netherlands, 2019; pp. 101–121. [CrossRef]
27. Laoufi, H.; Megherbi, Z.; Zeraibi, N.; Merzoug, A.; Ladmia, A. Selection of Sand Control Completion Techniques Using Machine Learning. In Proceedings of the International Geomechanics Symposium, Abu Dhabi, United Arab Emirates, 7–10 November 2022. [CrossRef]
28. Abdulkareem, N.M.; Abdulazeez, A.M. Machine Learning Classification Based on Radom Forest Algorithm: A Review. *Int. J. Sci. Bus.* **2021**, *5*, 128–142. [CrossRef]
29. Chellal, H.A.K.; Egenhoff, S.; Latrach, A.; Bakelli, O. Machine Learning Based Predictive Models for UCS and Young's Modulus of the Dakota Sand Using Schmidt Hammer Rebound. In Proceedings of the 57th U.S. Rock Mechanics/Geomechanics Symposium, Atlanta, GA, USA, 25–28 June 2023. [CrossRef]
30. Schapire, R.E. The Boosting Approach to Machine Learning: An Overview. In *Nonlinear Estimation and Classification*; Springer Science & Business Media: New York, NY, USA, 2003.
31. Natekin, A.; Knoll, A. Gradient boosting machines, a tutorial. *Front. Neurorobot.* **2013**, *7*, 21. [CrossRef] [PubMed]
32. Ouadi, H.; Mishani, S.; Rasouli, V. Applications of Underbalanced Fishbone Drilling for Improved Recovery and Reduced Carbon Footprint in Unconventional Plays. *Pet. Petrochem. Eng. J.* **2023**, *7*, 1–15. [CrossRef]
33. Ouadi, H.; Laalam, A.; Hassan, A.; Chemmakh, A.; Rasouli, V.; Mahmoud, M. Design and Performance Analysis of Dry Gas Fishbone Wells for Lower Carbon Footprint. *Fuels* **2023**, *4*, 92–110. [CrossRef]
34. Latrach, A. Application of Deep Learning for Predictive Maintenance of Oilfield Equipment. Master's Thesis, Université M'hamed Bougara Boumerdès, Boumerdas, Algeria, 2020. [CrossRef]
35. Emon, M.U.; Keya, M.S.; Meghla, T.I.; Rahman, M.M.; Al Mamun, M.S.; Kaiser, M.S. Performance Analysis of Machine Learning Approaches in Stroke Prediction. In Proceedings of the 2020 4th International Conference on Electronics, Communication and Aerospace Technology (ICECA), Coimbatore, India, 5–7 November 2020.
36. Ahsan, M.; Mahmud, M.A.P.; Saha, P.K.; Gupta, K.D.; Siddique, Z. Effect of Data Scaling Methods on Machine Learning Algorithms and Model Performance. *Technologies* **2021**, *9*, 52. [CrossRef]
37. Khan, R.; Tariq, Z.; Abdulraheem, A. Machine Learning Derived Correlation to Determine Water Saturation in Complex Lithologies. In Proceedings of the SPE Kingdom of Saudi Arabia Annual Technical Symposium and Exhibition, Dammam, Saudi Arabia, 23–26 April 2018. Available online: http://onepetro.org/SPESATS/proceedings-pdf/18SATS/All-18SATS/SPE-192307-MS/1246152/spe-192307-ms.pdf/1 (accessed on 20 May 2023).
38. Hamada, G.; Elsakka, A.; Chaw, N. Artificial Neural Network (ANN) Prediction of Porosity and Water Saturation of Shaly Sandstone Reservoirs. *Adv. Appl. Sci. Res.* **2018**, *9*, 26–31. Available online: www.pelagiaresearchlibrary.com (accessed on 22 May 2023).
39. Ghalanlo, H.; Amirpour, M.; Ahmadi, S. Estimation of water by using radial based function artificial neural network in carbonate reservoir: A case study in Sarvak formation. *Petroleum* **2016**, *2*, 166–170. [CrossRef]

eng

Article

Exploring Neogene Marine Diatomites in Western Crete: A New Source-Rock Candidate with Hydrocarbon Generation Potential?

Dimosthenis Telemenis [1], Vagia-Ioanna Makri [1,2], Emmanouil Manoutsoglou [1] and Spyridon Bellas [2,*]

1 School of Mineral Resources Engineering, Technical University of Crete, 73100 Chania, Greece
2 Institute of GeoEnergy, Foundation for Research and Technology-Hellas (FORTH/IG), 73100 Chania, Greece
* Correspondence: spyrosbellas@ipr.forth.gr

Abstract: Diatomites are sedimentary rocks rich in siliceous algae, mainly diatoms, and are evident in Greece mainly in Neogene successions. Despite their significance and worldwide potential as source rocks for hydrocarbons, little is known about them in the wider area of Greece, limited in their biostratigraphic characteristics and industrial use. This study assesses for the first time the Neogene diatomites in Western Crete and focuses on their source rock quality and hydrocarbon generation potential on top of their stratigraphic characteristics. The studied synthetic outcrop is of the Miocene age and is located in the Apokoronas sedimentary basin, in the Chania province. It is subdivided into four subsections reaching heights of 13 m. It has a total documented, visible extend of 90–100 m and presents adequate thickness in relation to other reported diatomitic occurrences in Crete. A SEM study and bulk sampling of 28 samples has been carried out on this outcrop and geochemical analysis has been conducted by means of a Rock-Eval 6 pyrolysis to facilitate the understanding of hydrocarbon potential. Stratigraphic analysis supports the establishment of system tracts (ST), with transgressive ones (TST) illustrated by fining-upward sequences including highstands (HST). At the top, a final coarsening-upwards sequence suggests a regressive sequence (RST) most probably related to the Messinian Salinity Crisis (MSC) event. Total organic carbon (TOC, %wt.) values are found to reach 3.4% in the diatomites, while siltstone/mudstone interlayers encounter lower TOC (%wt.), yet with exceptions reaching TOC levels as high as the diatomaceous facies. Overall, Rock-Eval pyrolysis shows that organic matter from the studied cross sections is immature with the hydrocarbon generation potential ranging from poor to excellent. The kerogen type is proved to be type III with poor to almost good quality. This suggests the presence of a prolific diatomaceous source rock in Western Crete demonstrating a high significance for the offshore hydrocarbon exploration in the Eastern Mediterranean that could potentially be related to the offshore Western and Southern Crete E&P-awarded blocks.

Keywords: diatomites; hydrocarbons; organic geochemical analysis; source rock; Crete; Neogene; Miocene; stratigraphy; pre-evaporitic; eastern Mediterranean

Citation: Telemenis, D.; Makri, V.-I.; Manoutsoglou, E.; Bellas, S. Exploring Neogene Marine Diatomites in Western Crete: A New Source-Rock Candidate with Hydrocarbon Generation Potential? *Eng* **2023**, *4*, 285–300. https://doi.org/10.3390/eng4010017

Academic Editors: Reza Rezaee and Yujie Yuan

Received: 15 December 2022
Revised: 12 January 2023
Accepted: 16 January 2023
Published: 18 January 2023

1. Introduction

Diatomites are light, soft, easily powdered, fine-grained, consolidated, sedimentary rocks mostly composed of the siliceous frustules of diatoms, silicoflagellates, sponge spicules and/or radiolaria, well-organized in cyclic alternations (laminae) of light and dark color respectively, and usually connected to seasonal variability between winter and summer. Unconsolidated lake or marine deposits with no diagenesis may be called diatomaceous ooze [1]. They include both freshwater and marine habitats. Rich marine diatomaceous deposits, diatom blooms, are usually connected to the increased productivity of upwelling systems, perhaps the most well-known of them being the offshore area of Chile (i.e., [2]).

Diatomitic occurrences are widespread in the Mediterranean. One of the first distribution maps of Messinian laminated diatomaceous formations of the western Mediterranean was provided by [3] where the authors tried to connect them to the upwelling and high nutrient influx. [4] published a similar map for the entire Mediterranean siliceous deposits. Recently, thorough research of the parameters that govern the relationship between biogenic silica and organic matter (OM) preservation during the upper Miocene of northern Italy is given in [5].

Most publications refer to the diatomites either from a lithostratigraphic and biostratigraphic point of view, or as industrial material with variable usages, including environmental ones [6–9]. Among others, their significance relies on their high value in hydrocarbon exploration. The latter is supported mainly by their gas generation potential as suggested by several research works [10–12] as well as their prolific hydrocarbon source-rock character in many basins worldwide ([12–17]).

Diatomites have never been utilized as prolific source rocks for hydrocarbon Exploration and Production (E&P) in the Mediterranean Sea, despite the sector's strong activation during the last two decades. In this context, the present work investigates diatomites outcropping in NW Crete, Greece (Figure 1). The first lithostratigraphic subdivision of western Crete (Chania Province) was introduced by [18]. According to this author, the presently studied siliceous deposits resemble the Italian "tripoli" and can be placed in the upper Miocene Khairethiana formation; compare also [19]. Messinian upwelling systems that resulted in diatomitic strata accumulation were reported from only four locations in Greece, three in Crete (see review in [20,21]) and one in the Gavdos Islands [22–24], while six in total are reported from the eastern Mediterranean (including one in Cyprus and one in Turkey respectively [25]. Thorough research on the upper Miocene strata of Crete geochemical signatures was recently published by [26], but no reference to diatomitic deposits is made therein.

This paper focuses on the geochemical study of a newly reported [27] thick occurrence of diatomitic laminites from the western Crete Island in Greece. An open system pyrolysis of the samples was applied by means of Rock-Eval 6 (RE6). Stratigraphic data are also provided for the better understanding of the geological connection to the prevailing paleodepositional environments. This approach aims to uncover the potential of the diatomites as a new source rock for hydrocarbon generation in Greece and broadly in the eastern Mediterranean Sea. In the recent context of energy supply deficiency in Europe and worldwide, such discoveries may provide the solution at least for the transition period to fully renewables development which will for sure last for the next two to three decades.

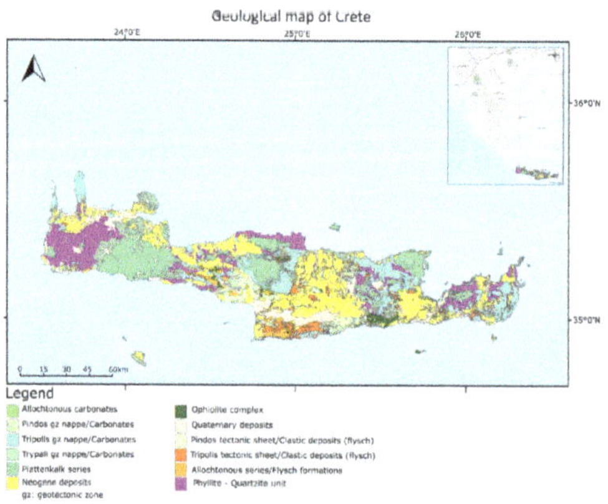

Figure 1. Geological map of Crete with the study outcrop indicated by a red box. Map modified after Decentralized Administration of Crete—IT and Communications Directorate (www.data.apdkritis.gov.gr, accessed on 11 November 2022) [28]. In yellow the main Neogene basins of Crete, bearing transgressive strata on the pre-Neogene basement.

2. Geological Setting

In the eastern Mediterranean Sea and particularly in Greece, a prominent geologic structure is the active collision of the European (Aegean microplate) and the African lithospheric plates. As the latter proceeds to the north, both the subduction and the roll-back effect contribute to a complex geodynamic regime, known as the Hellenic Arc and Trench System (HEAT) or the Hellenic Subduction Zone (HSZ), that is curvature-like and develops from the Adriatic/Ionian Seas to the northwest of Greece, down to the Crete Island to the south and further east to the Rhodes Island [29–32]. A number of mega-structures are directly related to this movement at the plate boundaries, namely the Aegean Volcanic Arc (i.e., Santorini volcanic island) to the north, an uplift (exhumation) resulting in the Crete island and subsequent neogene basin development [33], a backstop, the Mediterranean Ridge accretionary complex to the south [34] and deep offshore trenches mainly due to strike-slip accommodation to the west–southwest (e.g., Matapan) [35] and south–southeast of Crete (e.g., Pliny, Strabo and Ptolemy, [36,37]).

The modern situation of the Aegean, based mostly on offshore data, Plio-Quaternary Extension and strike-slip kinematics that led to the present structural framework was given by [38] and the references therein. Based on satellite-geodetic and earthquake analysis data a thorough reconstruction with two models for the past 1 and 5 Myr of the crustal rotation in the Aegean is recently provided by [39,40].

A structural fragmentation of the exhumed pre-Neogene basement (including mostly HP-LT metamorphic rocks) in multi-blocks led to the development of the Cretan basins, where late middle-to-upper Miocene and Pliocene clastic and carbonate sediments accumulated [41–43]. The initially N–S oriented extension was followed by the E–W extension in the course of Miocene, while the tilting of Crete to the N–NW prevailed in the Pliocene [33]. Beginning of subsidence and sedimentation is assigned an age of 10.8 Ma [44]. According to the later authors, an age of ~9.6 Ma may be attributed to the development of the present-day shape (horst-like) of Crete Island, although the rough relief today was largely shaped in the Pleistocene and Holocene [21]. Refs. [45,46] highlighted the Holocene paleoshoreline uplift of SW Crete due to normal faulting as a result of major earthquakes.

In western Crete and specifically in the Chania Province, the Neogene deposits present with a large geographical extension. Eleven (11) lithostratigraphic formations (Fms.) were recognized by [18], some of them passing laterally to one another. Two main Fms. groups were separated, based on the Gypsum presence stratigraphic level (documenting the Messinian Salinity Crisis evaporites, MSC). According to [18], these are the Ayios Georgios and Kissamou Fms. (below Gypsum), unconformably resting on the pre-Neogene, followed upwards by the Akrotiri, Khairethiana and Tavronitis Fms. (above Gypsum level). Following [33,41] in turn, there was a recognition of six (6) lithostratigraphic groups of Cretan formations. Four (4) of them belong to Miocene (Prina, Tefelion, Vrysses and Hellinikon groups) and two (Finikia and Aghia Galini groups) are of the Pliocene age.

All these lithostratigraphic schemes based on the facies differentiation assumption may clearly be the depositional result of two distinct and complete sedimentation cycles (transgression, highstand, regression), separated by the MSC regressional event; the first event of the Miocene age (lasting from the Tortonian to Messinian) and the second one of the Pliocene age (uppermost Zanclean to early Piacenzian) [20,47,48]. The later authors recognized a third sedimentation cycle as well, a Pleistocenic one, with terraces locally raised in some cases at 200 m altimeter as a result of the combination of high rates of tectonic uplift and enormous sea-level changes affecting various levels of existing deposits.

The diatomitic deposits that are presently studied represent a new occurrence of fine laminites in Crete [27]. The studied outcrop represents a synthetic one, subdivided into three main subsections. It is located in the Apokoronou District of Chania Prefecture, directly to the south of Soudha Bay, the main port of western Crete. It is widely (visibly) and linearly extended along the main street of Kalyves village in a west–east direction, but sparse occurrences of it can be followed up to at least 1 km away (to the west) of the main outcrop location. According to the map of [18] it is postulated that it belongs to the Khairethiana Fm. Although the later author placed it in the Pliocene (above Gypsum level), this formation's deposits were later on assigned a Miocene age based on planktonic Foraminifera biostratigraphy [49], a fact which fits well with our observations.

3. Materials and Methods

3.1. Lithostratigraphic Logging, Sampling Procedure and Microscopy

As mentioned above, the studied material comes from a synthetic outcrop, where Miocenic laminites are well exposed in the Kalyves village, Apokoronou District of Chania Prefecture. The methodology applied included field work for the identification of the specific rock type that we were interested in (i.e., diatomites), detailed section logging, combined with lateral equivalents in order to manage the strata continuity and appropriate sampling. Scanning Electron Microscopy (SEM) used on a set of samples to elaborate on the type (diatomites or not) and preservational status and structure of laminites.

Bulk sampling of twenty-eight (28) samples was carried out in three locations (Figure 2). It is an undisturbed profile, comprising a depositional sequence that measures more than a hundred meters in horizontal length and tens of meters in thickness. Layers are horizontally stratified, therefore following the lateral extension of it was not such a problem. The outcrop is subdivided into four (4) subsections, due to difficulties in access and sampling by heavy vegetation and steepness of it: the lower part, represented by subsections i and ia (twenty one (21) samples: 1, 1b to 9; plus the ex2 and 10 to 15), the intermediate part, represented by subsection b (three (3) samples: 20 to 22) and the upper part, subsection c (four (4) samples: 23 to 26). To control the lateral continuity, additional side sampling was undertaken as well (four (4) samples: 16 to 19). Sixteen (16) in total samples can be described as diatomites.

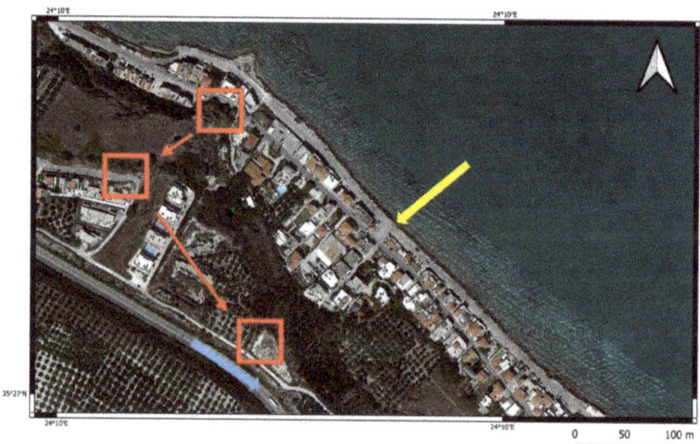

Figure 2. The studied synthetic outcrop in Kalyves village (west Crete, Greece, yellow arrow). Red boxes and relevant arrows point to the sampling route followed from the basal part of the profile (subsection i, ia; near shore), to the intermedium part of it (subsection b), and further to the top (subsection c; box to the right). The grey line crossing the map is the national road driving from Chania (to the left, or west) to Rethymnon city (to the right, east, light blue arrow).

SEM samples used both as raw rock material (rock fragments) and as processed samples. Six (6) samples in total were processed for SEM imaging, representing all four subsections of the studied outcrop (4, 8, 13, 21, 25) and a side one (16). Preparation included fragmentation of the rocks, treatment with light H_2O_2 (5%) for 24 h, and with deionized water for another day, followed by dispersion of the material onto the traditional SEM stubs and subjected to graphite coating procedure (film thickness of ~25 nm). Gold spattering procedure applied only in one sample (2022-4-6-CK-2(16)-ITE). Microphotographs were kindly undertaken in the Jeol JSM-IT500 SEM of the Hellenic Authority for Geological and Mining Research, under beam voltage of 20 kV and in low vacuum of 30–50 Pa, out of two samples that experimentally proceeded under ultra-high vacuum (samples 2022-4-6-CK-2(16)-ITE and 2022-4-6-CK-1(13)-ITE).

3.2. Rock Eval 6 Pyrolysis

Rock-Eval 6 (RE6) pyrolysis was performed on all the 28 outcrop samples using a Rock Eval 6® anhydrous open-system pyrolyser (Vinci Technologies, France) [50]. The basic setup/bulk-rock method for the organic matter analysis [51,52] was applied to characterize the potential source rock samples and assess their hydrocarbon generation potential. The analysis provided measurements from the sequential pyrolysis and oxidation of ca. 50–60 mg of finely grounded (<250 μm) sample amount, according to [53]. The resulting volatile hydrocarbons from pyrolysis were detected and quantified using a flame ionization detector (FID) which was also used to quantify the CO and CO_2 gases resulting from both pyrolysis and oxidation. Pyrolysis followed an isothermal trend from 300 °C where it stayed for 3 min and a subsequent 25 °C/min ramp until 650 °C. Subsequently the oxidation was carried out from 300 to 850 °C with a heating rate of 20 °C/min [53].

Based on this analysis, the following parameters were determined; The free hydrocarbons in the sample from the intensity of the S1 peak (S1; mg hydrocarbon per g rock), the hydrocarbons formed by the thermal cracking of kerogen determined from the intensity of the S2 peak (S2, mg HC/g rock), the total organic carbon (TOC, %wt.), the temperature Tmax (°C) at which the maximum number of hydrocarbons is generated. On top of these, the carbon monoxide (CO) and carbon dioxide (CO_2) released during the thermal decomposition of OM (S3CO and S3CO_2 peaks) were also determined after pyrolysis and oxidation, along with the inorganic carbon content (pyroMINC and oxiMINC peaks).

4. Results

4.1. Lithostratigraphy

The lower subsections (i and ia) display the base of the whole outcrop (Figure 3). Section i is approximately 2.90 m thick, followed by section ia which accounts for 2.70 m. The lowermost part of section i was grey/blueish, thinly to medium bedded, locally bioturbated, fine sand-to-siltstones moderately cemented, and reached up to 1.80 m. The following-upwards part is 1.10 m consisting entirely of whitish, brittle, silty diatomites (Figure 3, right part –i-, lower yellow horizon), well organized in fine (thickly laminated to very thinly bedded) laminae alternations (varve-like). Section ia consists of similar alternations, wherein the bottom part (1.35 m) contains mostly silty mudstones and an underlying of 0.95 m of well-laminated (Figure 3, right upper part –ia- yellow horizon) white-to-slightly-light grey diatomaceous deposits. The section "closes" at the top via a 0.4 m thick sandstone layer, which is partly oxidized.

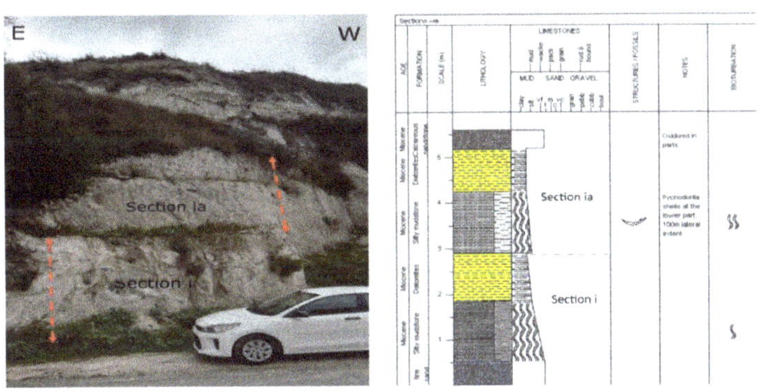

Figure 3. The lower (basal) part of the Kalyves synthetic outcrop. To the left is the field photograph of subsections i and ia, which are depicted with the red dashed lines. To the right is the profile logging. In yellow colour the laminated diatomite packages.

Section b (Figure 4) represents the middle package of diatomites and the upwards continuation of the lower part (sections i, ia). The whole profile falls inside a construction site, so unfortunately the available data will be lacking for future scientific work. The sampled sector is approximately three (3) meters thick, while the whole subsection has a thickness of about 7.60 m, extending a few meters further down but impossible to sample. A blueish-to-light grey silty mudstone comprises the bottom part, gradually followed upwards by pure, light, whitish, very friable diatomites (Figure 4, upper part in yellow). In one sample near the top, marine fish-fauna rests were found, but due to friability they were not possible to be identified, although this indicated that a deep marine paleoenvironment persisted in the area during the upper Miocene.

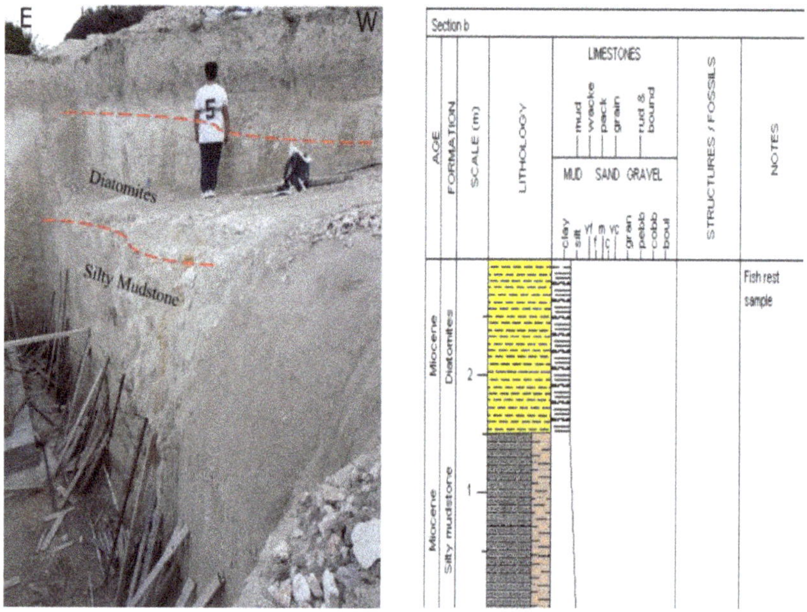

Figure 4. To the left, the construction site of studied subprofile b. Thinly-to-medium bedded silty mudstone at the base, passes gradually to the diatomitic laminites at the top (man for scale). To the right, the relevant lithostratigraphic log. In yellow color the diatomites.

Section c (Figure 5) represents the upper (top) package of silty mudstones and diatomites, comprising a total thickness of 3.60 m. In this part, yellowish silty mudstones underlie the diatomitic beds (Figure 5, middle part in yellow). The section closes at the top with three (3) medium-to-thickly bedded, indurated sandy mudstone beds, showing signs of oxidation and representing the closure of the studied synthetic outcrop whilst being covered by recent soil, bearing a root system of natural plants. A rapid uplift probably occurred in the area with subsequent erosion that did not enable the deposition of a normal regressive sequence to be developed.

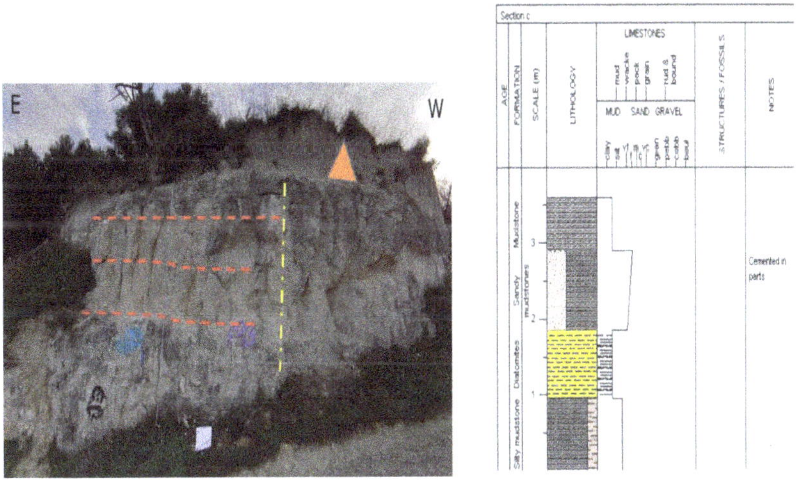

Figure 5. To the left is the topmost subsection of the studied synthetic outcrop of diatomites. The dashed red lines depict the sandy mudstone beds overlying the diatomitic horizons. Geological hammer for scale. To the right the subsection lithostratigraphic logging with the diatomites in yellow.

4.2. Sequence Stratigraphy

For a comprehensive review and standardization of the sequence stratigraphy methodology, the reader is referred to [54] and the references therein. The total synthetic outcrop currently studied is a depositional package belonging to the first Neogene sedimentation cycle of western Crete according to [20,41,47,48] of the Upper Miocene age. Between the basal part of it (subsection i) and the top (Section c), a cycling of deepening-to-shallowing strata is observed. The lower part of the subsection i (Figure 3) represents a transgressive, deepening deposit (TST), probably as a response to changes in the accommodation space (as a result of both the tectonic subsidence of north Cretan subbasins and sea-level rise). The absolute base of it is not visible, since no pre-Neogene basement occurs in the close neighborhood, therefore no real sequence boundary (unconformity) can be drawn. The diatomite strata that follow the up section consisting of fine, varve-like laminae seem to reflect an extra deepening, indicating a more distal setting than the lower part [55,56]. Similar siliceous sedimentary facies are usually attributed to basinal facies [57]. As the SEM study provided, the diatoms are dominated by circular forms (Figure 6), thus supporting the distal setting of deposition. Subsection ia (Figure 3, upper part) is lithostratigraphically a repeat of i, indicating firstly a slight shallowing (relative sea-level fall) followed again by a deepening and highstand, documented by the upper diatomites deposition (Figure 3, in yellow; Figure 7) (TST/HST) and closing with a coarsening upward sedimentation, i.e., the oxidized sandstone pointing to a mid-scale regression.

The middle part, section b (Figure 4), shows the same sedimentary pattern with i, where a fining-upward deposition prevails (the silty mudstones are overlain with a thick diatomitic package of almost equal thickness). It seems that cyclicity pertains as it is also documented in the third uppermost section c. Obviously we deal with parasequences [55] within the main depositional sequence. Section b is mostly attributable to relative small-scale sea-level changes (rise and highstand, TST to HST) and not to major structural rearrangements.

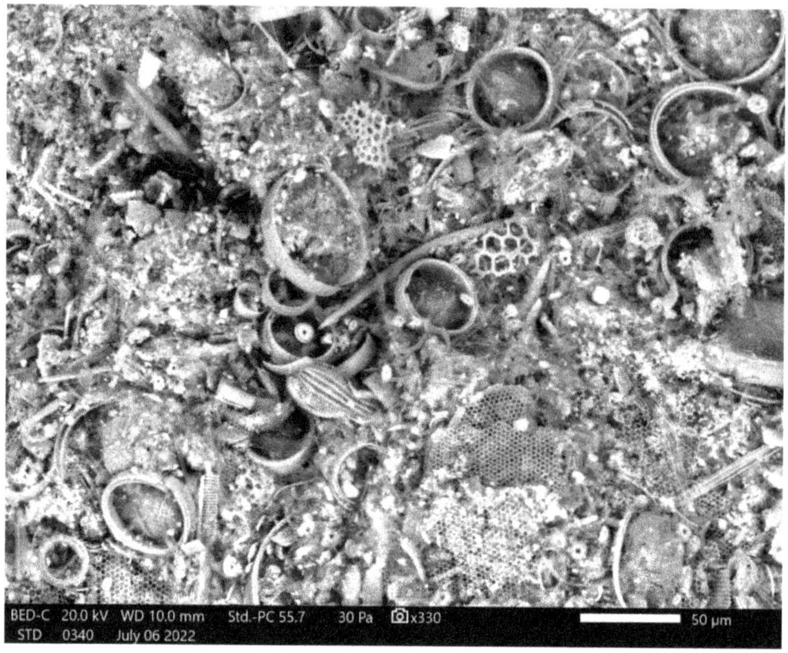

Figure 6. SEM picture of a diatomite rock fragment. Abundant, moderately preserved, marine circular diatoms (planktic forms), together with rare elongated (benthic forms) diatom frustules (A single *Diploneis bombus* can be recognized) and sponge spicules occur. Calcareous nannofossils are also to be observed (possible *Calcidiscus macintyrei*).

The top, section c, differs from the previous ones, and although it starts again at the base with silty mudstones (TST), the diatomitic package that overlies them is thinner, and a cycling of indurated sandy/silty mudstones alternations that seal the section overlies the diatomites. Oxidation traces in the top beds and the clear coarsening clearly point to the shallowing of the depositional environment. Subsequently, we interpret section c as a shallowing-upward parasequence, where the initial sea-level rise was followed by a relative fall (TST to RST). Deposition of the fine-grained facies such as diatomites represents the period of the sea-level rise where the basin reached its maximum depth. The coarse-grain package of the top could represent deposits that may be analogous to the MSC carbonate pattern that prevailed along the rim of the paleobasin by the end of the Messinian [56].

4.3. Rock-Eval Pyrolysis

Samples from a Neogene depositional succession have been characterized using RE6. Sample lithologies consist of mainly diatomites, mudstone and siltstone. According to RE6 pyrolysis the parameters regarding the thermal maturity and generation potential of the analyzed samples vary across the studied locations and across lithologies as they develop upwards. It should be noted that out of the complete set of twenty-eight (28) samples analyzed, the twenty-one (21) were found adequate for further characterization. This was linked to the mineral matrix effect [53,58,59] occurring when S2 is lower than 0.2 mg HC/g rock and the TOC content is below 0.5%wt., as well as to the oxidation of organic matter during outcrop weathering [58]. Thus, results from the RE6 pyrolysis include only the representative samples that are presented in Table 1 and plotted in the following figures.

Figure 7. Fine-laminae setting framework of undisturbed, well developed diatomites. One Euro coin for scale. Light and dark laminar scale alternations are clearly to be seen (compare the upper two dotted lines in red), developed over the lower bed of a fine sandy mudstone (red dotted line below the one Euro coin). The red dotted lines below and above the coin represent the interval transition between the fine sandy mudstone and the diatomites. Red double-arrow presents the diatomites interval.

Table 1. Sample information (section, lithology) and basic Rock-Eval Pyrolysis data.

Sample/Section	Lithology	S1	S2	S3	Tmax (°C)	TOC (%wt.)	HI	OI
smpl_1i	Fine sands silts	0.4	11.51	3.77	414	3.37	342	112
smpl_1Bi	Fine sands silts	0.28	9.41	3.52	413	3.27	288	108
smpl_2i	Fine sands silts	0.02	0.51	2.11	425	0.73	70	289
smpl_3i	Fine sands silts	0.06	1.84	3.15	425	1.24	148	254
smpl_4i	Diatomites	0.33	10.27	4.14	418	3.33	308	124
smpl_5i	Diatomites	0.27	8.56	3.84	418	3.05	281	126
ex_2i	Diatomites	0.31	8.84	3.51	418	3.02	293	116
smpl_6i	Diatomites	0.27	6.35	3.13	420	2.29	277	137
smpl_7i	Diatomites	0.27	6.52	2.91	417	2.45	266	119
smpl_8i	Diatomites	0.46	8.4	3.48	412	2.77	303	126
smpl_9i	Diatomites	0.25	6.02	2.94	421	2.19	275	134
smpl_10ia	Mudstones silts	0.04	1.61	1.62	415	0.77	209	210
smpl_11ia	Mudstones silts	0.07	2.62	2.47	424	1.28	205	193
smpl_12ia	Mudstones silts	0.17	5.02	4.62	418	2.14	235	216
smpl_13ia	Diatomites	0.09	2.24	2.17	416	1.23	182	176
smpl_14ia	Diatomites	0.04	0.51	1.59	428	0.47	109	338
smpl_15ia	Diatomites	0.05	0.88	1.84	423	0.71	124	259
smpl_16ia	Diatomite muds	0.38	9.25	3.36	419	3.29	281	102
smpl_17ia	Mudstones	0.08	0.38	4.65	416	0.88	43	528
smpl_18ia	Mudstones	0.02	0.15	4.37	409	0.57	26	767
smpl_21b	Diatomites	0.02	0.28	1.73	400	0.57	49	304

Acronyms used: TOC (total organic carbon), HI (hydrogen index), OI (oxygen index), Tmax (temperature of the maximum rate of hydrocarbon generation). S1, S2, S3, HI and OI in (mg/g).

4.4. Type and Thermal Maturity of Organic Matter

Based on the RE6 pyrolysis data, kerogen was classified using the data distribution across the studied subsections. The resulting HI (mg HC/g TOC) and OI (mg CO_2/g TOC) suggest values ranging between 40–308 mg HC/g TOC and 116–340 mg CO_2/g TOC respectively for the diatomitic samples, and 26–342 mg HC/g TOC and 108–767 mg CO_2/g TOC respectively for the rest lithologies. As illustrated in Figure 8 (left diagram), the analyzed samples correspond a gas-prone type III kerogen to an inert type IV, with most diatomitic samples obtaining a solid type III response.

Overall, the level of thermal maturity documented from the evaluated samples is low. This can be seen in Figure 8 (right diagram). More specifically, Tmax reaches a maximum level of 428 °C (Figure 9) for the studied samples of all sections suggesting kerogen immaturity [60–64]

4.5. Source Rock Potential

The hydrocarbon generation potential varies across the studied subsections. Section i (lower part of the outcrop) points towards a good to excellent source-rock potential with S2 varying between 6 and 10.27 mg HC/g rock and the TOC between 2.19 and 3.33 %wt. for the diatomitic samples (Figure 10). On the contrary, section b illustrates a fair hydrocarbon-generating potential, yet with a source rock quality not as high as in section i, with 0.28 mg HC/g rock S2 and 0.57 %wt. TOC content (Figure 10).

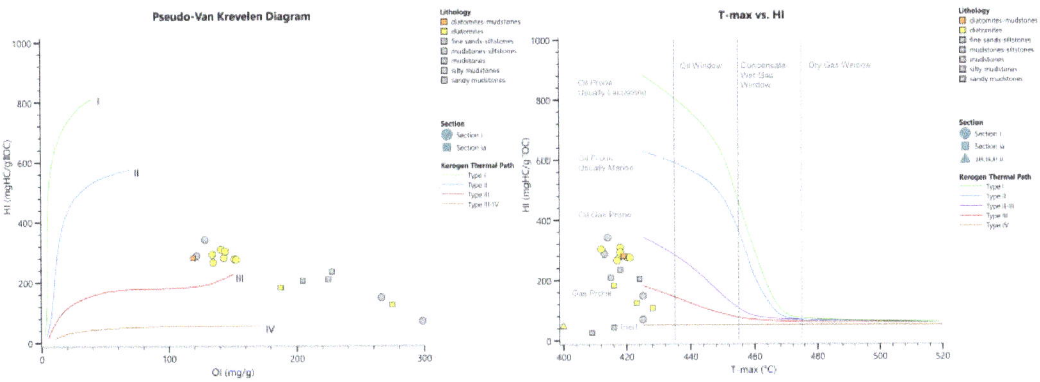

Figure 8. Pseudo Van Krevelen diagram, modified from Peters (1986), for the kerogen classification of the studied samples (**left**); HI (mg HC/g TOC) versus Tmax (°T) showing idealized kerogen types and thermal maturity of the samples separated by dashed lines according to Peters and Cassa (1994) and Baskin (1997) (**right**).

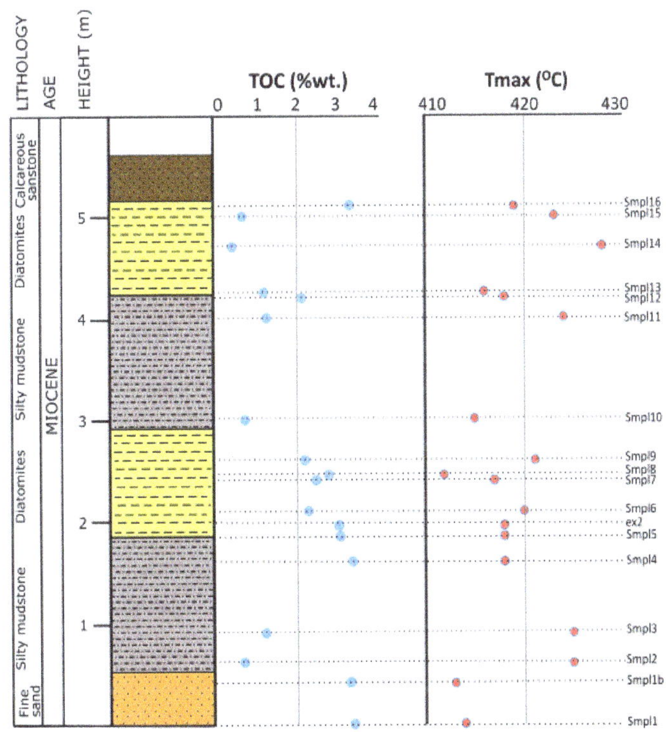

Figure 9. Lithological log of subsections i and ia (0–2.9 m height and 2.9 m-top respectively) along with the corresponding TOC (%wt.) (blue dots) and Tmax (°C) values (red dots). Note the high TOC values in the "basal" diatomites (samples 5 to 9, blue dots in the yellow interval).

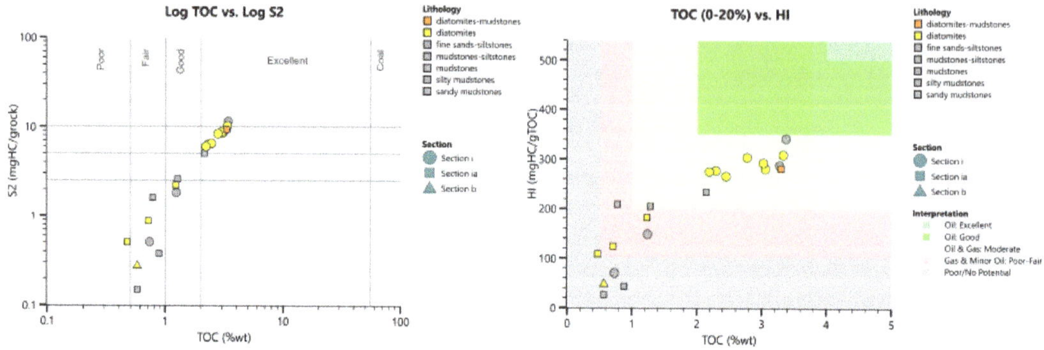

Figure 10. Logarithmic plot of TOC (%wt.) versus S2 (mg HC/g rock) illustrating the source-rock potential (**left**); TOC (%wt.) versus HI (mg HC/g TOC) indicating the source-rock potential and the kerogen type (**right**).

It is evident that diatomite samples (in yellow) obtain the highest TOC values corresponding to the highest HI values (Figure 10; right). This supports the overall frame of mainly-moderate kerogen potential, with evident differentiation between the subsections. Finally, the sum of free and potentially generated hydrocarbons (PP, S1 and S2) ranges between 0.17 and 11.91 mg HC/g rock in the studied samples. These fluctuations depict a poor-to-very good generation potential with the diatomites being clearly distinct from the rest samples.

5. Discussion

The lithostratigraphic results show that each subsection can be representative of a generally fining-upward depositional cycle, with the deeper being characterized by siliceous, fine-layered, pure diatomic packages ranging in thickness between 0.80 m (section c) and 1.5 m (section b). Coarser material closes each cycle at the top, indicating eustatic control of the deposition. Scanning Electron Microscopy gave evidence for the kind of laminites, proving that they are pure diatomites (Figure 6) with medium-to-well preservation of the frustules, since both valves are easily visible. Diagenesis seems not to have affected our material. No centrifugation was applied, trying not to bias the original depositional framework. The dominance of circular forms (centrales, diatoms) that the studied samples consist of reflects the basinal geological setting where they were accumulated. The results of this study suggest that deposition of the diatomites in western Crete is most likely connected to an early Messinian upwelling system that was active prior to the MSC onset and are in good agreement with Phase 1 of [65], (~7.2 to 6.9 Ma), where open marine conditions prevailed at the onset of the Messinian cooling.

The sequence-stratigraphic approach of the synthetic outcrop revealed that deposition occurred during the first sedimentation cycle of the Upper Miocene age. The main transgression (TST) developed as a response to tectonic movement (subsidence, basins development) and probably sea-level rise, but the four subsections studied provide evidence of a cyclicity that was controlled more by the sea-level oscillations, relevant to climate changes, than by tectonic movements. Diatomitic occurrences are connected to deepening (highstands in the transgressional regime, HST), while the coarsening of depositional material by the tops of sections represent sea-level fall and regression (RST).

The geochemical analysis of the samples along the outcrop indicated a poor (upper subsection) to very good (lower and middle subsection) hydrocarbon generation potential. Combining these results with the thickness and the geographical extending of the occurrence, it is clear that the diatomites may act as a possible gas/oil-gas prone source rock in the wider area. Further, the analysis documented high TOC values (ranging up to 3.4%wt.), a fact that makes the studied and other possible equivalent successions interesting candidates for further investigation. A kerogen type III (with a few samples falling between II and III) is mainly recorded by the analyzed set of samples. Additionally, although all samples were proved thermally immature, their hydrocarbon potential is leaning between fair to excellent and the quality of the source rock lies around intermediate levels.

6. Summary and Conclusions

Until presently, sub-salt, mixed carbonate/siliciclastic deposits from the upper Miocene age have been investigated from central Crete and Gavdos Isls., providing poor-to-fair and/or good in a few cases for hydrocarbon generation potential [24,26,66]. Diatomites in turn have long been studied as biostratigraphic, paleoecologic indicators and as industrial materials. There are only a few, but significant works that have investigated these siliceous sedimentary rocks as potential sources for hydrocarbon generation and subsequent exploration in various basins worldwide. The latter application is examined in the present work, within the context of the eastern Mediterranean Sea and the last decades' developments for the sufficient energy supply of Europe.

In this context, and as Western Crete diatomitic deposits have not yet been geochemically investigated, this study focuses on a combined stratigraphic and geochemical analysis of a synthetic outcrop, located in the Apokoronas Neogene basin (Chania province, Greece). The studied succession is a new diatomitic occurrence of the Messinian age, and can be subdivided into four subsections. The present set of samples were investigated and analyzed by means of the RE6 pyrolysis mainly in order to gain an insight into the potential of the diatomitic beds as source rocks for hydrocarbon generation. Additionally, lithostratigraphy, SEM and sequence stratigraphy applied for a better understanding of the depositional regime. Following outcomes emerged from the current work:

1. The studied strata represent pre-evaporitic, marine deposition during the Upper Miocene.
2. Initial transgression was both tectonically and climatically controlled, while later development within the basin was mostly controlled by cyclic, eustatic sea-level oscillations. Transgressive, Highstand and Regressive System Tracks were recorded along the outcrop.
3. Medium-to-well preserved diatomites deposited in distal settings, during deepening of the basin highstands.
4. Source rock quality of the studied diatomitic deposits is proved to be around intermediate levels.
5. Hydrocarbon-generating potential lies between fair to excellent.
6. The potentially generated hydrocarbons of the analyzed diatomitic samples reach values of 10.6 mg/g.
7. Both the stratigraphical data and source rock potential illustrate a new promising case for the offshore exploration of the Crete Island and possibly for the eastern Mediterranean.

Author Contributions: Conceptualization, S.B. and D.T.; methodology, S.B., D.T., V.-I.M.; software, V.-I.M.; validation, S.B., D.T., V.-I.M. and E.M.; formal analysis, D.T. and V.-I.M.; investigation, S.B. and D.T.; resources, S.B.; data curation, D.T., S.B. and V.-I.M.; writing—original draft preparation, D.T., S.B.; writing—review and editing, D.T., S.B., V.-I.M. and E.M.; visualization, D.T., S.B, V.M; supervision, S.B.; project administration, S.B.; funding acquisition, D.T., S.B. All authors have read and agreed to the published version of the manuscript.

Funding: This research received no external funding.

Institutional Review Board Statement: Not applicable.

Informed Consent Statement: Not applicable.

Data Availability Statement: The data used in this work is available on request.

Acknowledgments: N. Pasadakis (Institute of Geoenergy/FORTH) is acknowledged for his support throughout the project.

Conflicts of Interest: The authors declare no conflict of interest.

References

1. Zahajská, P.; Opfergelt, S.; Fritz, S.C.; Stadmark, J.; Conley, D.J. What Is Diatomite? *Quat. Res.* **2020**, *96*, 48–52. [CrossRef]
2. Pinochet, A.; Garces-Vargas, J.; Lara, C.; Olguin, F. Seasonal Variability of Upwelling off Central-Southern Chile. *Remote Sens.* **2019**, *11*, 1737. [CrossRef]
3. Moissette, P.; Bernard, C.; Martin, J. Saint Upwelling and Benthic Communities in the Messinian of Western Mediterranean. *Paleontol. I Evol.* **1992**, *24–25*, 245–254.
4. Pellegrino, L.; Dela Pierre, F.; Jordan, R.W.; Abe, K.; Mikami, Y.; Natalicchio, M.; Gennari, R.; Lozar, F.; Carnevale, G. The Upper Miocene Diatomaceous Sediments of the Northernmost Mediterranean Region: A Lamina-Scale Investigation of an Overlooked Palaeoceanographic Archive. *Sedimentology* **2020**, *67*, 3389–3421. [CrossRef]
5. Pellegrino, L.; Natalicchio, M.; Birgel, D.; Pastero, L.; Carnevale, G.; Jordan, R.W.; Peckmann, J.; Zanellato, N.; Dela Pierre, F. From Biogenic Silica and Organic Matter to Authigenic Clays and Dolomite: Insights from Messinian (Upper Miocene) Sediments of the Northern Mediterranean. *Sedimentology* **2022**, 1–33. [CrossRef]
6. Cameron, N.G. DIATOMS. In *Encyclopedia of Quaternary Science*, 2nd ed.; Elias, S.A., Mock, C.J., Eds.; Elsevier: Amsterdam, The Netherlands, 2013; pp. 522–525, ISBN 978-0-444-53642-6.
7. Flower, R.J. DIATOM METHODS | Diatomites: Their formation, distribution, and uses. In *Encyclopedia of Quaternary Science*, 2nd ed.; Elias, S.A., Mock, C.J., Eds.; Elsevier: Amsterdam, The Netherlands, 2013; pp. 501–506, ISBN 978-0-444-53642-6.
8. Stamatakis, M.; Stamatakis, G. The Use of Diatomaceous Rocks of Greek Origin As Absorbents of Olive-Oil Wasters. *Bull. Geol. Soc. Greece* **2017**, *43*, 2739. [CrossRef]
9. Stefanou, E.; Kantiranis, N.; Chatzicharalambous, K.; Mytiglaki, C.; Stamatakis, M.; Georgiadis, G. Diatomaceous Silica in Environmental Applications: A Case Study from the Lacustrine Deposit of Limnos Island, Aegean Sea, Greece. *Minerals* **2022**, *12*, 523. [CrossRef]
10. Aoyagi, K.; Omokawa, M. Neogene Diatoms as the Important Source of Petroleum in Japan. *J. Pet. Sci. Eng.* **1992**, *7*, 247–262. [CrossRef]
11. Aoyagi, K.; Omokawa, M. Diagenesis of Neogene Diatoms and Their Importance as a Source of Petroleum in Japan. *Isl. Arc* **1993**, *2*, 273–279. [CrossRef]
12. Tulan, E.; Radl, M.S.; Sachsenhofer, R.F.; Tari, G.; Witkowski, J. Hydrocarbon Source Rock Potential of Miocene Diatomaceous Sequences in Szurdokpüspöki (Hungary) and Parisdorf/Limberg (Austria). *Austrian J. Earth Sci.* **2020**, *113*, 24–42. [CrossRef]
13. Jirman, P.; Geršlová, E.; Bubík, M.; Sachsenhofer, R.F.; Bechtel, A.; Więcław, D. Depositional Environment and Hydrocarbon Potential of the Oligocene Menilite Formation in the Western Carpathians: A Case Study from the Loučka Section (Czech Republic). *Mar. Pet. Geol.* **2019**, *107*, 334–350. [CrossRef]
14. Martizzi, P.; Chiyonobu, S.; Hibi, Y.; Yamato, H.; Arato, H. Middle–Late Miocene Paleoenvironment of the Japan Sea Inferred by Sedimentological and Geochemical Characterization of Coeval Sedimentary Rocks. *Mar. Pet. Geol.* **2021**, *128*, 105059. [CrossRef]
15. Mayer, J.; Rupprecht, B.; Sachsenhofer, R.; Tari, G.; Bechtel, A.; Coric, S.; Siedl, W.; Kosi, W.; Floodpage, J. Source Potential and Depositional Environment of Oligocene and Miocene Rocks Offshore Bulgaria. *Geol. Soc. Lond. Spec. Publ.* **2017**, *464*, 307–328. [CrossRef]
16. Sachsenhofer, R.F.; Popov, S.V.; Coric, S.; Mayer, J.; Misch, D.; Morton, M.T.; Pupp, M.; Rauball, J.; Tari, G. Paratethyan petroleum source rocks: An overview. *J. Pet. Geol.* **2018**, *41*, 219–245. [CrossRef]
17. Sachsenhofer, R.F.; Popov, S.V.; Bechtel, A.; Coric, S.; Francu, J.; Gratzer, R.; Grunert, P.; Kotarba, M.; Mayer, J.; Pupp, M.; et al. Oligocene and Lower Miocene Source Rocks in the Paratethys: Palaeogeographical and Stratigraphic Controls. *Geol. Soc. Spec. Publ.* **2018**, *464*, 267–306. [CrossRef]
18. Freudenthal, T. Stratigraphy of Neogene Deposits in the Khania Province, Crete, with Special Reference to Foraminifera of the Family Planorbulinidae and the Genus Heterostegina. *Utr. Micropaleontol. Bull.* **1969**, *1*, 1–208.
19. Zachariasse, W.J.; Lourens, L.J. About the Age and Depositional Depth of the Sediments with Reported Bipedal Footprints at Trachilos (NW Crete, Greece). *Sci. Rep.* **2022**, *12*, 18471. [CrossRef] [PubMed]
20. Keupp, H.; Bellas, S. Neogene Development of the Sedimentary Basins of NW Crete Island, Chania Prefecture, South Aegean Arc System (Greece). *Berl. Geowiss. Abh.* **2000**, *34*, 3–117.
21. Zachariasse, W.J.; Kontakiotis, G.; Lourens, L.J.; Antonarakou, A. The Messinian of Agios Myron (Crete, Greece): A Key to Better Understanding of Diatomite Formation on Gavdos (South of Crete). *Palaeogeogr. Palaeoclimatol. Palaeoecol.* **2021**, *581*, 110633. [CrossRef]
22. Drinia, H.; Antonarakou, A.; Tsaparas, N.; Dermitzakis, M.D.; Kontakiotis, G. Foraminiferal record of environmental changes: Preevaporitic diatomaceous sediments from Gavdos island, southernGreece. *Bull. Geol. Soc. Greece* **2018**, *36*, 782. [CrossRef]
23. Frydas, D. Siliceous Phytoplankton Assemblages and Biostratigraphy of the Pre-Evaporite Messinian Diatomites on Gavdos Island, Greece. *Rev. Micropaleontol.* **2006**, *49*, 86–96. [CrossRef]
24. Pyliotis, I.; Hamilaki, E.; Pasadakis, N.; Manoutsoglou, E. Comparative Evaluation of Rock-Eval and Elemental Analysis to Determine Organic Carbon Content in Sediment Samples. *Bull. Geol. Soc. Greece* **2013**, *47*, 862–870. [CrossRef]
25. Pellegrino, L.; Dela Pierre, F.; Natalicchio, M.; Carnevale, G. The Messinian Diatomite Deposition in the Mediterranean Region and Its Relationships to the Global Silica Cycle. *Earth Sci. Rev.* **2018**, *178*, 154–176. [CrossRef]

26. Maravelis, A.G.; Kontakiotis, G.; Bellas, S.; Antonarakou, A.; Botziolis, C.; Janjuhah, H.T.; Makri, P.; Moissette, P.; Cornée, J.J.; Pasadakis, N.; et al. Organic Geochemical Signatures of the Upper Miocene (Tortonian—Messinian) Sedimentary Succession Onshore Crete Island, Greece: Implications for Hydrocarbon Prospectivity. *J. Mar. Sci. Eng.* **2022**, *10*, 1323. [CrossRef]
27. Telemenis, D.; Makridou, Z.; Bellas, S. New Evidence of Diatomitic Occurrences in Western Crete, Greece; A Preliminary Stratigraphic and Geochemical Approach and Its Implications. *Bull. Geol. Soc. Greece* **2022**, *9*, 81–82.
28. Decentralized Administration of Crete—IT and Communications Directorate Geological Map of Crete. Available online: https://www.apdkritis.gov.gr/ (accessed on 13 December 2022).
29. McKenzie, D. The Plate Tectonics of the Mediterranean Region. *Nature* **1970**, *226*, 239–243. [CrossRef] [PubMed]
30. Makris, J. The Crust and Upper Mantle of Aegean Region from Deep Seismic Soundings. *Tectonophysics* **1978**, *46*, 269–284. [CrossRef]
31. Le Pichon, X.; Angelier, J. The hellenic arc and trench system: A key to the neotectonic evolution of the eastern mediterranean area. *Tectonophysics* **1979**, *60*, 1–42. [CrossRef]
32. Angelier, J.; Lybéris, N.; Le Pichon, X.; Barrier, E.; Huchon, P. The Tectonic Development of the Hellenic Arc and the Sea of Crete: A Synthesis. *Tectonophysics* **1982**, *86*, 159–163, 165, 167–196. [CrossRef]
33. van Hinsbergen, D.J.J.; Meulenkamp, J.E. Neogene Supradetachment Basin Development on Crete (Greece) during Exhumation of the South Aegean Core Complex. *Basin Res.* **2006**, *18*, 103–124. [CrossRef]
34. Kopf, A. The Mediterranean Ridge: A Mass Balance across the Fastest Growing Accretionary Complex on Earth. *J. Geophys. Res.* **2003**, *108*, 2372. [CrossRef]
35. Le Quellec, P.; Mascle, J.; Got, H.; Vittori, J. Seismic Structure of Southwestern Peloponnesus Continental Margin1. *AAPG Bull.* **1980**, *64*, 242–263. [CrossRef]
36. Jongsma, D. Pliny and Strabo Trenches, South of the Hellenic Arc. *Geol. Soc. Am. Bull.* **1977**, *88*, 797–805. [CrossRef]
37. Peters, J.M.; Troelstra, S.R.; van Harten, D. Late Neogene and Quaternary Vertical Movements in Eastern Crete and Their Regional Significance. *J. Geol. Soc. London.* **1985**, *142*, 501–513. [CrossRef]
38. Sakellariou, D.; Tsampouraki-Kraounaki, K. Chapter 14—Plio-Quaternary extension and Strike-Slip tectonics in the Aegean. In *Transform Plate Boundaries and Fracture Zones*; Duarte, J.C., Ed.; Elsevier: Amsterdam, The Netherlands, 2019; pp. 339–374, ISBN 978-0-12-812064-4.
39. Lazos, I.; Sboras, S.; Chousianitis, K.; Kondopoulou, D.; Pikridas, C.; Bitharis, S.; Pavlides, S. Temporal Evolution of Crustal Rotation in the Aegean Region Based on Primary Geodetically-Derived Results and Palaeomagnetism. *Acta Geod. Geophys.* **2022**, *57*, 317–334. [CrossRef]
40. Sboras, S.; Lazos, I.; Bitharis, S.; Pikridas, C.; Galanakis, D.; Fotiou, A.; Chatzipetros, A.; Pavlides, S. Source Modelling and Stress Transfer Scenarios of the October 30, 2020 Samos Earthquake: Seismotectonic Implications. *Turk. J. Earth Sci.* **2021**, *30*, 699–717. [CrossRef]
41. Meulenkamp, J.E. Field Guide to the Neogene of Crete. *Publ. Dep. Geol. Paleontol. Univ. Athens* **1979**, *32*, 1–32.
42. Meulenkamp, J.E.; Wortel, M.J.R.; Van Wamel, W.A.; Spakman, W.; Strating, E.H. On the Hellenic Subduction Zone and the Geodynamic Evolution of Crete since the Late Middle Miocene. *Tectonophysics* **1988**, *146*, 203–215. [CrossRef]
43. Meulenkamp, J.E.; Van Der Zwaan, G.J.; Van Wamel, W.A. On Late Miocene to Recent Vertical Motions in the Cretan Segment of the Hellenic Arc. *Tectonophysics* **1994**, *234*, 53–72. [CrossRef]
44. Zachariasse, W.J.; van Hinsbergen, D.J.J.; Fortuin, A.R. Formation and Fragmentation of a Late Miocene Supradetachment Basin in Central Crete: Implications for Exhumation Mechanisms of High-Pressure Rocks in the Aegean Forearc. *Basin Res.* **2011**, *23*, 678–701. [CrossRef]
45. Mouslopoulou, V.; Begg, J.; Nicol, A.; Oncken, O.; Prior, C. Formation of Late Quaternary Paleoshorelines in Crete, Eastern Mediterranean. *Earth Planet. Sci. Lett.* **2015**, *431*, 294–307. [CrossRef]
46. Ott, R.F.; Wegmann, K.W.; Gallen, S.F.; Pazzaglia, F.J.; Brandon, M.T.; Ueda, K.; Fassoulas, C. Reassessing Eastern Mediterranean Tectonics and Earthquake Hazard from the 365 CE Earthquake. *AGU Adv.* **2021**, *2*, e2020AV000315. [CrossRef]
47. Frydas, D.; Keupp, H. Biostratigraphical Results in Late Neogene Deposits of NW Crete, Greece, Based on Calcareous Nannofossils. *Berl. geowiss. Abh.* **1996**, *E18*, 169–189.
48. Kontopoulos, N.; Zelilidis, A.; Frydas, D. Late Neogene Sedimentary and Tectonostratigraphic Evolution of Northwestern Crete Island, Greece. *Neues Jahrb. Fur Geol. Und Palaontologie—Abhandlungen* **1996**, *202*, 287–311. [CrossRef]
49. Zachariasse, W. Planktonic Foraminiferal Biostratigraphy of the Late Neogene of Crete (Greece). *Utr. Micropaleontol. Bull.* **1975**, *11*, 1–171.
50. Lafargue, E.; Marquis, F.; Pillot, D. Rock-Eval 6 Applications in Hydrocarbon Exploration, Production, and Soil Contamination Studies. *Oil Gas Sci. Technol.* **1998**, *53*, 421–437. [CrossRef]
51. Behar, F.; Beaumont, V.; De, H.L. Technologie Rock-Eval 6: Performances et Développements. *Oil Gas Sci. Technol.* **2001**, *56*, 111–134. [CrossRef]
52. Disnar, J.R.; Guillet, B.; Keravis, D.; Di-Giovanni, C.; Sebag, D. Soil Organic Matter (SOM) Characterization by Rock-Eval Pyrolysis: Scope and Limitations. *Org. Geochem.* **2003**, *34*, 327–343. [CrossRef]
53. Yang, S.; Horsfield, B. Critical Review of the Uncertainty of Tmax in Revealing the Thermal Maturity of Organic Matter in Sedimentary Rocks. *Int. J. Coal Geol.* **2020**, *225*, 103500. [CrossRef]

54. Catuneanu, O. Sequence stratigraphy: Guidelines for a standard methodology. In *Stratigraphy & Timescales*; Academic Press: Cambridge, MA, USA, 2017; Volume 2, pp. 1–57.
55. Clark, M.S. Sequence stratigraphy of an interbedded biogenic-clastic reservoir, Belridge diatomite at lost hills field, San Joaquin basin, California. *Geol. Soc. Am. Abstr. Programs* **2001**, *33*, 35.
56. Goldstein, R.H.; Franseen, E.K. Carbonate Play Models from Miocene Outcrops, Western Mediterranean: Part 2-Stratigraphic and Diagenetic Plays. AAPG Geoscience Technology Workshop. *AAPG* **2019**. [CrossRef]
57. Behl, R.J. Monterey Formation Overview and Context for New Research. *AAPG* **2022**. [CrossRef]
58. Peters, K.E. Guidelines for Evaluating Petroleum Source Rock Using Programmed Pyrolysis. *AAPG Bull.* **1986**, *70*, 3.
59. Reynolds, J.G.; Burnham, A.K. Comparison of Kinetic Analysis of Source Rocks and Kerogen Concentrates. *Org. Geochem.* **1995**, *23*, 11–19. [CrossRef]
60. Peters, K.E.; Cassa, M.R. Applied source rock geochemistry. In *The Petroleum System: From Source to Trap*; Magoon, L.B., Dow, W.G., Eds.; AAPG Memoir: Boulder, CO, USA; Toulsa, OK, USA, 1994; Volume 60; pp. 93–120.
61. Espitalie, J.; Marquis, F.; Barsony, I. Geochemical logging. In *Analytical Pyrolysis—Technics and applications*; Voorhees, K.J., Ed.; Butterworth: Boston, MA, USA, 1984; pp. 276–304. [CrossRef]
62. Baskin, D. Atomic H/C Ratio of Kerogen as an Estimate of Thermal Maturity and Organic Matter Conversion. *AAPG Bull.* **1997**, *81*, 1437–1450. [CrossRef]
63. Killops, S.; Killops, V. *An Introduction to Organic Geochemistry*; Wiley: Hoboken, NJ, USA, 2013; pp. 1–408.
64. Dellisanti, F.; Pini, G.A.; Baudin, F. Use of T Max as a Thermal Maturity Indicator in Orogenic Successions and Comparison with Clay Mineral Evolution. *Clay Miner.* **2010**, *45*, 115–130. [CrossRef]
65. Kontakiotis, G.; Butiseacă, G.A.; Antonarakou, A.; Agiadi, K.; Zarkogiannis, S.D.; Krsnik, E.; Besiou, E.; Zachariasse, W.J.; Lourens, L.; Thivaiou, D.; et al. Hypersalinity Accompanies Tectonic Restriction in the Eastern Mediterranean Prior to the Messinian Salinity Crisis. *Palaeogeogr. Palaeoclimatol. Palaeoecol.* **2022**, *592*, 110903. [CrossRef]
66. Kontakiotis, G.; Karakitsios, V.; Cornée, J.-J.; Moissette, P.; Zarkogiannis, S.D.; Pasadakis, N.; Koskeridou, E.; Manoutsoglou, E.; Drinia, H.; Antonarakou, A. Preliminary Results Based on Geochemical Sedimentary Constraints on the Hydrocarbon Potential and Depositional Environment of a Messinian Sub-Salt Mixed Siliciclastic-Carbonate Succession Onshore Crete (Plouti Section, Eastern Mediterranean). *Mediterr. Geosci. Rev.* **2020**, *2*, 247–265. [CrossRef]

eng

MDPI

Article

Reservoir Adaptability Evaluation and Application Technology of Carbon Quantum Dot Fluorescent Tracer

Jinjian Chen [1,2], Jianxin Liu [1,2,*], Jijian Dai [3], Bo Lin [3], Chunyu Gao [1,2] and Ci Wang [1,2]

[1] School of Petroleum Engineering, Yangtze University, Wuhan 430100, China
[2] Key Laboratory of Drilling and Production Engineering for Oil and Gas, Wuhan 430100, China
[3] Oil Production Plant No. 1, Sinopec Henan Oilfield Branch, Nanyang 474780, China
* Correspondence: liujianxin@yangtzeu.edu.cn

Abstract: This study investigates the application of carbon quantum dots as tracers in inter-well connectivity monitoring. A new laboratory-made water-soluble carbon quantum dot fluorescent tracer (CQD-W) was studied using 3D fluorescence characterization, structural characterization, reservoir suitability evaluation, and core flow experiments. The experimental results showed that CQD-W has a size of about 2 nm, a minimum detection limit of 10^{-2} mg·L^{-1}. It has good stability when the salinity is 200,000 mg·L^{-1}, the concentration of Ca^{2+} is 1000 mg·L^{-1}, the pH value is 1–9, and the temperature is 80 °C. Because CQD-W contains many functional groups, such as carboxyl and hydroxyl, it shows good water solubility and has a negative surface charge. In the process of formation flow, CQD-W has a small adsorption amount, high tracer resolution, and excellent injectivity and mobility, meaning it is less likely to cause reservoir damage. Through the study of this method, the application field of carbon quantum dots is broadened, and it is proved that the CQD-W fluorescent tracer has a high potential for application in the oil industry, laying the foundation for the popularization of this technology.

Keywords: carbon quantum dots; fluorescent tracers; nanoparticles; reservoir adaptation

Citation: Chen, J.; Liu, J.; Dai, J.; Lin, B.; Gao, C.; Wang, C. Reservoir Adaptability Evaluation and Application Technology of Carbon Quantum Dot Fluorescent Tracer. *Eng* **2023**, *4*, 703–718. https://doi.org/10.3390/eng4010042

Academic Editors: Reza Rezaee and Yujie Yuan

Received: 24 December 2022
Revised: 15 February 2023
Accepted: 20 February 2023
Published: 22 February 2023

1. Introduction

In oil and gas development, it is important to obtain inter-well parameters. Tracer technology is considered an effective way to obtain subsurface information [1,2]. The tracers currently used are primarily classified as radioactive tracers and non-radioactive tracers [3,4]. Radioactive tracers are primarily represented by tritium and tritium compounds that are used in small quantities, are easy to detect, and have a minimum detection limit of 10^{-5} mg·L^{-1}. However, they are limited due to their radioactivity, the complexity of operation, hazards to staff, and environmental safety [5]. The non-radioactive tracers are primarily thiocyanates (SCN^-), halogenated ions (Cl^-, Br^-), fluorescent dyes, halogenated hydrocarbons, and alcohols. They have been commonly used as inter-well tracers due to their wide variety and low toxicity. However, these tracers also have disadvantages, such as poor adaptability and selectivity, easy interaction with minerals, and adsorption [6–9].

Carbon quantum dots are a new class of carbon nanomaterial with fluorescent properties [10,11]. Due to their advantages in water solubility, chemical inertness, low toxicity, easy functionalization, and resistance to photobleaching, they are primarily used in biomedical imaging [12–17], photocatalysis [16–19], and water treatment [20,21]. These nanomaterials consist of a carbon core (internal carbon atoms) and surface functional groups. The carbon core is a skeleton composed of sp^2 heterogeneous graphitic microcrystalline carbon and sp^3 heterogeneous amorphous carbon. Its surface usually shows abundant oxygen-containing functional groups, such as hydroxyl and carboxyl groups [22–29]. Therefore, the carbon quantum dots exhibit good stability and water solubility. In addition, a rigid planar structure is formed in the center of the carbon nucleus. The energy trap formed on the surface can trap additional electrons to ensure that it fluoresces strongly when excited [25,30].

Related studies in the biomedical field have shown that carbon quantum dots exhibit strong mobility and have more advantages in terms of luminescence intensity and stability than conventional organic dyes [31,32]. In addition, these nanomaterials inherit the biocompatibility of carbon materials and can meet the needs of real-time monitoring of complex biological environments [33,34].

In recent years, related research scholars started to study the application of carbon quantum dots as tracers in oil fields. Murugesan et al. [2] synthesized a nitrogen-doped carbon quantum dot using an electrochemical method and preliminarily verified the potential of carbon quantum dots as a tracer through a core flow experiment. Ma et al. [1] synthesized five carbon quantum dots using a hydrothermal method. He conducted temperature and acid-base resistance tests as well as core flow experiments. He confirmed that the adsorption of carbon quantum dots on the rock surface determines whether they can be effective as tracers. Hu et al. [35] synthesized a carbon quantum dot and studied its flow performance in sandstone cores. The results showed that the carbon quantum dots could detect the oil saturation of unknown sandstone cores. Shi et al. [25] synthesized a high fluorescence yield quantum dot using a hydrothermal method and confirmed that carbon quantum dots have good stability, mobility, and low adsorption with different minerals through reservoir adaptability evaluation experiments and core flow experiments.

Previous studies on applying carbon quantum dots as tracers in the petroleum industry have focused on the reservoir suitability study and oil saturation determination of carbon quantum dots. In this paper, the team has designed a Si-doped CQD-W fluorescent nanotracer, conducted preliminary research on its three-dimensional fluorescence characteristics, structural characteristics, and reservoir adaptability, and verified the injectivity and mobility of CQD-W through single-tube core flow experiments. Finally, a double-tube core flow experiment study was conducted to confirm whether carbon quantum dots have the ability to distinguish dominant channels. This latter experiment is expected to be applied to oil field development.

2. Materials and Methods

2.1. Experimental Materials

Main experimental reagents used are showned in Table 1.

Table 1. Main experimental reagents.

Reagent Name	Purity	Manufacturers
Sodium chloride (NaCl)	Analysis of pure	National Pharmaceutical Reagent
Calcium chloride ($CaCl_2$)	Analysis of pure	Sinopharm Reagent
Disodium hydrogen phosphate (Na_2HPO_4)	Analysis of pure	National Pharmaceutical Reagent
37% Hydrochloric acid (HCI)	Analysis of pure	National Pharmaceutical Reagent
Citric acid anhydrous ($C_6H_8O_7$)	Analysis of pure	National Pharmaceutical Reagent
Potassium bromide (KBr)	Analysis of pure	National Pharmaceutical Reagent

CQD-W is homemade in the laboratory; quartz sand is purchased from Jiangxi Shanggao Huasil Mining Co.; Artificial Sandstone Cores (Φ2.5 × 10) A (611 mD) and B (795 mD).

2.2. Experimental Methods

2.2.1. Method of Infrared Spectroscopy Test

6 mg of CQD-W and 600 mg of KBr were taken and ground, and 10 Mpa pressure was applied, pressing the materials into thin slices. The treated samples were tested by infrared spectroscopy using a Fourier transform infrared spectrometer (Vector-33, Bruker Instruments, Germany) [1,35].

2.2.2. Method of ζ-Potential Test and Particle Size Characterization

A ζ-potential cuvette was loaded with 1 $mg \cdot L^{-1}$ CQD-W solution 0.5 mL and the cuvette was placed into a nanoparticle size ζ-potential analyzer (Litesizer 500, Anton Paar,

Austria) for ζ-potential testing. The CQD-W particle size testing was performed similarly, with a quartz cuvette chosen for the test [36].

2.2.3. Method of Fluorescence Characteristics Test

The concentration of 1 mg·L^{-1} CQD-W solution was configured. 1 mL of the solution was taken in a quartz fluorescent cuvette and the quartz fluorescent cuvette was put into a fluorescence spectrophotometer (F97pro, Shanghai Prism, China) for three-dimensional fluorescence scanning with a slit width of 10 and a gain of 9 (the fluorescence intensity tests were all tested under this parameter).

The standard concentration solutions of 0~20 mg·L^{-1} were configured. The quartz fluorescent cuvettes with a standard concentration of CQD-W solutions were put into a fluorescence spectrophotometer (F97pro, Shanghai Prism, China) at the optimal excitation wavelength, emission wavelength, and slit width to measure the fluorescence intensity. Finally, the relationship between CQD-W and fluorescence intensity was obtained.

2.2.4. Method of Reservoir Suitability Evaluation

The CQD-W mixed solutions with concentrations of 0~200,000 mg·L^{-1} NaCl of 0.1 mg·L^{-1} CQD-W, 0~1000 mg·L^{-1} $CaCl_2$ of 0.1 mg·L^{-1} CQD-W, and 0.1 mg·L^{-1} CQD-W at pH = 1~9 were prepared and tested for CQD-W fluorescence intensity by diluting 10 times with pure water.

CQD-W mixed solutions with concentrations of 0.1 mg·L^{-1} CQD-W and 5000 mg·L^{-1} NaCl were prepared, placed in small sample bottles, sealed and heated in an electric thermostatic water bath (DK-8AX, Shanghai Heheng Instruments, China). Every 24 h, a portion of the mixed solution was removed and diluted 10 times using pure water for the CQD-W fluorescence intensity test (repeated for one week).

$$\delta = \frac{x_1 - x_2}{x_1} \tag{1}$$

where: δ—relative error.

x_1—control CQD-W concentration.

x_2—CQD-W concentration in the experimental group.

The CQD-W mixed solution with concentrations of 0.1 mg·L^{-1}, 5000 mg·L^{-1} NaCl was prepared, and the 0.1 mg·L^{-1} CQD-W solution and liquid paraffin were mixed according to the volume ratio (20:20) with magnetic stirring for 5 h. Then, the solution was transferred to a partition funnel and left for 2 h. After being completely layered, the aqueous/oil phase fraction was removed for fluorescence intensity test analysis [37].

2.2.5. Static Adsorption Experiments

CQD-W solution of 0.1 mg·L^{-1} was added to a 250 mL reagent bottle with quartz sand (60–120 mesh) at a solid-liquid ratio of 1:4 (mass ratio). Then the bottle was placed in a 50 °C digital water bath shaker (SHA-CA, Changzhou Jintan Kexing Instruments) for 2 h. Every 2 h, all the supernatant samples were removed and separated by centrifugation at 4000 r/min for 10 min in a benchtop high-speed centrifuge (TG16, Shanghai Lu Xiangyi Centrifuge Instruments Co. Ltd.). Part of the supernatant was removed to determine its fluorescence intensity, then new quartz sand was added according to the solid-liquid ratio of 1:4 for adsorption. This process was repeated 8 times [38].

$$\Gamma = \frac{(C_0 - C_1)V}{m} \tag{2}$$

where: Γ—apparent adsorption of CQD-W per unit mass of quartz sand, ug/g.

C_0—concentration of CQD-W before adsorption, mg/L.

C_1—concentration of CQD-W after adsorption, mg/L.

V—the volume of CQD-W added, mL.

m—single quartz sand mass, g.

2.2.6. Core Flow Experiments

The multi-functional core repelling device (Jiangsu Lianyou Research Instruments) was installed. The core A (porosity 25.1%) was saturated with deionized water before the experiment and injected with a mixture of 5000 mg·L^{-1} NaCl and 1 mg·L^{-1} CQD-W solution at a constant flow rate of 1 mL/min at standard atmospheric pressure and a temperature of 25 °C. The brine concentration of the extracted fluid and the concentration of the CQD-W solution were measured periodically for a certain PV. The concentration output curves of NaCl and CQD-W were summarized. The schematic diagram of the single-tube repulsion device is shown in Figure 1.

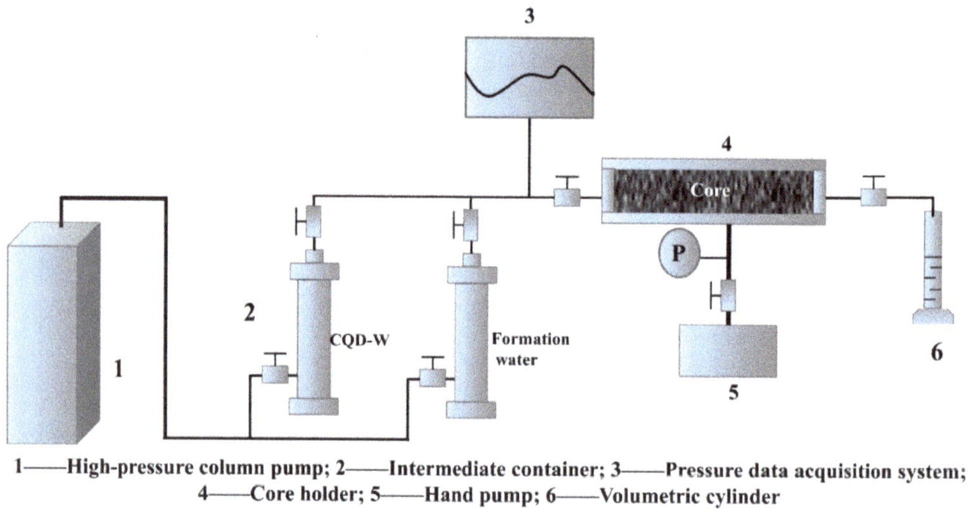

Figure 1. Schematic diagram of core flow device.

The experimental method for the double-tube core flow is similar, except that an additional group of core B (porosity 30.4%) is connected in parallel for the experiment (cores A and B are saturated with 5000 mg·L^{-1} NaCl before the experiment), and 0.1 PV 5000 mg·L^{-1} NaCl mixed with 5 mg·L^{-1} CQD-W is injected at a constant flow rate of 2 mL/min. The concentration is detected as above.

3. Results and Discussion

3.1. Characterization Analysis of CQD-W

3.1.1. Infrared Spectral Analysis of CQD-W

As shown in Figure 2, the Fourier transform infrared (FTIR) spectra indicated CQD-W exhibits abundant hydrophilic groups on its surface, including O-H (3600–3100 cm^{-1}), C=C (1525.63 cm^{-1}), C=O (1639.03 cm^{-1}) and Si-O (1208.87 cm^{-1}), thus promoting a good solubility in water. In addition, the O-H vibrational band, about 3437.14 cm^{-1}, is broad peaked, indicating that the CQD-W surface has multiple structures of hydroxyl groups, thus leading to high polarity and hydrophilicity [35,39,40]. These functional groups make the carbon quantum dots scalable and their surface tunable for a variety of applications [7,41,42].

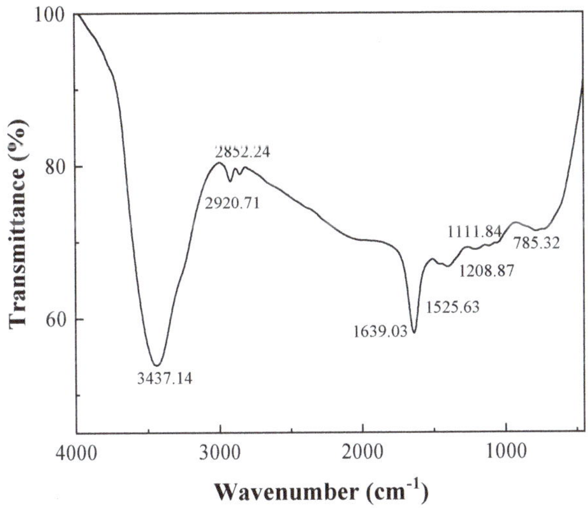

Figure 2. Fourier Transform Infrared (FTIR) Spectroscopy of CQD-W.

3.1.2. ζ-Potential Analysis of CQD-W

As shown in Figure 3, the average ζ-potential of CQD-W is 0.3 mV, primarily due to the presence of Si-O and carboxyl groups' in the CQD-W structure. In general, the carboxyl group exists in the dissociated state (-COO-). The covalent bond between Si-O is very polar, and it will attract many hydrogen ions in water to form Si-OH, that will be ionized under certain conditions [43–47].

$$Si-OH \rightleftharpoons Si-O^- + H^+ \quad (3)$$

$$Si-OH + OH^- \rightleftharpoons Si-O^- + H_2O \quad (4)$$

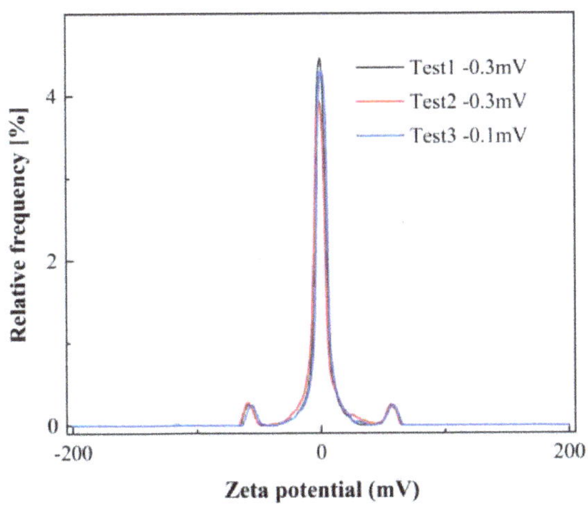

Figure 3. ζ-potential distribution of CQD-W.

3.1.3. Particle Size Test Analysis of CQD-W

As shown in Figure 4, CQD-W has a very narrow strong peak at 0~2 nm, indicating that CQD-W has good particle size and particle size homogeneity (particle size is about

1 nm). The size of CQD-W microspheres needs to be adapted to the fluid flow in the formation. According to the equation of the hydrodynamic radius of the pore throat [48]:

$$R = \frac{d}{4} = \frac{r}{2} = \frac{1}{2}\left(\frac{8K}{\phi}\right)^{0.5} \quad (5)$$

where: R—the hydrodynamic diameter of the orifice throat.

K—permeability.

ϕ—Take the value of 10%.

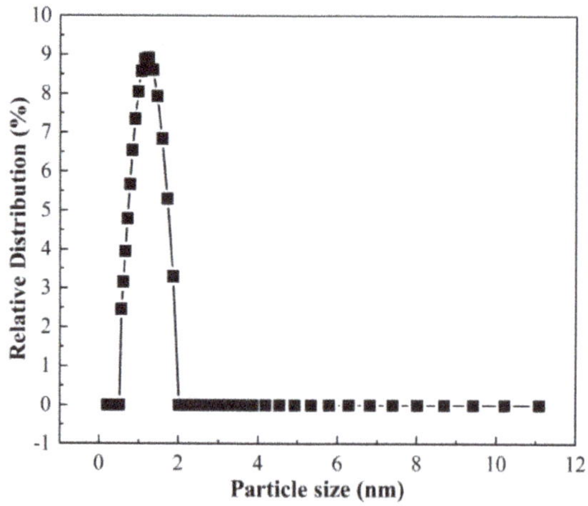

Figure 4. Particle size distribution (intensity) diagram of CQD-W.

To ensure that the tracer particles can pass through the formation pore throat smoothly, the radius of the tracer microspheres should be no larger than the hydrodynamic radius of the pore throat. For size 10 nm particles, this generally applies to reservoirs with permeability above 0.005 mD.

3.2. *Analysis of CQD-W Fluorescence Characteristics*

3.2.1. Three-Dimensional Fluorescence Characterization of CQD-W

As shown in Figure 5, CQD-W appears clear and transparent under natural light conditions and exhibits a strong fluorescence emission band (green fluorescence) under UV light irradiation.

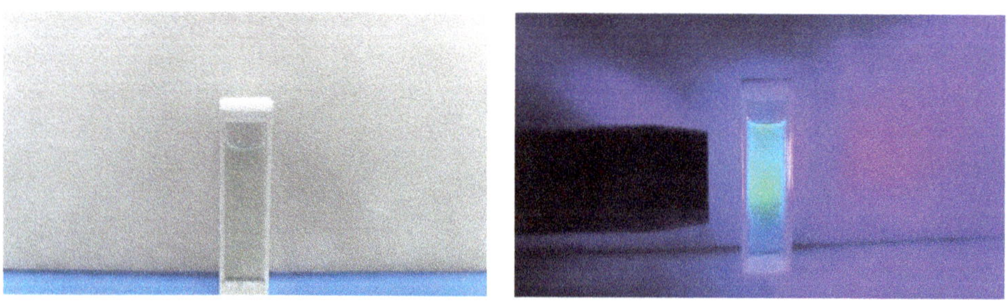

Figure 5. Color development of CQD-W under natural light (**left**) and UV light (**right**) irradiation.

As shown in Figure 6, the three-dimensional fluorescence spectra showed that the fluorescence waveform of CQD-W had a single fine shape and obvious fluorescence characteristics with good discrimination. The best excitation wavelength was 491.1 nm and the best emission wavelength was 518.9 nm.

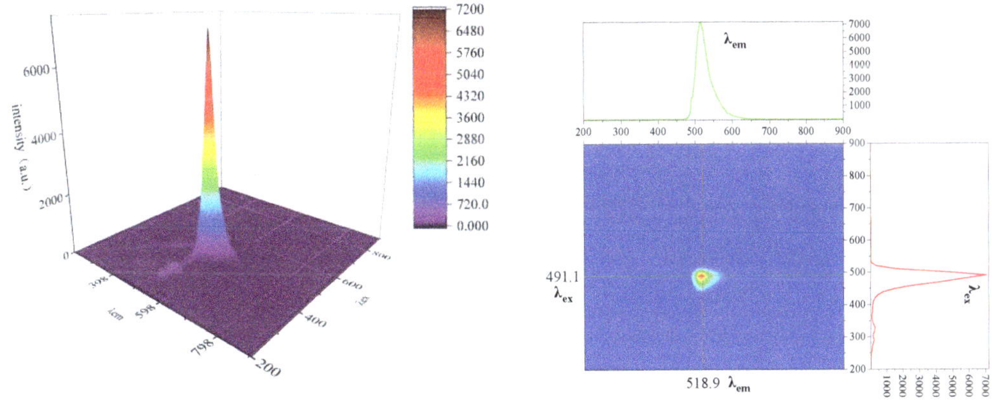

Figure 6. 3D scanning situation of CQD-W (**left**) top view profile of CQD-W (**right**).

3.2.2. Standard Curve of CQD-W

As shown in Figure 7, the concentration of CQD-W is no longer linear above 3 mg·L^{-1}. This is primarily because when the CQD-W concentration is too high due to factors such as internal filtration and energy transfer, causing the fluorescence intensity and concentration are not stable, and fluorescence quenching phenomena such as linearity and spectral distortion [49–52]. In the concentration range of 0 mg·L^{-1} to ~0.5 mg·L^{-1}, the tracer concentration has an excellent linear correlation with the fluorescence intensity (bandwidth 10, gain 9), which satisfies the Lang–Bier law. It can be used as a standard curve for concentration measurement [53,54], and its linearity equation is as follows.

$$I = -55.50 + 13697.96 \times C \ \left(R^2 = 0.998\right) \tag{6}$$

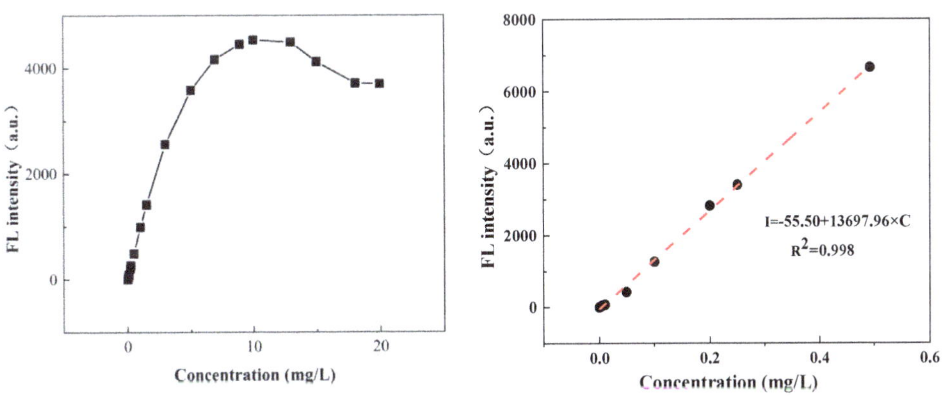

Figure 7. The concentration of CQD-W versus fluorescence intensity (bandwidth 10, gain 7).

3.3. Reservoir Suitability Analysis of CQD-W

3.3.1. Stability Analysis of CQD-W

Common ions in stratigraphic water may bind to the tracer and thus alter the fluorescence intensity of the tracer [55]. This paper examines the measurements of 0~200,000 mg·L^{-1} NaCl solution and 0~1000 mg·L^{-1} $CaCl_2$ solution in a calibrated 0.1 mg·L^{-1} tracer solution.

As shown in Table 2, the mineralization and Ca^{2+} had minimal interference with the concentration of the tracer and their concentration deviations were within 5%, indicating that the CQD-W tracer has good stability [25,35,56].

Table 2. Effect of stratigraphic ions on the stability of CQD-W.

The Concentration of NaCl (%)	The Concentration of CQD-W (mg/L)	Relative Error	The Concentration of $CaCl_2$ (mg/L)	The Concentration of CQD-W (mg/L)	Relative Error
Blank group	0.10083	0.00%	Blank group	0.10307	0.00%
0.5	0.10087	0.04%	50	0.10482	1.70%
5	0.10206	1.21%	100	0.10488	1.76%
10	0.10243	1.58%	500	0.10581	2.66%
20	0.10190	1.06%	1000	0.10693	3.75%

The concentration variation of CQD-W solution of 0.1 mg·L^{-1} in the temperature range of 30~80 °C is shown in Figure 8. The results showed that temperature and aging time influenced the measurement results, and its measurement error was basically within 10%. The maximum deviation did not exceed 8%, indicating that CQD-W has good reservoir adaptability [25,35].

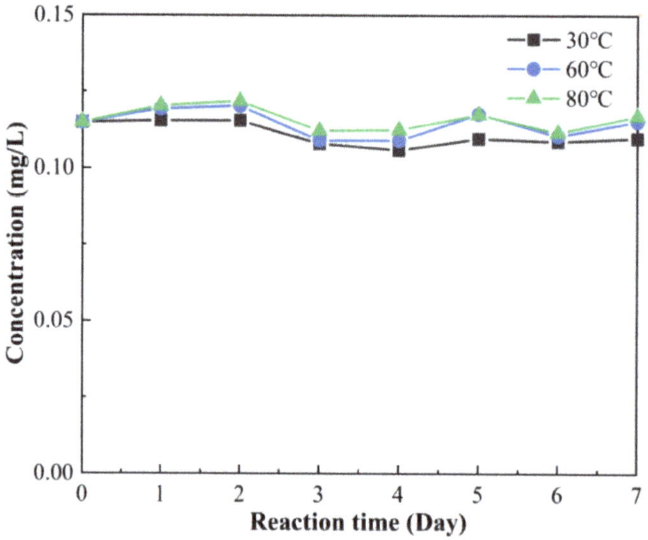

Figure 8. Changes in concentration of CQD-W after aging at different temperatures over a period of time.

As shown in Table 3, the pH value of the solution had little effect on the measurement results, and its concentration deviation was within 6%, indicating that the tracer has good pH stability [56].

Table 3. Effect of pH on concentration test of CQD-W.

pH	The Concentration of CQD-W (mg/L)	Relative Error
Blank group	0.10785	0.00%
1	0.10189	5.53%
3	0.10234	5.11%
5	0.10714	0.66%
7	0.10416	3.43%
9	0.10564	2.05%

3.3.2. Solubility of CQD-W in Oil and Water

Since formation fluids generally exist as multiphase flows with oil and water coexistence, the dissolution properties and partitioning ratio of tracer in oil and water phases significantly impact the separation methods and test results of samples [57,58]. As shown in Figure 9, the water phase part of the solution showed obvious fluorescence under UV irradiation after CQD-W and liquid wax were entirely mixed and left to stand. In contrast, the oil phase part did not show fluorescence characteristics.

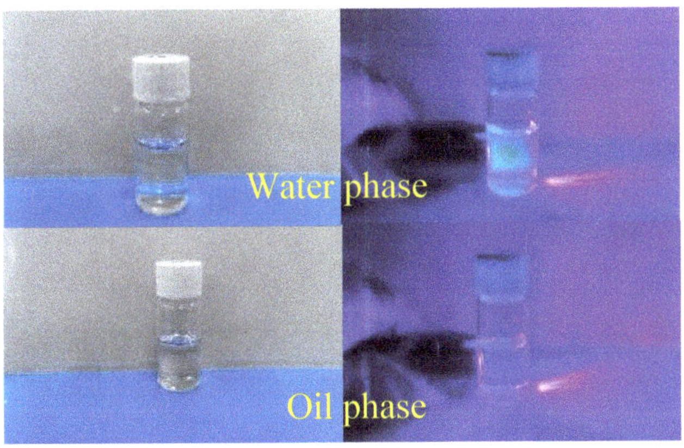

Figure 9. Solution after static stratification.

As shown in Figure 10, the fluorescence value of CQD-W changed very little after oil-water mixing, indicating that this CQD-W has good hydrophilic properties. It is because CQD-W contains many carboxyl and hydroxyl groups, that can easily form hydrogen bonds with water molecules and therefore has better water solubility [59,60]. In the actual testing process, only a simple oil-water separation of the extracted fluid is required for testing and analysis.

3.3.3. Static Adsorption Experiments of CQD-W

Tracer adsorption occurs when the tracer flows over the rock surface. This leads to tracer loss and may impact the tracer concentration detection during the actual detection process. As shown in Figure 11, the tracer concentration of CQD-W was 0.92 mg·L^{-1} after eight repeated adsorptions, with a fluctuation of 8.7% relative to the initial concentration and average adsorption of 0.0435 ug/g. This is because CQD-W is negatively charged with the formation rock and its adsorption is low under the effect of electrostatic repulsion [25].

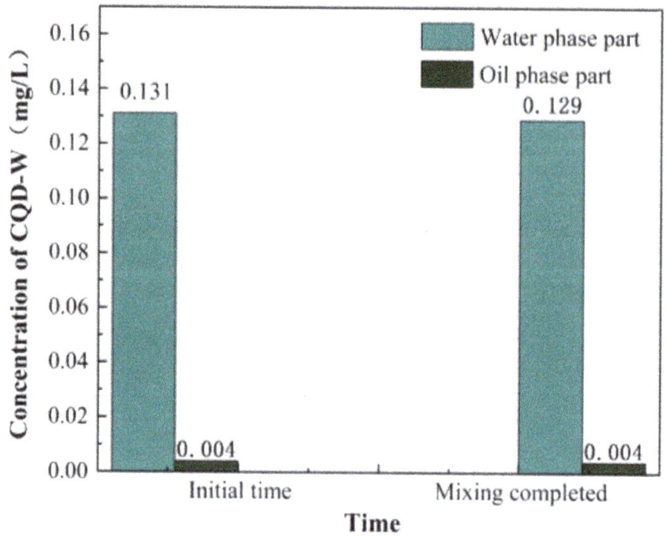

Figure 10. Change in concentration of CQD-W after static stratification of the solution.

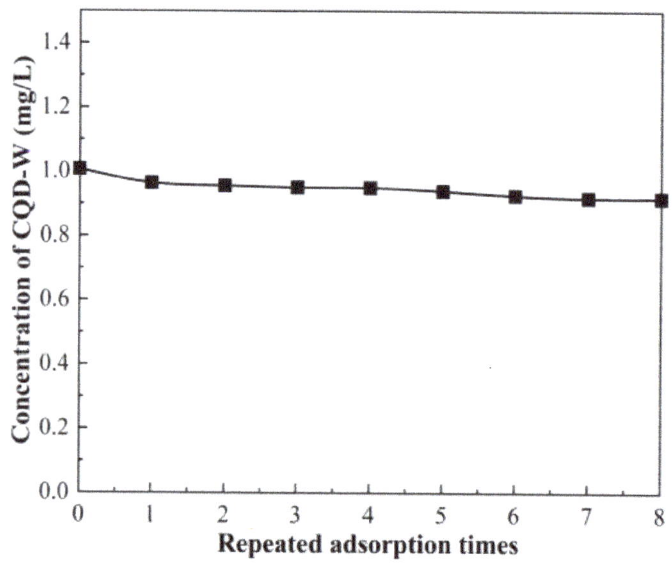

Figure 11. The concentration of CQD-W with the number of repetitions of adsorption.

3.4. *Core Flow Experiment*

3.4.1. Injectivity and Brine Flow Variability Analysis of CQD-W

As shown in Figure 12, in CQD-W and brine synchronous flow experiments, CQD-W and brine flow were consistent with each other and had good flow performance inside the core. No adsorption hysteresis was observed in the CQD-W concentration curve when compared with the brine flow curve. In the process of core flow, the pressure inside the core remained unchanged after injecting CQD-W and pure water. It indicates that the internal pore structure of the core was not changed after the flow of CQD-W, which is not easy to cause reservoir damage and other problems [25].

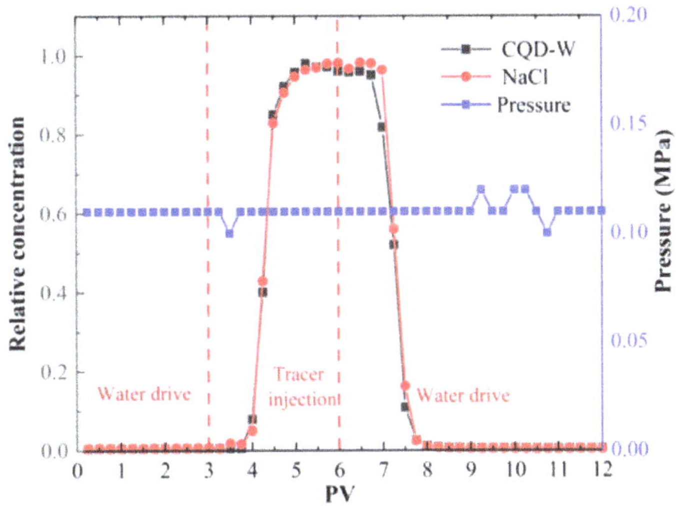

Figure 12. Relative concentration and internal pressure versus injection volume.

As shown in Figure 13, the cumulative recovery of NaCl reached 99.21%, and the cumulative recovery of CQD-W reached 96.83%. The two confirm each other, indicating that the adsorption between CQD-W and the core is minimal, primarily because of the repulsive effect between CQD-W and the negatively charged core [7,25].

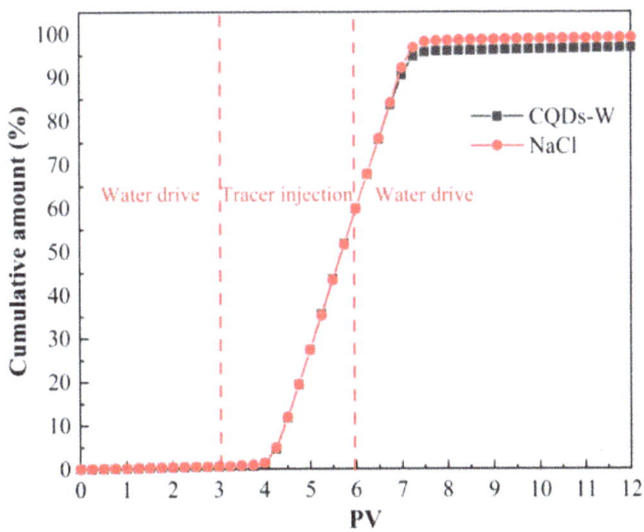

Figure 13. CQD-W cumulative volume versus injection volume.

3.4.2. Description of Cores by CQD-W

As shown in Figure 14, after injecting a 0.1 PV small segment plug CQD-W in two segments of the core with little difference in permeability, the slightly larger permeability core saw the CQD-W first. This is because the flow rate of the segment plug in the flow pipe increases after the permeability increases and the time to see the CQD-W becomes shorter. With the same plug length, the time difference between the leading and trailing edges of the plug to reach the production well becomes shorter. Therefore, the overall time to see the CQD-W becomes shorter. This is reflected in the shape of the curve, i.e., the bandwidth becomes narrower [61]. As a side effect, the CQD-W tracer can distinguish superior channels with higher permeability and has excellent tracer resolution.

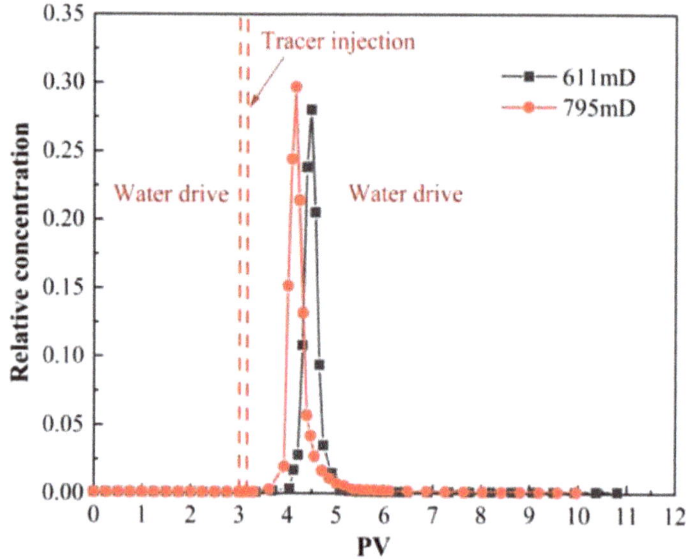

Figure 14. Relative concentration versus injection volume.

As shown in Figure 15, the cumulative recovery of low and high permeability cores reached 97.41% and 96.57%, respectively, indicating that the adsorption of CQD-W is low and meets the performance requirements for the tracer used.

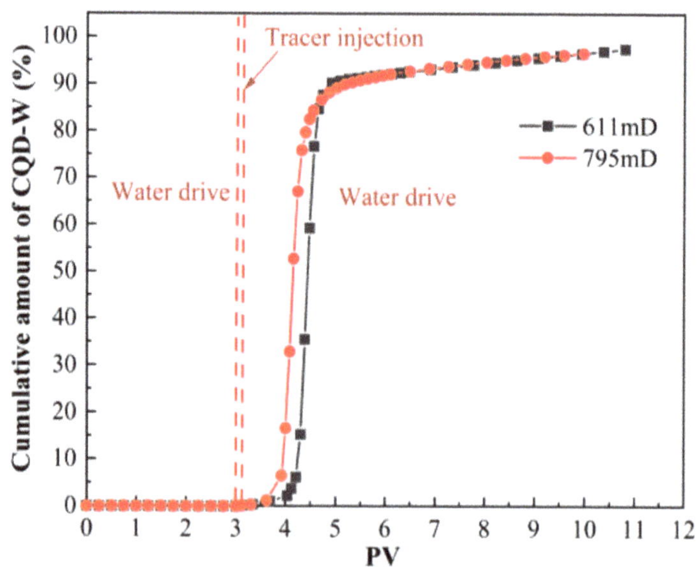

Figure 15. The cumulative volume of CQD-W versus injection volume.

As shown in Figure 16, the two sections of cores with little difference in permeability were subjected to core flow experiments. The shunt ratios of both in the pre-water drive, CQD-W injection, and subsequent water drive stages were maintained, indicating that the internal pore structure of the cores was less affected by the flow of CQD-W [62,63].

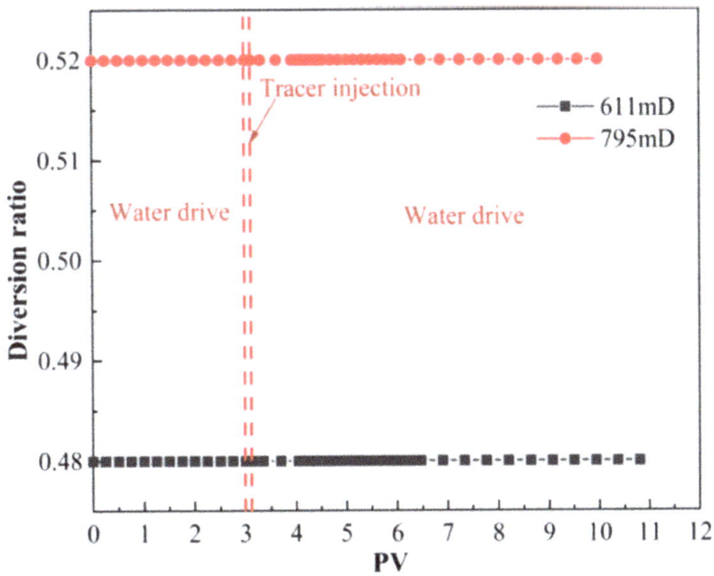

Figure 16. Relationship between shunt ratio and injection volume.

4. Conclusions

The surface of CQD-W is rich in polycarboxyl and siloxy groups with good water solubility, and CQD-W shows negative electrical properties in pure water; this makes CQD-W adsorption on the rock surface very low. The particle size characterization results show that the average particle size of CQD-Ws is 2 nm; this can meet the reservoir mobility and injectivity with permeability above 0.005 mD.

(1) CQD-W three-dimensional fluorescence spectrogram shows that the three-dimensional fluorescence peak is a single peak that has better differentiation in practical application detection. The minimum detection limit of CQD-W reaches 10^{-8}. CQD-W has the advantage of small dosage, convenient detection, and the luminescence situation of CQD-W is very little affected by temperature, mineralization, and pH.

(2) In the core flow experiment, after injecting a 0.1 PV small section plug CQD-W tracer in two sections of cores with little difference in permeability, the permeability of slightly larger cores saw the CQD-W first. This shows that CQD-W can distinguish the superior channel with higher permeability and has excellent tracer resolution. At the same time, CQD-W has an excellent flow ability and will not change the internal pore structure of the core during the flow, making it more difficult to produce reservoir damage. The cumulative recovery rate in the low and high permeability cores reached 97.41% and 96.57%, respectively, primarily because CQD-W has negative electricity and the negative core has a repulsive effect between the two.

(3) CQD-W testing and characterization, reservoir suitability analysis, and core flow experiments can verify that CQD-W meets the performance requirements of conventional tracers and, to a certain extent, proves the possibility of CQD-W's future application in oil field sites. It can also provide in-depth guidance on the preparation of target CQD-W materials and broaden the application fields of carbon quantum dots.

Author Contributions: Conceptualization, J.L.; data curation, J.C.; formal analysis, J.C. and C.W.; methodology, J.C. and C.G.; project administration, J.C. and J.D.; resources, J.L. and B.L.; supervision, J.L. and B.L.; writing—original draft, J.C.; writing—review and editing, J.L. All authors have read and agreed to the published version of the manuscript.

Funding: This research was supported by the Open Foundation of Hubei Key Laboratory of Oil and Gas Drilling and Production Engineering, Yangtze University (No. YQZC202105) and the Yangtze University 2021 Innovation and Entrepreneurship Training Program for College Students "Research on the Mechanism of Deep Drive Regulating by Resistant Polymer Microspheres" (Project No. Yz20211253).

Institutional Review Board Statement: Not applicable.

Informed Consent Statement: Not applicable.

Data Availability Statement: Not applicable.

Acknowledgments: We would like to thank the Key Laboratory of Drilling and Production Engineering for Oil and Gas of Yangtze University for their support. At the same time, I would like to thank my teacher, Jian-xin Liu, for his help.

Conflicts of Interest: The authors declare no conflict of interest.

References

1. Ma, J.; Zhang, P.; Tian, S.; Sheng, M.; Xu, Q.; Lu, Z. Carbon Dots as Fluorescent Sensitive Tracers in Reservoir Engineering. In Proceedings of the ARMA-CUPB Geothermal International Conference, Beijing, China, 5–8 August 2019.
2. Murugesan, S.; Kuznetsov, O.; Suresh, R.; Agrawal, D.; Monteiro, O.; Khabashesku, V.N. Carbon Quantum Dots Fluorescent Tracers for Production and Well Monitoring. In Proceedings of the SPE Annual Technical Conference and Exhibition, Dubai, United Arab Emirates, 26–28 September 2016.
3. Pu, C.; Jing, C.; He, Y.; Gu, X.; Zhang, Z.; Wei, J. Multistage interwell chemical tracing for step-by-step profile control of water channeling and flooding of fractured ultra-low permeability reservoirs. *Pet. Explor. Dev.* **2016**, *43*, 679–688. [CrossRef]
4. Torrealba, V.A.; Hoteit, H.; Chawathe, A. Improving Chemical-Enhanced-Oil-Recovery Simulations and Reducing Subsurface Uncertainty Using Downscaling Conditioned to Tracer Data. *SPE Reserv. Eval. Eng.* **2019**, *22*, 1426–1435. [CrossRef]
5. Hutchins, R.D.; Dovan, H.T.; Sandiford, B.B. Aqueous Tracers for Oilfield Applications. In Proceedings of the SPE International Symposium on Oilfield Chemistry, Anaheim, CA, USA, 20–22 February 1991.
6. Zhao, B.; Panthi, K.; Mohanty, K.K. Tracer eluting proppants for hydraulic fracture characterization. *J. Pet. Sci. Eng.* **2020**, *190*, 107048. [CrossRef]
7. Franco, C.A.; Candela, C.H.; Gallego, J.; Marin, J.; Patiño, L.E.; Ospina, N.; Patiño, E.; Molano, M.; Villamil, F.; Bernal, K.M.; et al. Easy and Rapid Synthesis of Carbon Quantum Dots from Mortiño (Vaccinium Meridionale Swartz) Extract for Use as Green Tracers in the Oil and Gas Industry: Lab-to-Field Trial Development in Colombia. *Ind. Eng. Chem. Res.* **2020**, *59*, 11359–11369. [CrossRef]
8. Sharma, A.N.; Luo, D.; Walter, M.T. Hydrological Tracers Using Nanobiotechnology: Proof of Concept. *Environ. Sci. Technol.* **2012**, *46*, 8928–8936. [CrossRef]
9. Maldonado-Domínguez, M.; González-Antonio, O.; Díaz-Paulino, L.E.; Cañas-Alonso, R.C.; Xochitiotzi-Flores, E.; Jiménez-García, C.A.; Almaraz-Girón, M.A.; Luna-Pabello, V.M.; Romero-Ávila, M.; Barragán-Aroche, F.; et al. Passive organic cationic tracers for carbonate formations: Development and performance at oil reservoir emulating conditions. *Fuel* **2021**, *304*, 121377. [CrossRef]
10. Xu, X.; Ray, R.; Gu, Y.; Ploehn, H.J.; Gearheart, L.; Raker, K.; Scrivens, W.A. Electrophoretic Analysis and Purification of Fluorescent Single-Walled Carbon Nanotube Fragments. *J. Am. Chem. Soc.* **2004**, *126*, 12736–12737. [CrossRef]
11. Xu, X.; Caswell, K.K.; Tucker, E.; Kabisatpathy, S.; Brodhacker, K.L.; Scrivens, W.A. Size and shape separation of gold nanoparticles with preparative gel electrophoresis. *J. Chromatogr. A* **2007**, *1167*, 35–41. [CrossRef]
12. Jana, J.; Lee, H.J.; Chung, J.S.; Kim, M.H.; Hur, S.H. Blue emitting nitrogen-doped carbon dots as a fluorescent probe for nitrite ion sensing and cell-imaging. *Anal. Chim. Acta* **2019**, *1079*, 212–219. [CrossRef]
13. Pichardo-Molina, J.L.; Cardoso-Avila, P.E.; Flores-Villavicencio, L.L.; Gomez-Ortiz, N.M.; Rodriguez-Rivera, M.A. Fluorescent carbon nanoparticles synthesized from bovine serum albumin nanoparticles. *Int. J. Biol. Macromol.* **2020**, *142*, 724–731. [CrossRef]
14. Wang, T.; Luo, H.; Jing, X.; Yang, J.; Huo, M.; Wang, Y. Synthesis of Fluorescent Carbon Dots and Their Application in Ascorbic Acid Detection. *Molecules* **2021**, *26*, 1246. [CrossRef] [PubMed]
15. Huang, S.-W.; Lin, Y.-F.; Li, Y.-X.; Hu, C.-C.; Chiu, T.-C. Synthesis of Fluorescent Carbon Dots as Selective and Sensitive Probes for Cupric Ions and Cell Imaging. *Molecules* **2019**, *24*, 1785. [CrossRef] [PubMed]
16. Tripathi, K.M.; Tran, T.S.; Tung, T.T.; Losic, D.; Kim, T. Water Soluble Fluorescent Carbon Nanodots from Biosource for Cells Imaging. *J. Nanomater.* **2017**, 7029731. [CrossRef]

17. Tripathi, K.M.; Tran, T.S.; Kim, Y.J.; Kim, T. Green Fluorescent Onion-Like Carbon Nanoparticles from Flaxseed Oil for Visible Light Induced Photocatalytic Applications and Label-Free Detection of Al(III) Ions. *ACS Sustain. Chem. Eng.* **2017**, *5*, 3982–3992. [CrossRef]
18. Li, L.; Zhao, F.; Zhang, T.; Lü, C. A facile method to prepare polymer functionalized carbon dots inspired by the mussel chemistry for LED application. *Dye. Pigment.* **2019**, *162*, 845–854. [CrossRef]
19. Sai, L.; Liu, S.; Qian, X.; Yu, Y.; Xu, X. Nontoxic fluorescent carbon nanodot serving as a light conversion material in plant for UV light utilization. *Colloids Surf. B: Biointerfaces* **2018**, *169*, 422–428. [CrossRef]
20. Nasrollahzadeh, M.; Sajjadi, M.; Iravani, S.; Varma, R.S. Carbon-based sustainable nanomaterials for water treatment: State-of-art and future perspectives. *Chemosphere* **2021**, *263*, 128005. [CrossRef]
21. Zhao, D.L.; Chung, T.-S. Applications of carbon quantum dots (CQDs) in membrane technologies: A review. *Water Res.* **2018**, *147*, 43–49. [CrossRef]
22. Singh, R.K.; Kumar, R.; Singh, D.P.; Savu, R.; Moshkalev, S.A. Progress in microwave-assisted synthesis of quantum dots (graphene/carbon/semiconducting) for bioapplications: A review. *Mater. Today Chem.* **2019**, *12*, 282–314. [CrossRef]
23. Langer, M.; Paloncýová, M.; Medved', M.; Pykal, M.; Nachtigallová, D.; Shi, B.; Aquino, A.J.A.; Lischka, H.; Otyepka, M. Progress and challenges in understanding of photoluminescence properties of carbon dots based on theoretical computations. *Appl. Mater. Today* **2021**, *22*, 100924. [CrossRef]
24. Sharma, S.; Umar, A.; Sood, S.; Mehta, S.K.; Kansal, S.K. Photoluminescent C-dots: An overview on the recent development in the synthesis, physiochemical properties and potential applications. *J. Alloy. Compd.* **2018**, *748*, 818–853. [CrossRef]
25. Shi, X.; Yan, L.; Fan, S.; Huang, Y.; Xu, H.; Lai, N. Indoor synthesis of carbon quantum dots and its potential applications study as tracers in oilfields. *J. Pet. Sci. Eng.* **2022**, *213*, 110325. [CrossRef]
26. Tang, X.; Yu, H.; Bui, B.; Wang, L.; Xing, C.; Wang, S.; Chen, M.; Hu, Z.; Chen, W. Nitrogen-doped fluorescence carbon dots as multi-mechanism detection for iodide and curcumin in biological and food samples. *Bioact. Mater.* **2021**, *6*, 1541–1554. [CrossRef]
27. Kumar, A.; Asu, S.; Mukherjee, P.; Singh, P.; Kumari, A.; Sahu, S.K. Single-step synthesis of N-doped carbon dots and applied for dopamine sensing, in vitro multicolor cellular imaging as well as fluorescent ink. *J. Photochem. Photobiol. A Chem.* **2021**, *406*, 113019. [CrossRef]
28. Zhu, P.; Tan, K.; Chen, Q.; Xiong, J.; Gao, L. Origins of Efficient Multiemission Luminescence in Carbon Dots. *Chem. Mater.* **2019**, *31*, 4732–4742. [CrossRef]
29. Sun, Z.; Li, X.; Wu, Y.; Wei, C.; Zeng, H. Origin of green luminescence in carbon quantum dots: Specific emission bands originate from oxidized carbon groups. *New J. Chem.* **2018**, *42*, 4603–4611. [CrossRef]
30. Bao, L.; Liu, C.; Zhang, Z.L.; Pang, D.W. Photoluminescence-tunable carbon nanodots: Surface-state energy-gap tuning. *Adv. Mater.* **2015**, *27*, 1663–1667. [CrossRef]
31. Myint, A.A.; Rhim, W.-K.; Nam, J.-M.; Kim, J.; Lee, Y.-W. Water-soluble, lignin-derived carbon dots with high fluorescent emissions and their applications in bioimaging. *J. Ind. Eng. Chem.* **2018**, *66*, 387–395. [CrossRef]
32. Zhong, Y.; Xue, F.; Wei, P.; Li, R.; Cao, C.; Yi, T.J.N. Water-soluble MoS_2 quantum dots for facile and sensitive fluorescence sensing of alkaline phosphatase activity in serum and live cells based on the inner filter effect. *Nanoscale* **2018**, *10*, 21298–21306. [CrossRef]
33. Yang, X.-C.; Li, Q.; Tang, M.; Yang, Y.-L.; Yang, W.; Hu, J.-F.; Pu, X.-L.; Liu, J.; Zhao, J.-T.; Zhang, Z.-J. One Stone, Two Birds: pH- and Temperature-Sensitive Nitrogen-Doped Carbon Dots for Multiple Anticounterfeiting and Multiple Cell Imaging. *ACS Appl. Mater. Interfaces* **2020**, *12*, 20849–20858. [CrossRef]
34. Panwar, N.; Soehartono, A.M.; Chan, K.K.; Zeng, S.; Xu, G.; Qu, J.; Coquet, P.; Yong, K.-T.; Chen, X. Nanocarbons for Biology and Medicine: Sensing, Imaging, and Drug Delivery. *Chem. Rev.* **2019**, *119*, 9559–9656. [CrossRef]
35. Hu, Z.; Gao, H.; Ramisetti, S.B.; Zhao, J.; Nourafkan, E.; Glover, P.W.J.; Wen, D. Carbon quantum dots with tracer-like breakthrough ability for reservoir characterization. *Sci. Total Environ.* **2019**, *669*, 579–589. [CrossRef]
36. Hassanzadeh, S.; Adolfsson, K.H.; Hakkarainen, M. Controlling the cooperative self-assembly of graphene oxide quantum dots in aqueous solutions. *RSC Advances*, 2015; 5, 57425–57432.
37. Jiang, H.-Y.; Qin, S.-Q.; Chen, J.-J.; Dai, J.-J.; Lin, B.; Gao, C.-Y.; Ruan, X.-X.; Yang, Y.-Q.; Liu, J.-X. Reservoir Adaptability Evaluation of Core-shell Fluorescent Nanotracers. *Contemporary Chemical Industry.* **2022**, *51*, 1572–1576.
38. Guo, Y.-J.; Liu, J.-X.; Zhang, X.-M.; Feng, R.-S.; Li, H.-B.; Zhang, J.; Lv, X.; Luo, P.-Y. Solution Property Investigation of Combination Flooding Systems Consisting of Gemini–Non-ionic Mixed Surfactant and Hydrophobically Associating Polyacrylamide for Enhanced Oil Recovery. *Energy Fuels* **2012**, *26*, 2116–2123. [CrossRef]
39. Ding, H.; Yu, S.-B.; Wei, J.-S.; Xiong, H.-M. Full-Color Light-Emitting Carbon Dots with a Surface-State-Controlled Luminescence Mechanism. *ACS Nano* **2016**, *10*, 484–491. [CrossRef]
40. Li, Y.-X.; Lee, J.-Y.; Lee, H.; Hu, C.-C.; Chiu, T.-C. Highly fluorescent nitrogen-doped carbon dots for selective and sensitive detection of Hg^{2+} and ClO^- ions and fluorescent ink. *J. Photochem. Photobiol. A Chem.* **2021**, *405*, 112931. [CrossRef]
41. Hwang, C.-C.; Wang, L.; Lu, W.; Ruan, G.; Kini, G.C.; Xiang, C.; Samuel, E.L.; Shi, W.; Kan, A.T.; Wong, M.S. Highly stable carbon nanoparticles designed for downhole hydrocarbon detection. *Energy Environ. Sci.* **2012**, *5*, 8304–8309. [CrossRef]
42. Hwang, C.-C.; Ruan, G.; Wang, L.; Zheng, H.; Samuel, E.L.; Xiang, C.; Lu, W.; Kasper, W.; Huang, K.; Peng, Z. Carbon-based nanoreporters designed for subsurface hydrogen sulfide detection. *ACS Appl. Mater. Interfaces* **2014**, *6*, 7652–7658. [CrossRef]
43. Kim, D.H.; Suh, J.H.; Ju, J.Y.; Yum, S.K.; Shin, C.S. Analysis of sisomicin binding sites in Micromonospora inyoensis cell wall. *FEMS Microbiol. Lett.* **1998**, *166*, 9–13. [CrossRef]

44. Yang, J.; Tan, Y.-Z.; Wang, Z.-H.; Shang, Y.-D.; Zhao, W.-B. Study on the coal dust surface characteristics and wetting mechanism. *J. China Coal Society.* **2007**, *7*, 737–740.
45. Liu, C.; Cao, X.; Chen, C.; Hu, Y. Study on the flotation performance of dodeca-tertiary amine series trap for quartz. *Min. Metall. Eng.* **2009**, *29*, 37–39.
46. Liao, M.; Huang, C.Y.; Xie, C.M. Effect of pH on the transport and morphology of cadmium in soil-water systems. *J. Environ. Sci.* **1999**, *19*, 81–86.
47. Wang, X.-X.; Luo, T.; Cheng, N.; Wang, H.-H.; Huang, J.; Li, Y.-J.; Ren, H.; Liu, X.; Yu, D.-H. Study on Liquid Paraffin Pickering Emulsion Stabilized with Fluorescent Carbon Quantum Dots s and Laponite. *J. Qilu Univ. Technology.* **2018**, *32*, 18–25.
48. Yang, S.L. *Petrophysics*; Petroleum Industry Press: Beijing, China, 2004.
49. Das, R.; Rajender, G.; Giri, P.K. Anomalous fluorescence enhancement and fluorescence quenching of graphene quantum dots by single walled carbon nanotubes. *Phys. Chem. Chem. Phys.* **2018**, *20*, 4527–4537. [CrossRef]
50. Luo, T.; Zhou, T.; Zhao, Y.; Liu, L.; Qu, J. Multiplexed fluorescence lifetime imaging by concentration-dependent quenching. *J. Mater. Chem. B* **2018**, *6*, 1912–1919. [CrossRef]
51. Sun, C.; Zhang, Y.; Sun, K.; Reckmeier, C.J.; Zhang, T.-Q.; Zhang, X.; Zhao, J.; Wu, C.; Yu, W.W.; Rogach, A.L. Combination of carbon dot and polymer dot phosphors for white light-emitting diodes. *Nanoscale* **2015**, *7*, 12045–12050. [CrossRef]
52. Lv, J.; Yong, K.L. Identification of sudden extinction phenomenon in three-dimensional fluorescence analysis of crude oil. *Anal. Lab.* **1998**, *17*, 28–31.
53. The applicability of Lambert-Beer's law. Documenta ophthalmologica. *Adv. Ophthalmol.* **1974**, *38*, 279–282.
54. Maentele, W.; Deniz, E. UV-VIS absorption spectroscopy: Lambert-Beer reloaded. *Spectrochim. Acta Part A—Mol. Biomol. Spectrosc.* **2017**, *173*, 965–968. [CrossRef]
55. Guo, Y.; Cao, F.; Qiu, P.; Wang, Z. Studies of the effect of halide ions on the fluorescence of quinine sulfate. *Luminescence* **2019**, *34*, 450–455. [CrossRef]
56. Gao, X.; Zhang, Y.; Fu, Z.; Cui, F. One step synthesis of ultra-high quantum yield fluorescent carbon dots for "on-off-on" detection of Hg(2+) and biothiols. *J. Fluoresc.* **2022**, *32*, 1921–1930. [CrossRef]
57. Hou, C. Detection methods for non-distributable tracer and distributable tracer concentrations. *Sci. Technol. Innov. Her.* **2011**, *21*, 116.
58. Galeev, R.I.; Bolotov, A.V.; Varfolomeev, M.A.; Mukhutdinova, A.R.; Smirnov, A.E.; Kornilov, A.V.; Kruglov, D.S.; Zhirov, A.V.; Sansiev, G.V.; Fedorchenko, G.D. New and simple methods of determination partition coefficient and degree hydrolysis of tracer for estimating residual oil saturation by SWCTT technologies. *Pet. Sci. Technol.* **2021**, *39*, 1043–1059. [CrossRef]
59. Yuan, Y.; Liu, R.; Wang, C.; Luo, J.; Liu, X. Synthesis of UV-curable acrylate polymer containing sulfonic groups for anti-fog coatings. *Prog. Org. Coat.* **2014**, *77*, 785–789. [CrossRef]
60. Lai, Y.; Tang, Y.; Gong, J.; Gong, D.; Chi, L.; Lin, C.; Chen, Z. Transparent superhydrophobic/superhydrophilic TiO_2-based coatings for self-cleaning and anti-fogging. *J. Mater. Chem.* **2012**, *22*, 7420–7426. [CrossRef]
61. Chen, G.; Lin, C.; Jiang, R.; Zhang, W.; Li, Q.; Li, G. Inter-well tracer interpretation model for fractured low-permeability reservoirs based on flowtube model. *Logging Technol.* **2018**, *42*, 200–204.
62. Lv, P.; Liu, Y.; Zhang, Y.; Li, Y.; Xia, H. Experimental research on novel oil displacement and profile control system for heterogeneous reservoir. *J. Pet. Explor. Prod. Technol.* **2020**, *10*, 481–485. [CrossRef]
63. Li, S.; Li, Z.; Li, B.F. Experimental study of the effect of permeability on foam diversion. *Pet. Sci. Technol.* **2012**, *30*, 1907–1919. [CrossRef]

eng

MDPI

Article

Aperture Design Optimization of Wire-Wrapped Screens for SAGD Production Wells

Jesus David Montero Pallares [1,2], Chenxi Wang [1,3], Mohammad Haftani [1] and Alireza Nouri [1,*]

1 Department of Civil and Environmental Engineering, University of Alberta, Edmonton, AB T6G 2E3, Canada
2 Gran Tierra Energy, Calgary, AB T2P 0R3, Canada
3 China National Offshore Oil Corporation (CNOOC) Research Institute Ltd., China National Offshore Oil Corporation, Beijing 100028, China
* Correspondence: anouri@ualberta.ca

Abstract: Wire-wrapped screens have been established as one of the primary sand control devices in Steam-Assisted Gravity Drainage (SAGD) wells due to the high open-to-flow area and superior plugging attributes. However, their design is still a point of interest for thermal operations. Generally, existing approaches rely on one or more particular points of reservoir sands' particle size distribution (PSD) and rules of thumb inferred from other devices like the slotted liners. This study used Sand Retention Testing (SRT) to analyze the performance of WWS under various testing conditions, which were neglected in the current design criteria. The experimental investigation leads to a set of graphical design criteria that provide an optimum aperture size window. The results show that the sand retention performance of WWS is highly dependent on the flow velocities of the wetting phase. Moreover, the testing showed satisfactory plugging performance of WWS even with narrow aperture sizes, proving a superior performance for low-quality oil sands.

Keywords: wire wrapped screen; design criteria; traffic light system; SAGD; sand control

Citation: Pallares, J.D.M.; Wang, C.; Haftani, M.; Nouri, A. Aperture Design Optimization of Wire-Wrapped Screens for SAGD Production Wells. *Eng* **2023**, *4*, 1058–1070. https://doi.org/10.3390/eng4020062

Academic Editors: Reza Rezaee and Yujie Yuan

Received: 1 March 2023
Revised: 27 March 2023
Accepted: 30 March 2023
Published: 6 April 2023

1. Introduction

SAGD is the primary technology to extract heavy oil from oil sands. The continuous steam injection from the upper well generates a growing steam chamber that contacts and transfers heat to the bitumen. Consequently, oil viscosity is reduced, which allows the gravity-assisted flow of melted bitumen and condensed steam toward the lower production well [1] (Figure 1a,b).

Oil sands are geologically young formations typically found at shallow depths (up to 450 m) [2]. The inherent loose characteristic of oil sands requires sand control methods to avoid damage in surface and downhole facilities [3]. Sand particles are erosive at high flow velocities [4], and their accumulation in horizontal wells can obstruct the free flow of production fluids, leading to remedial workover operations and high treatment costs [5]. Over the years, SAGD operators have employed Standalone Screens (SAS) to provide sand control and mechanical support to the wellbore. Slotted Liner (SL), Wire-Wrapped Screen (WWS), and Punched Screens (PS) are the most common sand control devices (SCD) in SAGD wells (Figure 1c,e).

The screen would form sand bridges over the slots, providing retention of sand grains [6]. The relatively low flow rates in SAGD wells prompted the operators to opt for SL as a low-cost option with robust mechanical integrity [7,8]. However, SL may exhibit severe plugging tendencies and corrosion issues, resulting in a gradual shift from SL to higher OFA screens, such as WWS and PS [9,10]. WWS consists of a continuous profiled wire wrapped onto a base pipe (Figure 1d). The wires typically have a trapezoid cross-section and are supported by ribs or rods in direct contact with a perforated pipe. The hole density on the base pipe affects the integrity and strength of the screen, which is out of the scope of the current study. WWS provides a high OFA, from 6% to 15%, compared to 2–3% for SL and 3–8% for PS [11].

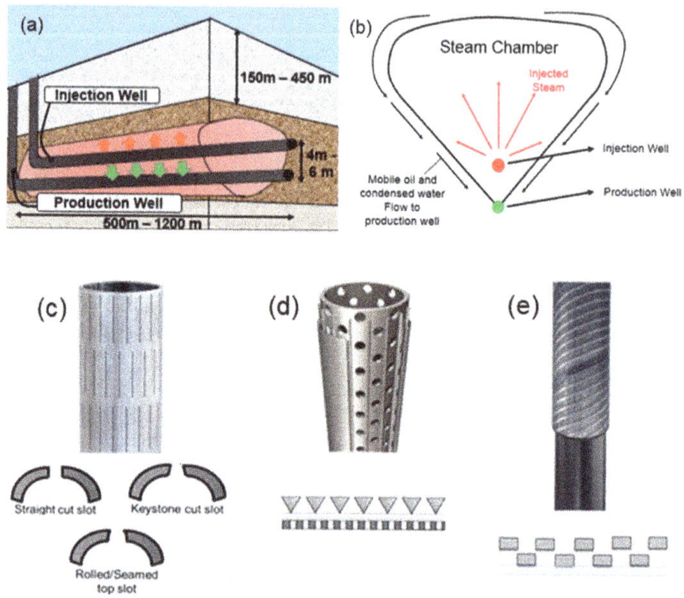

Figure 1. (**a**) A schematic of the SAGD process, (**b**) steam chamber, (**c**–**e**) the depiction of sand control devices and their respective slot geometry, (**c**) Slotted Liner, (**d**) Wire-Wrapped Screens, and (**e**) Punched Screens.

A sand control screen optimization effort minimizes sand production while keeping an acceptable flow performance throughout wellbore life [7]. Generally, the screen aperture is the crucial parameter in sanding [8,12], whereas OFA dominates the flow response of the device [13]. The lower the OFA, the faster a device can achieve severe plugging. WWS offers the benefit of high OFA and has proven more effective for low-quality sands with active clay components [14]. Increasing the slot density for SL generates a slight increase in OFA, while varying the aperture size in WWS can drastically escalate the OFA.

This study aims to elaborate a new set of criteria using the "Traffic Light System" (TLS) method to identify the optimum aperture window for the McMurray Formation PSD classes. The criteria employ the experimental data to discretize the performance into different production scenarios, accounting for sanding and flow performance. The operational procedure incorporates various flow velocities and fluid phases to gain an insight into the WWS performance.

2. Existing Aperture Design Criteria

Over time, several authors have tried to develop criteria for the sand control screen aperture selection. Coberly [15] introduced 2D10 as the maximum aperture at which stable bridges can form. D10 represents the sieve size that retains 10% of the material mass. Suman et al. [6] proposed a more conservative criterion (≤D10) but still recognized the importance of larger particles on bridge stability. Initial groundwater applications of WWS followed a sizing recommendation of D40 [16]. Subsequent criteria continued to rely on one single point of the PSD. For instance, Gillespie et al. [17] proposed 2D50 as the upper limit for WWS aperture size based on slurry SRT for different sand classes. Ballard and Beare [18] combined pre-packed and slurry SRT results to suggest aperture sizing on the D30 and stated that it provides better results than the D10 criteria. Likewise, Weatherford guidelines recommend D25 for WWS. Fattahpour et al. [8] presented a rule of thumb for SAGD wells after comparing experimental data with field applications, selecting WWS with apertures 0.004 to 0.008 inches less than the equivalent SL selection for the same sand. The existing aperture design criteria for WWS do not consider operational conditions, fluid properties, stress levels, or PSD shape.

3. Experimental Setup and Procedure

The investigation employs a pre-packed SRT that emulates the high-porosity zone formed around the sand control screen after borehole collapse. The experiment mimics the conditions around SAGD production wells using reasonable flow rates, fluid ratios, and production scenarios. All tests were conducted with 60 psi of axial stress to replicate the early stages of the near-wellbore region with relatively low effective stresses [19].

3.1. Experimental Apparatus

The pre-packed SRT equipment (Figure 2a) encompasses five units: (1) cell and accessories, (2) fluids injection unit, (3) data acquisition system, (4) collection and back-pressure units, and (5) load frame.

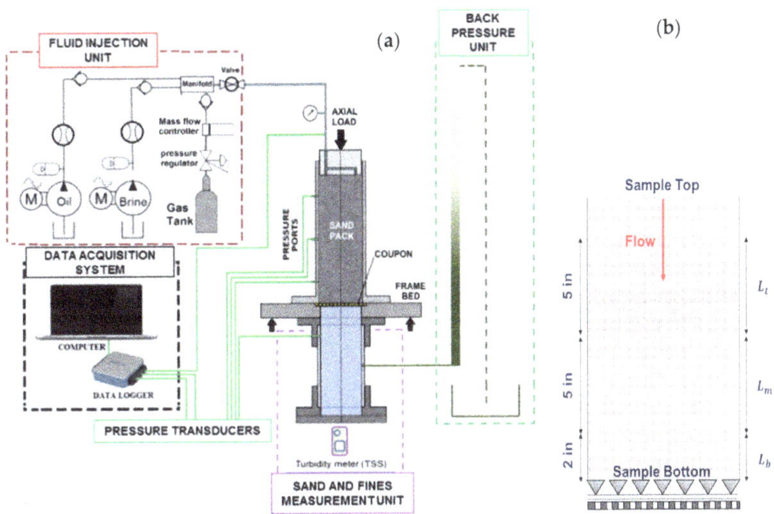

Figure 2. (**a**) Pre-packed SRT, schematic, (**b**) sand-pack sections. L_b, L_m, and L_t represent the length of near-screen, middle, and top intervals.

The core holder, 6 inches in diameter and 18.5 inches in length, accommodates the sand pack sitting over the sand control screen samples. Three connection points along the cell allow recording pressure drop evolution (Figure 2b) (differential transducers with 0.25% accuracy). The near-screen zone is defined as the 2-inch interval of sand above the screen. The other points are located 7 and 12 inches above the screen. Flow is injected from the top to the bottom of the sand pack and is then directed toward the back-pressure column, which provides 3 psi pressure at the sample bottom. Produced particles at each step are captured and accumulated in the sand trap.

3.2. Testing Materials

Several tests were performed on three representative PSD classes from the McMurray formation (DC-I to DC-III) to evaluate the response of WWS under different sand characteristics. DC-III is considered medium-coarse sand, while DC-II and DC-I are fine and very fine sands, respectively. The porosity of sand samples is about 35%. Synthetic sand mixtures using commercial sands, silts, and clays are employed to replicate the PSDs categorized by Abram and Cain [20]. Figure 3 compares commercial and actual formation sands. Table 1 describes each sample's detailed D-values and shape factors (uniformity coefficient and sorting coefficient). Mahmoudi et al. [21] showed that commercial mixtures also display similar strength properties and shape factors as the formation sands. Kaolinite was used as the clay mineral in the samples as it is the dominant clay in the McMurray Formation [21].

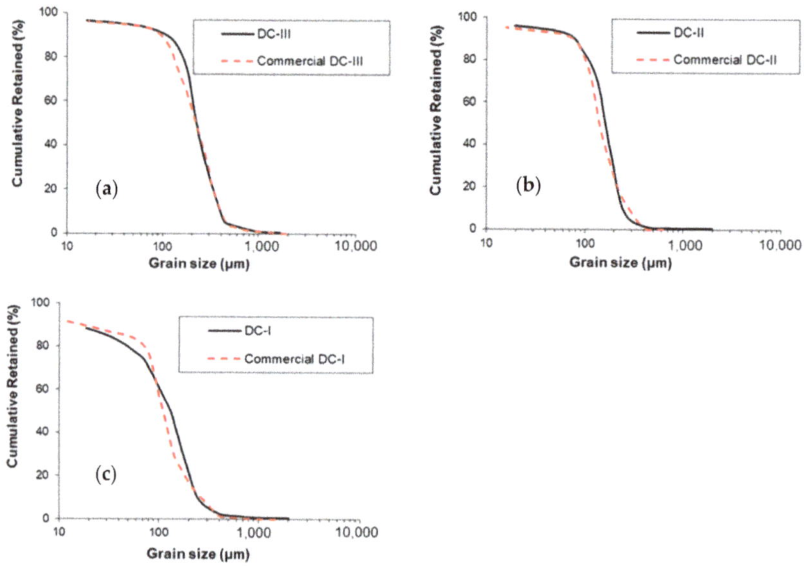

Figure 3. PSD of formation sands and corresponding synthetic samples; (**a**) DC-III, (**b**) DC-II, (**c**) DC-I.

Table 1. PSD Characteristics of the McMurray Formation Sands.

Sand	D90	D70	D50	D40	D10	Uniformity Coefficient	Sorting Coefficient
DC-I	25	80	135	147	232	5.9	9.3
DC-II	76	118	175	205	260	2.7	3.4
DC-III	110	187	215	264	341	2.4	3.1

Uniformity Coefficient (UC = D40/D90), Sorting Coefficient (SC = D10/D90)

Figure 4a introduces the test matrix that covers existing criteria to incorporate both sanding and plugging conditions. Figure 4b shows a schematic with the design parameters of WWS coupons with 6 inches diameter. WWS coupons correspond to 6-inch-diameter disks cut from the screen. The performance of each sand is evaluated for different aperture sizes.

Haftani et al. [22] summarized the reported pH values of produced water from several SAGD projects, indicating the pH range from 7.3 to 8.8 with a wide range of salinity values. In this study, 7.9 is selected as the pH of the brine phase in all tests. Na^+ and Cl^- are the dominant ions encountered in produced water from SAGD wells, and 400 ppm was found to be the lowest value in SAGD wells [22], representing the worst-case scenario for fines migration.

Mineral oil with 8-cp viscosity is used to emulate bitumen at downhole temperature conditions [7]. Nitrogen is the gas phase used to represent steam-breakthrough episodes.

Aperture Size	PSD		
0.006"	DC-I	DC-II	DC-III
0.010"	DC-I	DC-II	DC-III
0.014"	DC-I	DC-II	DC-III
0.018"	DC-I	DC-II	DC-III
0.022"	-	-	DC-III

(a)

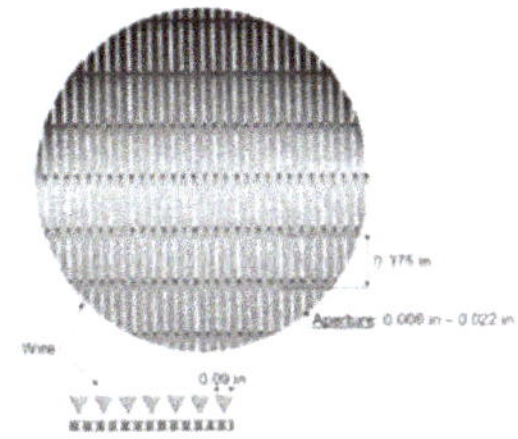

(b)

Figure 4. (**a**) Testing program and (**b**) schematic of WWS coupons with design specifications.

3.3. Testing Procedure

Sand preparation starts with the dry mixing of commercial sands, silts, and clay. Brine at 10% weight of the sand is then added and thoroughly mixed with the dry sand. The wet sand is packed inside the test cell following a layer-by-layer technique known as the moist tamping method [23] to ensure a uniform porosity and permeability distribution along the sand pack. Once the sand-pack reaches the top, the top platen is installed, and the load piston applies 60 psi of axial stress. Next, the sand pack is saturated from bottom to top at a low flow rate to avoid permeability damage in the sample.

The absolute permeability of the sample is measured before brine displacement by oil. Three differential readings across the sample confirm an even permeability. Oil is injected from the sample top towards the coupon at 1250 cc/h to displace the brine and reach irreducible water saturation, emulating the reservoir's initial saturation condition [7].

The test includes the injection of single-, two-, and multi-phase flow stages (Figure 5). The duration of each stage depends on the distribution of flowing phases and pressure stabilization to achieve a steady-state flow. Three flow rate levels are designed to account for scenarios like reservoir heterogeneity, aperture plugging, and non-uniform flow distribution. The maximum liquid flow rate (7200 cc/h) corresponds to 4000 bbl/day of liquid production from an 800-m SAGD production well equipped with a 7-inch screen, where only 15% of the well contributes to the flow. The medium- and minimum-flow rates correspond to scenarios where 50% and 70% of the well contribute to the flow, respectively.

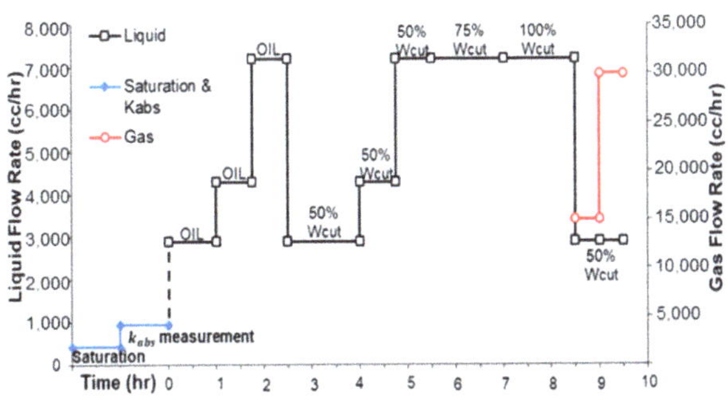

Figure 5. Fluid injection program.

The three single-phase oil stages in Figure 5 emulate sections of the well that produce high oil cuts. Next, the injection of brine and oil captures the changing conditions of SAGD wells, where different liquid rates and water cuts are experienced throughout the wellbore life [24]. The flow scheme includes a single-phase brine stage to account for high water cuts, representing the worst-case scenario for sand production and fines migration [25]. Lastly, two stages of co-injection gas (N_2), brine and oil, resemble the event of steam breakthrough. Although the precise influx conditions during such episodes are not known, the liquid rates are dropped to the initial level (total liquid rate of 2900 cc/h) since it is anticipated that high steam mobility may restrict liquid inflow.

4. Testing Program and Evaluation Method

Cumulative sand production and retained permeability are the sand retention and flow performance indicators, respectively. Produced sand is reported in pounds per square foot (lb/ft^2) of the coupon area. Retained permeability is evaluated after the last liquid stage (100% brine flow) as the ratio of effective permeability at the near-screen zone over initial effective permeability at irreducible oil conditions. The Wang et al. [26] method was used in this study to obtain the retained permeability under multi-phase conditions at residual oil saturation tests (Table 2).

Table 2. Relative permeability at irreducible oil conditions.

Sand Class	k_{abs} (md)	k_{rel} (Fraction) at S_{or}
DC-I	950	0.48
DC-II	1800	0.52
DC-III	2400	0.54

5. Testing Results

5.1. Sand Retention Performance

WWS's ability to retain the production of solid particles relies on the stability of particle bridges, which is mainly controlled by the ratio of the aperture size over the particle diameter [27], local fluid velocity [28], grain shape [29], water cut [25,30], and flowing phases [7].

This study established sand production limits at 0.12 lb/ft^2 for moderate sanding and 0.15 lb/ft^2 as the upper threshold for cumulative sanding. These limits correspond to limiting cumulative sand to less than 1% of the liner volume [31]. Similarly, Hodge et al. [32] correlated laboratory performances with field data and proposed a value of 0.12 lb/ft^2.

Figure 6 shows negligible sand production during the oil stages for all sand classes, which can be attributed to the strong capillarity bonding between grains. Even at high oil flow rates and wider slots, minimal sanding is observed. After the water breakthrough, more sand production is expected due to the reduction of the capillary bonding force. However, for 0.006″ and 0.010″ coupons in DC-III, there is minimal sanding even after water breakthrough, indicating stable sand bridges are formed on the small apertures that could provide a strong capability to prevent sand production. However, wider slots exhibit transient sanding. In transient sanding, some sanding is observed upon changing the flow rate or water cut, with the rate eventually declining until stable bridges are formed at constant flow rates.

Increasing flow rates induce higher pressure gradients through the sand bridges, and the drag forces can exceed the frictional resistance of the bridge. The impact of flow rate fluctuations is more substantial for wider slots, since greater aperture size to grain size results in weaker sand bridges.

Transient sanding was also observed during the co-injection of brine, oil, and gas, but at more intense levels. Interestingly, gas–liquid flow can destabilize the bridges, even over narrow apertures such as 0.006″. Wider slots (>0.014″) displayed significant sand production. In cases like the 0.018″ test for DC-I, bridge stability is never achieved, and particles are continuously produced, known as continuous sanding.

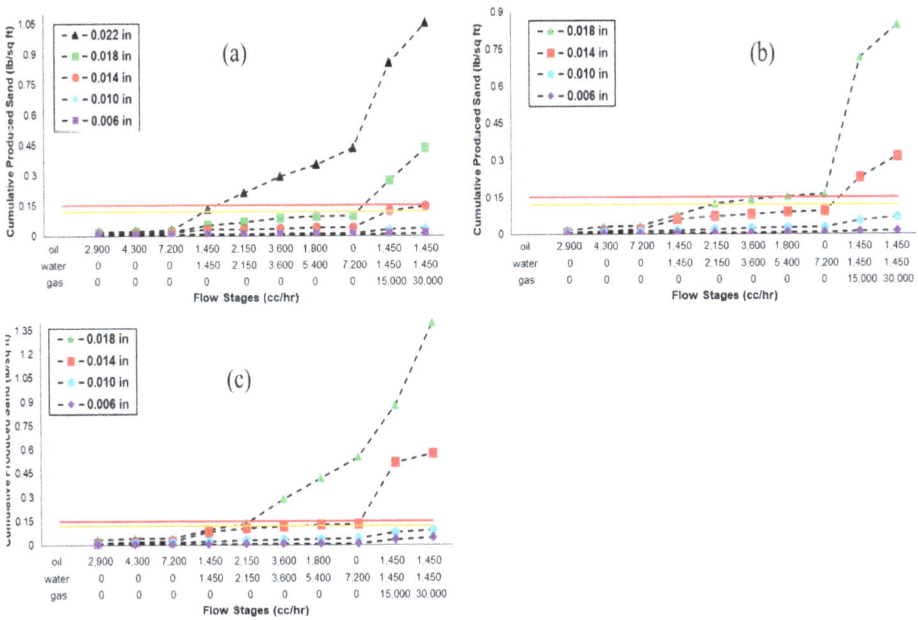

Figure 6. Cumulative sand production; (**a**) DC-III, (**b**) DC-II, (**c**) DC-I. Red and yellow horizontal lines represent the sanding limits of 0.15 lb/ft^2 and 0.12 lb/ft^2, respectively.

5.2. Flow Performance

The retained permeability normalizes the final permeability of the near-screen zones against the initial permeability. This parameter allows for a better comparison of the flow performance of different sand control designs. Effective retained permeability determination uses the relative permeability values ($k_{rw@Sor}$) from Table 2 and the pressure drop reading from the last liquid stage:

$$k_{ret} = \frac{k_{w,bottom}}{k_{abs} k_{rw\ @S_{or}}}$$

$$k_{w,bottom} = \frac{q_w \mu_w L_b}{\Delta P_b A}$$

where $k_{rw@Sor}$ is the relative permeability to water at residual oil saturation, $k_{w,bottom}$ is the final effective permeability after the last liquid stage, and k_{ret} is the retained permeability. ΔP_b is the pressure drop at the bottom section of the sand-pack. q_w and μ_w represent the water flow rate and viscosity, respectively.

Burton and Hodge [33] analytically found that a negligible impact on productivity occurs for retained screen permeability above 20% due to the high permeability of the screens compared to that of a porous medium. Later, retained permeability of 50% in the near-wellbore region was proposed to account for other formation damage sources [12,32]. This study considers 50% and 70% for the marginal and acceptable limits of retained permeability, respectively.

Figure 7 displays the retained permeability as a function of aperture size for the three sand classes. As expected, the fines can easily be dislodged from the sand-pack for the wider aperture size, resulting in minimal plugging. Remarkably, WWSs provide retained permeability values above 50%, even for narrow apertures such as 0.006" in low-quality sand, i.e., DC-I. In DC-III, increasing the aperture size beyond 0.018" does not further improve skin but results in higher sanding levels. Flow performance decreases from DC-III to DC-I, as narrow pore throats are prone to plugging.

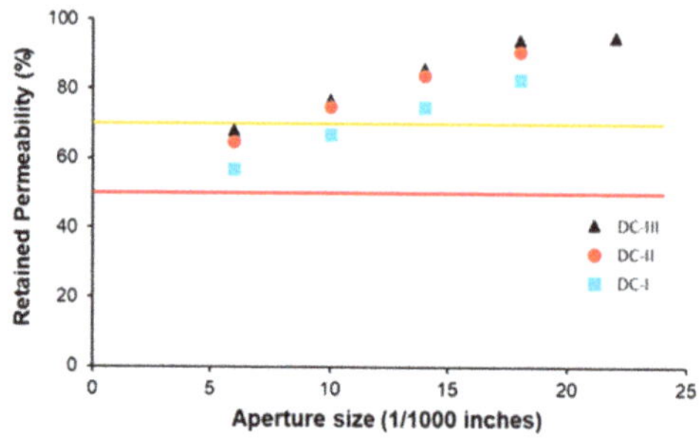

Figure 7. Retained permeability for different aperture sizes and sand classes. The red and yellow lines represent the retained permeability limits of 50% and 70%, respectively.

6. Design Criteria for WWS

Flow and sand production performances obtained with the SRT are combined to elaborate a set of graphical design criteria for WWS. The proposed criteria specify an aperture size window that keeps the produced sand within an acceptable limit while minimizing plugging potentials and maintaining wellbore productivity. Unlike previous criteria, the criteria introduced here differentiate production scenarios to evaluate their influence on the safe aperture zone. The optimum aperture window provides an upper bound and a lower bound. The upper bound is governed by sand production performance, while the flow performance dominates the lower bound.

6.1. The Traffic Light System (TLS)

The TLS is a graphical approach to rank the performance of different aperture sizes for specific sand classes proposed by Wang et al. [34] for slotted liners. This paper uses the same procedure for WWS. The performance is presented on an axial representation of the PSD (Figure 8) that contains representative D-values along the axis (red lines) as well as the aperture sizes implemented in the testing (dashed lines). The performance limits are used to categorize the response of WWS as acceptable, marginal, and unacceptable for sand retention and flow performance. Table 3 presents the color definitions for each indicator.

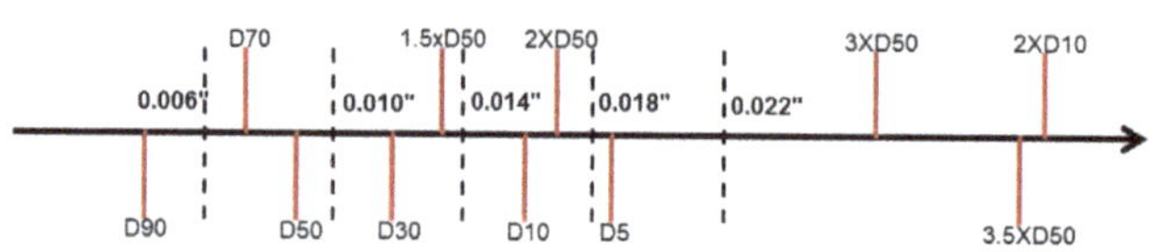

Figure 8. Example of linear x-axis representation of the PSD for DC-III.

Table 3. TLS color code.

Color Symbol and Description	Sand Retention Performance for Cumulative Sand Production	Flow Performance by Retained Permeability
Unacceptable	More than 0.15 lb/ft^2	Retained permeability less than 50%
Marginal	Between 0.12–0.15 lb/ft^2	Retained permeability between 50–70%
Acceptable	Less than 0.12 lb/ft^2	Retained permeability less than 50%

6.2. Design Criteria

The criteria differentiate two production scenarios, including regular and aggressive SAGD conditions. Experimental results up to the last liquid stage represent a regular SAGD case where a well exhibits changes in flow rate and water cuts throughout its life. The aggressive flow, defined as the flow during the three-phase flow condition, represents the steam breakthrough (see Figure 5). SAGD operators strive to identify "hot spots" and control these events due to their impact on the steam-chamber growth efficiency and risks of liner damage.

Figure 9a,b show an example of the axes for sand retention and flow performance for DC-III, respectively, under regular conditions. Combining both performance indicators generates the optimum aperture window (Figure 9c). The same method is used to create the final aperture size window for DC-II and DC-I under regular SAGD conditions (Figure 10). The safe aperture window narrows down from DC-III to DC-I. Fine sands produce more solids, requiring the implementation of smaller apertures. However, smaller apertures promote pore and slot plugging, reducing the range of optimum sizes.

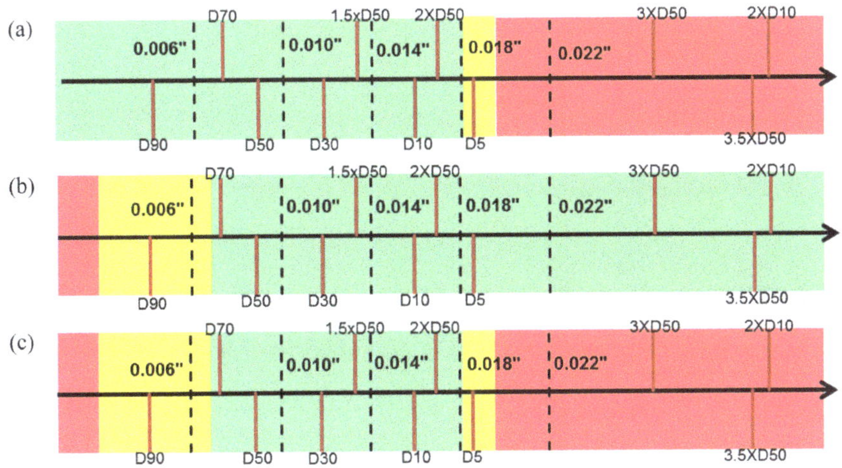

Figure 9. Aperture window for DC-III for regular SAGD conditions (red: unacceptable, yellow: marginal, green: acceptable); (**a**) sanding performance, (**b**) flow performance, (**c**) overall design window.

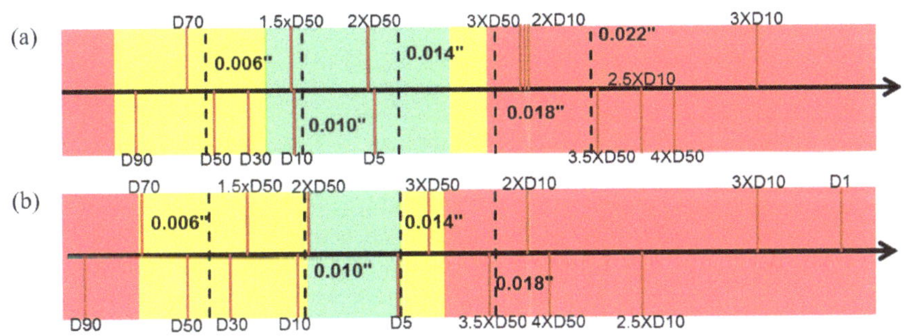

Figure 10. Overall aperture window in regular SAGD conditions (red: unacceptable, yellow: marginal, green: acceptable): (**a**) DC-II, (**b**) DC-I.

Figure 11 presents the TLS design criteria for the samples in aggressive conditions (during steam influx). By comparing the TLS criteria for regular and aggressive flow situations, it is evident that the aperture window shrinks for the aggressive condition. The reason is high levels of sanding due to strong drag forces during gas flow, shifting the upper bound to the left.

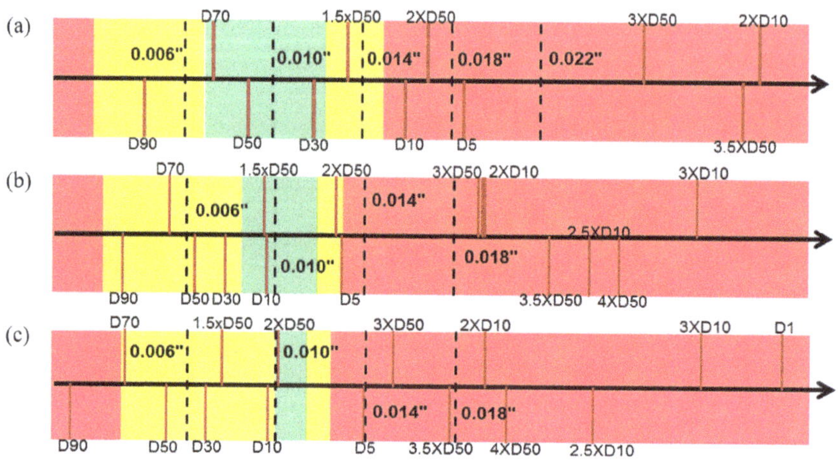

Figure 11. Overall aperture window in aggressive SAGD conditions (red: unacceptable, yellow: marginal, green: acceptable); (**a**) DC-III, (**b**) DC-II, (**c**) DC-I.

In summary, the high OFA of WWS results in low pressure gradients and high fines discharge that reduce pore plugging potential. As a benefit of high OFA, WWS allows the selection of narrow apertures that can handle a wide range of PSDs. However, for coarser sands, selecting narrow apertures may diminish the advantages of high OFA. Furthermore, the TLS shows how the adequacy of current criteria (i.e., D10, 2D50) is highly dependent on production scenarios, and that such simple criteria do not work consistently for all PSDs. Therefore, the graphical design criteria are created for each sand class, accounting for the entire shape of the PSD curve and the operating conditions.

7. Discussion

This paper presents the performance of WWS through an experimental study by including several key parameters, such as flow rate, water cut, PSD, and aperture size. Based on the sanding and flow performance, the aperture design criteria are generated for different SAGD operation conditions, which could provide some guidance to the field engineers in selecting the proper aperture size.

It should be mentioned here that there are more influential factors in the testing design that are not investigated in this study, such as temperature, paraffin deposition, etc. [35,36]. Thus, additional testing scenarios and field data are required to validate and calibrate the proposed aperture design criteria.

8. Conclusions

This research introduces design criteria for WWS in SAGD applications, considering the role of operational scenarios, flow rate, PSD, and aperture size on the screen performance. The results show that production scenarios strongly influence sanding intensity; wider slots exhibit extreme sanding levels during steam-breakthrough stages but respond reasonably well during liquid stages. Drag forces play a critical factor in disrupting particle bridges. A positive conclusion for WWS is that the retained permeability values stay above acceptable limits (50–70%), even for finer sands. WWS displays a substantial ability to release fines.

Sanding and flow performance results are combined to determine optimum-aperture windows for three sand classes using the TLS approach, incorporating PSD and production scenarios. Aggressive conditions shrink the safe-aperture window compared to normal SAGD conditions, which signifies the influence of operational practices in the performance of screens. Elevated flow rates and steam production significantly impact sanding and plugging and must be considered in aperture size design. For instance, for low sub-cool levels, the risk for the steam influx is more eminent, and the effect of aggressive conditions would affect the aperture sizing selection.

Author Contributions: J.D.M.P. (Conceptualization, methodology, validation, investigation, writing—original draft preparation); C.W. (methodology, validation, investigation, writing—original draft preparation); M.H. (investigation, resources, writing—review and editing, visualization); A.N. (validation, writing—review and editing, funding acquisition, supervision). All authors have read and agreed to the published version of the manuscript.

Funding: This research was funded by RGL Reservoir Management (now Variperm Energy Services) and The Natural Sciences and Engineering Research Council of Canada (NSERC) through their CRD program-Grant # NSERC CRDPJ 488486.

Institutional Review Board Statement: Not applicable.

Informed Consent Statement: Not applicable.

Data Availability Statement: Further supporting data can be found at: https://era.library.ualberta.ca/items/6898d3ef-ce7a-42f9-ba87-45389c013c3d.

Acknowledgments: The authors acknowledge the support and feedback from colleagues in the sand control lab during this research.

Conflicts of Interest: The authors declare no conflict of interest.

Abbreviations

OFA	Open-to-Flow Area
PS	Punched Screens
PSD	Particle Size Distribution
SAGD	Steam Assisted Gravity Drainage
SAS	Standalone Screens
SCD	Sand Control Devices
SL	Slotted Liner
SRT	Sand Retention Test
TLS	Traffic Light System
WWS	Wire-Wrapped Screen

References

1. Butler, R.M. Gravity Drainage to Horizontal Wells. *J. Can. Pet. Technol.* **1992**, *31*, 31–37. [CrossRef]
2. Collins, P.M. Geomechanical Effects on the SAGD Process. *SPE Reserv. Eval. Eng.* **2007**, *10*, 367–375. [CrossRef]
3. Han, G.; Dusseault, M. Quantitative Analysis of Mechanisms for Water-Related Sand Production. In Proceedings of the International Symposium and Exhibition on Formation Damage Control, Lafayette, LA, USA, 20–21 February 2002.
4. Ali, M.A.; Al-Hamad, K.; Al-Haddad, A.; AlKholosy, S.; Abu Sennah, H.; Sanyal, T.; Aniel, J. Laboratory Challenges of Sand Production in Unconsolidated Cores. In Proceedings of the SPE Kuwait International Petroleum Conference and Exhibition, Kuwait City, Kuwait, 10–12 December 2012.
5. Oyeneyin, B. Integrated Sand Management for Effective Hydrocarbon Flow Assurance, Chapter 6: Sand Control Completion Strategy. In *Developments in Petroleum Science Series*; Elsevier: Amsterdam, The Netherlands, 2015.
6. Suman, G.O.; Ellis, R.C.; Snyder, R.E. *Sand Control Handbook*, 2nd ed.; Gulf Publishing Company, Book Division: Houston, TX, USA, 1983.
7. Bennion, D.B.; Gupta, S.; Gittins, S.; Hollies, D. Protocols for Slotted Liner Design for Optimum SAGD Operation. *J. Can. Pet. Technol.* **2009**, *48*, 21–26. [CrossRef]

8. Fattahpour, V.; Mahmoudi, M.; Wang, C.; Kotb, O.; Roostaei, M.; Nouri, A.; Fermaniuk, B.; Sauve, A.; Sutton, C. Comparative Study on the Performance of Different Standalone Sand Control Screens. In Proceedings of the SPE International Conference and Exhibition on Formation Damage Control, Lafayette, LA, USA, 7–9 February 2018.
9. Romanova, U.; Ma, T. An Investigation on the Plugging Mechanisms in a Slotted Liner from the Steam-Assisted Gravity Operations. In Proceedings of the SPE European Formation Damage Conference & Exhibition, Noordwijk, The Netherlands, 3–4 June 2013.
10. Montero Pallares, J.D.; Wang, C.; Haftani, M.; Nouri, A. Experimental Correlations for the Performance and Aperture Selection of Wire-Wrapped Screens in Steam-Assisted Gravity Drainage Production Wells. *SPE Prod. Oper.* **2020**, *35*, 537–548. [CrossRef]
11. Matanovic, D.; Cikes, M.; Moslavac, B. *Sand Control in Well Construction and Operation*, 1st ed.; Springer Science & Business Media: Berlin/Heidelberg, Germany, 2012.
12. Markestad, P.; Christie, O.; Espedal, A.; Rorvik, O. Selection of Screen Slot Width to Prevent Plugging and Sand Production. In Proceedings of the SPE Formation Damage Control Symposium, Lafayette, LA, USA, 14–15 February 1996.
13. Kaiser, T.M.V.; Wilson, S.; Venning, L.A. Inflow Analysis and Optimization of Slotted Liners. *SPE Drill Complet.* **2002**, *17*, 200–209. [CrossRef]
14. Romanova, U.G.; Ma, T.; Piwowar, M.; Strom, R.; Stepic, J. Thermal Formation Damage and Relative Permeability of Oil Sands of the Lower Cretaceous Formations in Western Canada. In Proceedings of the SPE Canada Heavy Oil Technical Conference, Calgary, AB, Canada, 9–11 June 2015.
15. Coberly, C.J. Selection of Screen Openings for Unconsolidated Sands. In Proceedings of the Drilling and Production Practice, New York, NY, USA, 1 January 1937.
16. Driscoll, F.G. *Groundwater and Wells*, 2nd ed.; Johnson Filtration Systems Inc.: St. Paul, MN, USA, 1986.
17. Gillespie, G.; Deem, C.K.; Malbrel, C. Screen Selection for Sand Control Based on Laboratory Tests. In Proceedings of the SPE Asia Pacific Oil and Gas Conference and Exhibition, Brisbane, Australia, 17–19 October 2000.
18. Ballard, T.; Beare, S. An Investigation of Sand Retention Testing with a View to Developing Better Guidelines for Screen Selection. In Proceedings of the SPE International Symposium on Formation Damage Control, Lafayette, LA, USA, 23–24 February 2012.
19. Guo, Y.; Roostaei, M.; Nouri, A.; Fattahpour, V.; Mahmoudi, M.; Jung, H. Effect of Stress Build-Up around Standalone Screens on the Screen Performance in SAGD Wells. *J. Pet. Sci. Eng.* **2018**, *171*, 325–339. [CrossRef]
20. Abram, M.; Cain, G. Particle-Size Analysis for the Pike 1 Project, McMurray Formation. *J. Can. Pet. Technol.* **2014**, *53*, 339–354. [CrossRef]
21. Mahmoudi, M.; Fattahpour, V.; Nouri, A.; Rasoul, S.; Yao, T.; Baudet, B.A.; Leitch, M.; Soroush, M. An Investigation into the Use of Commercial Sands and Fines to Replicate Oil Sands for Large-Scale Sand Control Testing. In Proceedings of the SPE Thermal Well Integrity and Design Symposium, Banff, AB, Canada, 28 November–1 December 2016.
22. Haftani, M.; Wang, C.; Montero, J.; Mahmoudi, M.; Fattahpour, V.; Nouri, A. An Investigation into the Effect of Brine Salinity on Fines Migration in SAGD Operations. In Proceedings of the SPE Western Regional Meeting, San Jose, CA, USA, 23–26 April 2019.
23. Ladd, R.S. Preparing Test Specimens Using Undercompaction. *Geotech. Test. J.* **1978**, *1*, 16–23.
24. Stahl, R.M.; Smith, J.D.; Hobbs, S.; Clark, C.M. Application of Intelligent Well Technology to a SAGD Producer: Firebag Field Trial. In Proceedings of the SPE Heavy Oil Conference-Canada, Calgary, AB, Canada, 10–12 June 2014.
25. Wu, B.; Tan, C.P. Effect of Water-cut on Sandstone Strength and Implications in Sand Production Prediction. In Proceedings of the DC Rocks 2001, The 38th U.S. Symposium on Rock Mechanics (USRMS), Washington, DC, USA, 7–10 July 2001.
26. Wang, C.; Montero Pallares, J.D.; Haftani, M.; Nouri, A. Developing a Methodology to Characterize Formation Damage (Pore Plugging) due to Fines Migration in Sand Control Tests. *J. Pet. Sci. Eng.* **2020**, *186*, 106793. [CrossRef]
27. Chanpura, R.A.; Mondal, S.; Sharma, M.M.; Andrews, J.S.; Mathisen, A.M.; Martin, F.; Marpaung, F.; Ayoub, J.A.; Parlar, M. Unraveling the Myths Associated with Selection of Standalone Screens and a New Methodology for Sand Control Applications. *SPE Drill. Complet.* **2013**, *28*, 227–236. [CrossRef]
28. Muecke, T.W. Formation Fines and Factors Controlling Their Movement in Porous Media. *J. Pet. Technol.* **1979**, *31*, 144–150. [CrossRef]
29. Santamarina, C. Soil Behavior at the Microscale: Particle Forces. In Proceedings of the ASCE Symposium on Soil Behavior and Soft Ground Construction, in Honor of Charles, C. Ladd, Cambridge, MA, USA, 5–6 October 2001.
30. Vaziri, H.; Barree, B.; Xiao, Y.; Ian, P.; Kutas, M. What is the Magic of Water in Producing Sand? In Proceedings of the SPE Annual Technical Conference and Exhibition, San Antonio, TX, USA, 29 September–2 October 2002.
31. Montero, J.; Chissonde, S.; Kotb, O.; Wang, C.; Roostaei, M.; Nouri, A.; Mahmoudi, M.; Fattahpour, V. A Critical Review of Sand Control Evaluation Testing for SAGD Applications. In Proceedings of the SPE Canada Heavy Oil Technical Conference, Calgary, AB, Canada, 13–14 March 2018.
32. Hodge, R.M.; Burton, R.C.; Constien, V.; Skidmore, V. An Evaluation Method for Screen-only and Gravel-pack Completions. In Proceedings of the International Symposium and Exhibition on Formation Damage Control, Lafayette, LA, USA, 10–12 February 2002.
33. Burton, R.C.; Hodge, R.M. The Impact of Formation Damage and Completion Impairment on Horizontal Well Productivity. In Proceedings of the SPE Annual Technical Conference and Exhibition, New Orleans, LA, USA, 27–30 September 1998.
34. Wang, C.; Pang, Y.; Mahmoudi, M.; Haftani, M.; Salimi, M.; Fattahpour, V.; Nouri, A. A Set of Graphical Design Criteria for Slotted Liners in Steam Assisted Gravity Drainage Production Wells. *J. Pet. Sci. Eng.* **2020**, *185*, 106608. [CrossRef]

35. Ilyushin, Y.V.; Fetisov, V. Experience of Virtual Commissioning of a Process Control System for the Production of High-Paraffin Oil. *Sci. Rep.* **2022**, *12*, 18415. [CrossRef] [PubMed]
36. Lira-Galeana, C.; Hammami, A. Chapter 21 Wax Precipitation from Petroleum Fluids: A Review. In *Developments in Petroleum Science*; Elsevier: Amsterdam, The Netherlands, 2000; Volume 40, pp. 557–608.

eng

MDPI

Article

Offset Well Design Optimization Using a Surrogate Model and Metaheuristic Algorithms: A Bakken Case Study

Ahmed Merzoug * **and Vamegh Rasouli**

Department of Petroleum Engineering, University of North Dakota, Grand Forks, ND 58202, USA; vamegh.rasouli@und.edu
* Correspondence: ahmed.merzoug@und.edu; Tel.: +1-701-215-5991

Abstract: Fracture-driven interaction FDI (colloquially called "Frac-hit") is the interference of fractures between two or more wells. This interference can have a significant impact on well production, depending on the unconventional play of interest (which can be positive or negative). In this work, the surrogate model was used along with metaheuristic optimization algorithms to optimize the completion design for a case study in the Bakken. A numerical model was built in a physics-based simulator that combines hydraulic fracturing, geomechanics, and reservoir numerical modeling as a continuous simulation. The stress was estimated using the anisotropic extended Eaton method. The fractures were calibrated using Microseismic Depletion Delineation (MDD) and microseismic events. The reservoir model was calibrated to 10 years of production data and bottom hole pressure by adjusting relative permeability curves. The stress changes due to depletion were calibrated using recorded pressure data from MDD and FDI. Once the model was calibrated, sensitivity analysis was run on the injected volumes, the number of clusters, the spacing between clusters, and the spacing between wells using Sobol and Latin Hypercube sampling. The results were used to build a surrogate model using an artificial neural network. The coefficient of correlation was in the order of 0.96 for both training and testing. The surrogate model was used to construct a net present value model for the whole system, which was then optimized using the Grey Wolf algorithm and the Particle Swarm Optimization algorithm, and the optimum design was reported. The optimum design is a combination of wider well spacing (1320 ft), tighter cluster spacing (22 ft), high injection volume (1950 STB/cluster), and a low cluster number per stage (seven clusters). This study suggests an optimum design for a horizontal well in the Bakken drilled next to a well that has been producing for ten years. The design can be deployed in new wells that are drilled next to depleted wells to optimize the system's oil production.

Keywords: surrogate modeling; optimization; hydraulic fracturing; fracture-driven interaction

Citation: Merzoug, A.; Rasouli, V. Offset Well Design Optimization Using a Surrogate Model and Metaheuristic Algorithms: A Bakken Case Study. *Eng* **2023**, *4*, 1290–1305. https://doi.org/10.3390/eng4020075

Academic Editor: George Z. Papageorgiou

Received: 15 March 2023
Revised: 3 April 2023
Accepted: 4 April 2023
Published: 23 April 2023

1. Introduction

Hydraulic fracturing and horizontal drilling have unlocked huge amounts of oil and gas reserves. This technique allows operators to extract resources from low permeability rocks [1,2]. In the US, operators have been drilling wells on 640 acres to hold the lease before it expires in one to three years. This is called hold by production (HBP). Operators then come to perform infill drilling to increase production. The primary well is colloquially called "parent well", while the offset wells are called colloquially "child well". If the wells are drilled at the same time, they are colloquially called sibling wells. The problem in these configurations is the interference between the fractures of the offset wells and the primary well. This interference is called Fracture Driven Interaction, FDI, colloquially known as "Frac-Hit".

Ajani and Kelkar [3] and Daneshy et al. [4] were the first to present the concept of FDI. Several factors can promote the occurrence of FDI. These factors are reported in a thorough critical review by Gupta et al. [5]. Several statistical studies quantified this effect in terms of productivity gains and production losses [6–8]. Table 1 summarizes the effect of FDI on production in several plays in the US.

Table 1. Effect of fractured child wells due to the production from the parent well (modified from Miller et al. [7]).

	Bakken	Eagle Ford	Haynesville	Woodford	Niobrara
Positive	50%	24%	58%	4%	6%
No Change	35%	36%	24%	4%	38%
Negative	15%	41%	19%	64%	56%

Cozby and Sharma [6] reported that the wells' spacing and produced volumes significantly affected the FDI intensity. They also reported that early wells in the Bakken exhibited positive behavior through FDI, mainly because these wells have been understimulated. This work mainly focused on the optimization of infill well completions that were drilled next to early wells in the Bakken.

Depletion is one of the many reasons that promotes the occurrence of FDI. Depletion is associated with poroelastic stress changes around the depleted primary well. This causes an asymmetric growth of the fractures from the offset wells towards the path of least resistance. The effect of depletion on FDI has been reported and studied by many researchers [3,9–13].

Many researchers have conducted numerical modeling studies to understand the effect of FDI on the performance of primary and offset wells. Kumar et al. [14] investigated the relationship between production losses, the diffusivity coefficient, and the well spacing. Rezaei et al. [15] simulated stress changes due to depletion and their effect on fracture propagation and reorientation from offset wells under different stress contrasts. Cipolla et al. [16] and Fowler et al. [9] modeled stress changes in a field-scale laboratory study. Morsy et al. [17] reported a numerical simulation workflow to optimize offset well design through sensitivity analysis in the Wolfcamp formation. Cai and Taleghani [18] worked on the theory developed by Daneshy [19] using numerical modeling for the interpretation of pressure response in primary wells during the stimulation of offset wells. Ratcliff et al. [13] modeled the increase in water production due to FDI in the Meramec of the STACK play. Their work modeled production losses and remediation techniques for conductivity damage.

Hydraulic Fracture (HF) treatment design optimization has been a topic of interest for both academia and the industry. Yu and Sepehrnoori [20] attempted to optimize the net present value (NPV) using numerical modeling and a response surface methodology (RSM) algorithm. They optimized the half-length, fracture spacing, and well spacing. Ma et al. [21] used reservoir simulation to optimize the design of HF stages. They used a covariance matrix adaptation evolution strategy, a genetic algorithm (GA), and simulations with perturbation stochastic approximation. Plaksina and Gildin [22] and Rahmanifard and Rahmanifard [23] attempted to optimize the number of stages, location, and fracture half-length using multi-objective optimization using numerical and analytical models. Cipolla et al. [24] used a fully integrated numerical workflow to optimize the Utica Shale's well spacing and perf cluster spacing. Kim and Choe [25] used proxy models through linear regression to optimize the HF design. The previous research was conducted using reservoir numerical simulation, and the science behind fracture propagation was ignored. Wang and Olson [26] developed a numerical three-dimensional fracture propagation code and optimized the fracture effective contact area using the generic algorithm (GA) and the pattern search algorithm (PSA). Their research focused only on the fracture contact area, which can sometimes be economically overstimulated. Garcia Ferrer et al. [27] suggested a workflow using numerical simulation to optimize child wells in a depleted environment. Wang et al. [28] used a data-driven approach in the Montney to maximize the productivity of child wells. Kang et al. [29] implemented a workflow for automated history matching and optimization for hydraulic fracturing design; in their study, they used a different optimization algorithm (COBYLA) and kriging as their proxy models.

Pudugramam et al. [30] optimized the well spacing and completion design for Hydraulic Fracturing Test Site 2 to maximize the net present value and return on investment.

In this work, a fully physics-based model was built and calibrated using data from logs, cores, microseismics, and production. The stress changes were estimated from and calibrated to microseismic depletion delineation (MDD). Once the model was built, a sensitivity analysis was run to build a proxy model. The simulation runs were structured using Sobol and Latin Hypercube sampling approaches. The proxy model was built using an artificial neural network (ANN) with an R_2 score of the order of 0.96. The proxy model was used to build a net present value function. Metaheuristic algorithms were used to optimize the net present value and find the optimum completion design for Bakken wells drilled next to a pre-depleted primary well.

2. Problem Statement

The setup of this project in the Bakken is part of a multi-year journey of data acquisition projects conducted by the operator to improve the reliability of the physics-based model. The project is called Red Sky (RS). The aim of the project was to understand well-to-well connectivity, drainage areas, and HF geometry. The setup is composed of a horizontal well (H1) and two vertical wells (V1 and V2) at a distance of 1000 ft and 1200 ft from the lateral of well H1 (see Figure 1). Table 2 summarizes the timeline for the setup.

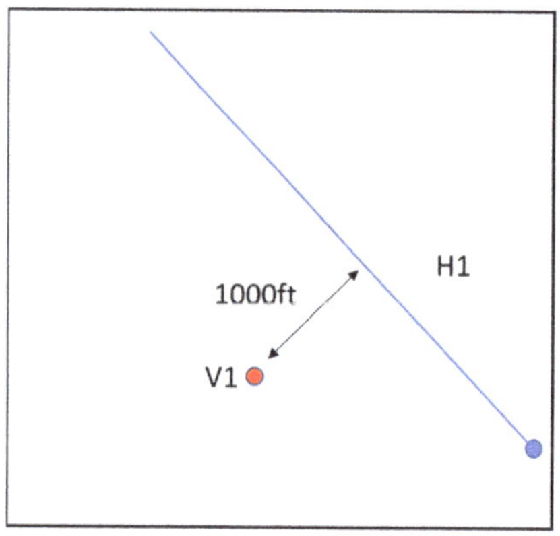

Figure 1. Schematic of the well location.

Table 2. Setup timeline.

Production/Testing Event	Start Date	End Date
Hydraulic Fracturing Treatment in H1	29 September 2005	29 September 2005
H1 Primary Production	29 September 2005	8 October 2016
Shut-in H1	8 October 2016	5 January 2017
V1 DFIT #1	7 November 2016	7 November 2016
H1 MDD	9 December 2016	11 December 2016
V1 Fracture Stimulation	11 December 2016	11 December 2016
H1 Back to Production	5 January 2017	3 May 2017

The completion design for H1 consists of one stage that was hydraulically fractured in one go using ball sealers and six diversion stages. The stage consists of 33 clusters separated by 134 ft. The pumping treatment had a very small total injected volume of 4600 bbl of XL fluid pumped at a rate of 60 bpm and a total of 420,000 lb of 20/40 sand. The radioactive

(RA) tracer showed very good coverage of proppant in the 33 clusters. The well produced 110,000 bbl prior to microseismic depletion delineation (MDD).

The MDD was first suggested by Dohmen et al. [31]. It was mainly used to map the depleted area (an approximation to fracture geometry) and measure stress changes. More details about the interpretation can be found in Cipolla et al. [16].

The time-lapse DFITs were interpreted by Cipolla et al. [16] using the compliance method with an estimated closure stress of 7150 psi and a pore pressure of 5800 psi in the Middle Bakken Formation. Note that in the model calibration, only the MDD, DFIT #1, and V1 fracture treatment responses were matched. The reason for doing this is that calibrating the pressure response from the MDD is sufficient for stress change calibration because it represents an average value for all H1. In contrast, the DFITs represent local measurements that the heterogeneity of the formation can cause. Another difference between this work and the work of Cipolla et al. [16] and Fowler et al. [9] is that the contribution of the Three Forks is considered in the modeling. The numerical modeling section will discuss more details about the modeling process.

3. Numerical Modeling

The numerical model for this study was built based on data from Cipolla et al. [16] and Fowler et al. [9] with some minor changes. The numerical simulation was run in a fully integrated physics-based simulator. The numerical simulator incorporates modeling hydraulic fracturing with reservoir and geomechanics components integrated. This approach was adopted to build a representative model of hydraulic fracture propagation and calibrate the stress changes due to depletion and reservoir performance. The numerical simulator used has independent meshes between the fracture and the reservoir grid. The two components are connected using a 1D submesh to capture the fluid flow between the rock matrix grid and the fracture grid. This approach overcomes previously used approaches because it couples most of the physics in the system without ignoring parameters when going from hydraulic fracturing simulation to reservoir simulation. The numerical model provides more confidence in the generated results because all aspects have been constrained in one way or another. Figure 2 illustrates the workflow used for building the numerical model.

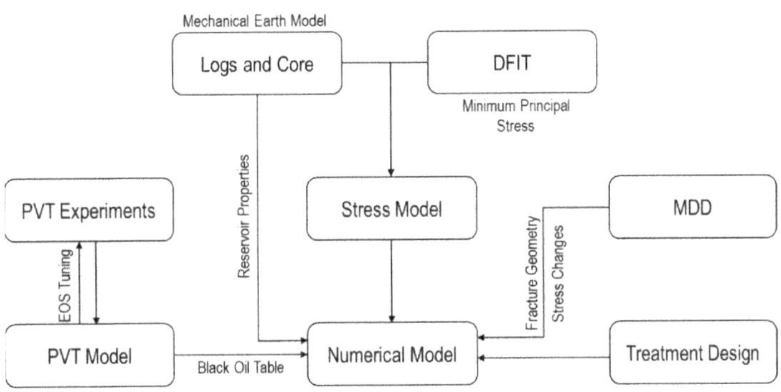

Figure 2. Workflow for building a numerical model.

The stress model was estimated using the transverse isotropic (TI) stress model (see Appendix A). The Diagnostic Fracture Injection Tests (DFIT) were interpreted using the compliance method as reported in Cipolla et al. [16] and Fowler et al. [9]. The MDD, along with radioactive tracer data, were used to calibrate fracture geometry. The MDD pressure response was used to calibrate stress changes in the formation. The PVT model was calibrated using Constant Composition Expansion and Differential Liberation data in a PVT numerical simulator. A black oil PVT was generated and used as an input for the numerical model. Reservoir and rock properties were taken from data reported by Cipolla et al. [16].

The fracture geometry from the horizontal well was estimated from MDD. The mapping results show events from the lower Three Forks to the Lodgepole with a height growth of approximately 250 ft. The lateral growth is averaged at 500 ft, with a dominant fracture at the heel of the well at 1000 ft. The recorded microseismic events belong only to the first 2000 ft of the lateral. This suggests that the toe of the well was not drained effectively. The drainage can be explained by pressure losses along the lateral, which minimize drainage at the toe of the well. The low drainage did not alter the stress state at the toe of the well. This is supported by the effectiveness of the treatment shown in the radioactive tracers. Figure 3 illustrates the recorded microseismic events. In the modeling process, both RA and microseismic data were combined to model and calibrate the lateral extent of the fractures using a fracture toughness of 2000 psi·in$^{0.5}$. The height was estimated and calibrated by building a mechanical earth model (MEM) and a DFIT (see Appendix A).

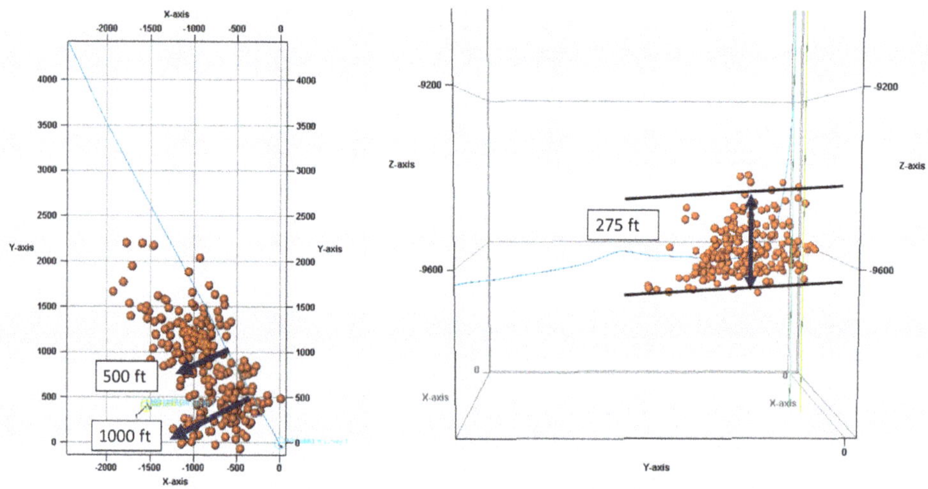

Figure 3. Microseismic Depletion Delineation recorded event and estimated fracture geometry.

The geomodel was built using data from logs, cores, and data reported by Cipolla et al. [16]. The rock properties are summarized in Table 3. The production is assumed to be only from the middle Bakken (MB) (divided into four by Cipolla et al. [16]) and Three Forks (TF) Formations, whereas the permeability in the Lodgepole, Upper, and Lower Bakken Shales was assumed to be zero [32,33]. The production history was matched by adjusting the matrix relative permeability curves using a total liquid production constraint to ensure mass conservation. Figure 4 reports the adjusted relative permeability curves from Cho et al. [34] and Cipolla et al. [16]. The history matching results are reported in Figure 5. The grid sizes are reported as follows:

- A fracture grid size of 80 ft was used. This value is acceptable for the range of application of the fracture propagation algorithm [35];
- A geomodel with 5635 ft height, 15,000 ft length, and 1280 ft height was built;
- A logarithmic grid length in the Shmax direction was used to account for the sensitivity analysis fracture geometry.

Table 3. Reservoir rock properties used in the geomodel.

Formation	Permeability	Porosity	Water Saturation
MB1	0.004	6.5	42
MB2	0.0026	7	36
MB3	0.0026	5.5	42
MB4	0.001	5.5	40
TF	0.0015	5	40

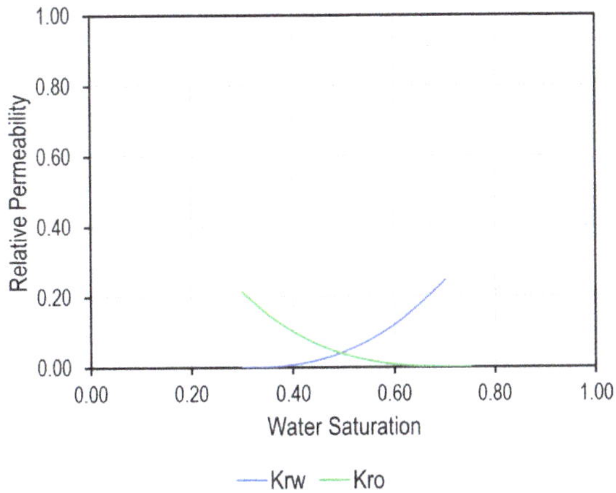

Figure 4. Relative permeability curves used in the history match.

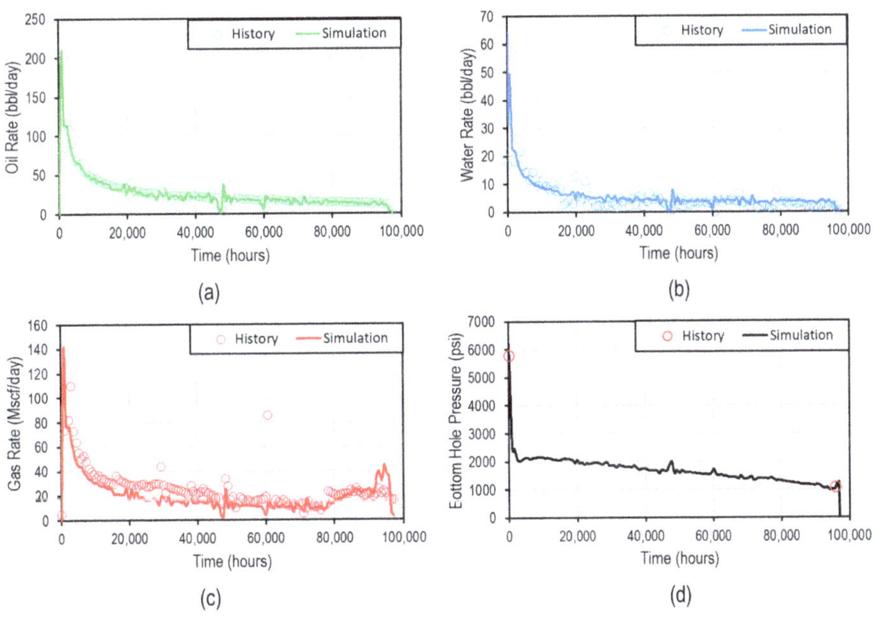

Figure 5. Production and bottom-hole pressure match. (**a**) oil rate match, (**b**) water rate match, (**c**) gas rate match, and (**d**) bottom hole pressure match.

Stress changes due to depletion were measured through MDD and hydraulic fracturing treatment at well V1. The pressure of the MDD stabilizes at 6000 psi. This pressure is close to the magnitude of FDI, with a magnitude of 5100 psi. The magnitude of the FDI pressure is related to the location of the occurrence of the fracture interaction. Communication can take several forms. Direct connection between fractures from the offset well and the primary well, drainage area overlap, and connection through fractures [5]. Daneshy [36] laid the groundwork for the interpretation of the pressure response; however, the interpretation process is still subjective, and the explanations can vary. According to the theory of poroelasticity, stress changes are dependent on the Biot coefficient and the poroelasticity coefficient [37]. The stress changes can be estimated using the following equation [37].

$$\Delta S_{hmin} = \alpha\left(\frac{1-2\nu}{1-\nu}\right) \tag{1}$$

To match the pressure change, the Biot coefficient was adjusted following the process described in Fowler et al. [9]. The calibration value of the Biot coefficient was 0.7. This value is different from the study by Fowler et al. [9] for the same case study, where they reported a value of 0.34. The discrepancy in the results is due to the different approaches used for modeling fractures in well H1. In their work, the approach was to implement pre-existing fractures that force the simulator to have uniform drainage along the fractures. In the current work, the fractures were modeled by creating lower drainage towards the tip of the fracture. These effects change the MDD modeling response in the two cases. Figure 6 illustrates the bottom hole pressure match for the MDD part and the pressure response due to the fracture interaction from the offset well V1. The Biot coefficient of 0.7 was sufficient to match both fracture interaction and the MDD. This match gives more confidence to our model in terms of poroelastic stress change. Figure 7 illustrates the poroelastic stress changes and the FDI from well V1. Note that the fracture from well V1 is connected to the fracture from the horizontal well.

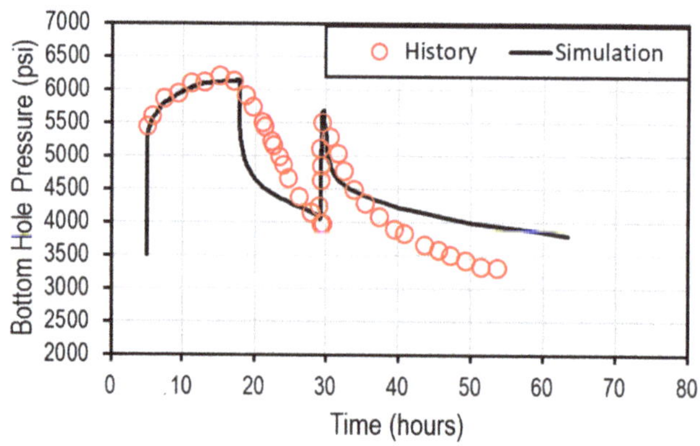

Figure 6. MDD pressure response and fracture-driven interaction match.

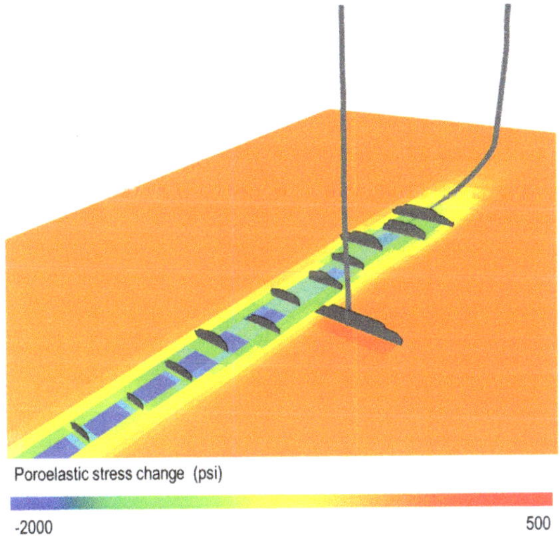

Figure 7. Predicted poroelastic stress changes and fracture driven interaction.

4. Optimization Formulation

The optimization process requires a mathematical formulation for the problem's so-called objective function. This step can be addressed by building a proxy model. Proxy models are objective-oriented; thus, a good understanding of the problem and control variables is necessary before embarking on the modeling process. In this work, the objective function is the net present value (NPV). It is expressed as follows:

$$\text{NPV} = f(IV, CN, CS, WS) \times O_p - SV_p \times IV \times TCN - NPT_C \times SN \quad (2)$$

where:

IV: Injected volume per cluster;
CN: Number of clusters per stage;
CS: Cluster spacing;
WS: Well spacing;
O_p: Oil price;
SV_p: Fracture volume price per barrel;
TCN: Total clusters number;
NPT_C: Non-productive time cost (time between stages);
SN: Stages number.

The function $f(IV, CN, CS, WS)$ was built through proxy modeling using neural networks. The ANN model was used because the problem is an interpolation one. Thus, it would be fine to overfit the parameters in the solution domain. The design parameters and their range are summarized in Table 4.

To ensure that the pumping schedule used is realistic, the treatment design was obtained from Cipolla et al. [38]. The design was for 14 clusters (with two perforations each) in one stage with a total injected volume of 7000 bbl. This treatment was upscaled to the sensitivity analysis intended by keeping the pumping rates and the proppant concentrations constant and only changing the time to adjust the total volume of injected fluid. This approach will automatically upscale the total volume and the injected proppant. Figure 8 illustrates the treatment design pumped in the Bakken. The scaling process was calculated as follows:

$$ratio = \frac{IV \times CN}{7000} \quad (3)$$

The ratio is then multiplied by the time intervals of the design depicted in Figure 8 to scale the total volume of fluid and total proppant.

Table 4. Design parameters for the hydraulic fracturing job.

Parameter	Unit	Range
Injected volume per cluster	bbl/cluster	400–2000
Number of clusters	/	7–20
Spacing between the clusters	ft	15–50
Well spacing	ft	440; 660; 880; 1320
Treatment design		High-Viscosity Friction Reducer (Proppant size: 50% 100 mesh and 50% 40/70 mesh)

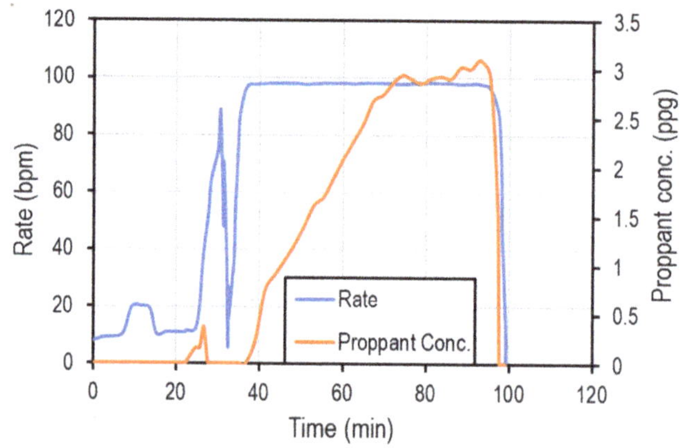

Figure 8. Hydraulic fracturing treatment schedule for one stage.

The neural network used had an architecture of seven layers. The input layer has four inputs, followed by five hidden layers, the first three of which each have six neurons. The fourth layer had four neurons, followed by a three neuron layer, and finally an output layer. The activation function was ReLu, and the optimizer was Adams. A total of 1000 iterations were performed during the training. These parameters were selected based on trial and error.

Before any training, the data was normalized as follows to reduce the space of solutions for the Adam optimizer [32]. The following equation was used:

$$X_{Normalized} = \frac{X-\mu}{\sigma} \quad (4)$$

where $X_{Normalized}$ is the normalized data, X is the data, μ is the average, and σ is the standard deviation.

The error was quantified using the root mean square error (RMSE) as follows:

$$\text{RMSE} = \sqrt{\frac{\sum_{i=1}^{n}\left(Y_i^{proxy} - Y_i^{simulation}\right)^2}{n}} \quad (5)$$

where n is the number of scenarios and Y_i is the prediction result for the proxy model or simulation.

Two metaheuristic algorithms were used to optimize the completion design in this study. These algorithms were selected because they are gradient-free algorithms. The advantage of incorporating these algorithms into an optimization process is that they help to find a global optimum due to their stochastic search patterns, which prevent local optimum issues [39]. Both algorithms are discussed in the following section.

4.1. Grey Wolf Optimization Algorithm

Mirjalili et al. [40] introduced the algorithm Grey Wolf Optimization, which is inspired by the Grey Wolf's behavior in nature. The algorithm has been applied in the petroleum engineering literature [40–44].

The wolves act in a hierarchical order. First, the alpha wolf is the leader of the pack. The alpha wolf is responsible for making decisions about hunting prey. Second, the beta wolves are subordinate to the alpha wolves in making and enforcing the decision. These wolves are the leaders of the pack. Next, the delta wolves are subordinate to the alpha and beta wolves. The lowest level of the hierarchy are the omega wolves [40].

The hunting process follows the hierarchy of the wolves. This process was modeled mathematically as the Grey Wolf algorithm to optimize processes and objective functions. The algorithm proceeds as follows [41]:

- Generate a random set of solutions (can be bounded) to represent the location of wolves in the solution space;
- Evaluate the locations according to the cost function (NPV for this work);
- Rank the solutions and assign them according to the hierarchy of the wolves α, β, δ, and the rest of the solutions to γ;
- Update the location of the wolves according to the best solution (assumed to be alpha);
- Repeat the process until the maximum number of iterations is reached.

4.2. Particle Swarm Optimization Algorithm

The Particle Swarm Optimization algorithm mimics the natural behavior of insects, birds, and fish. It was first suggested by Kennedy and Eberhart [45]. Each solution represents the position of a particle. The positions of each particle are updated from one iteration to another. These solutions are characterized by a position and a velocity [46]. Three parameters mainly control the algorithm [47]:

- *Cognitive*: serves as the memory of the particle; ensures that the particle moves towards the best values; and limits the step size in the search and convergence process;
- *Social*: determines the step size while converging to the swarm's best solution;
- *Inertia*: control the speed of convergence and encourage exploration of new solutions.

The algorithm is executed as follows [48]:

- Initialize the PSO parameters;
- Generate primary swarm positions;
- Evaluate the fitness of each position using the objective function (NPV);
- Record the best position for each particle along with the global best particle;
- Update the position and the velocity of the particles until the maximum number of iterations is reached.

5. Sensitivity Analysis

Different sampling approaches have been used to reduce the total number of simulations needed to build a proxy model. Eighty different treatment designs were generated using Latin Hypercube sampling [49] and Sobol Sequence sampling [50]. The combination of these approaches allows the generation of spaced data points that efficiently represent the space of solutions [51]. The computation is run on the cloud. A restart file was used to reduce the computation time. Modeling multistage hydraulic fracturing and production for all the wells can be time-consuming; thus, only one stage of the well in each scenario was simulated. The results were then upscaled to the full well, assuming each stage has an equal contribution to the total production of the well.

Note that local grid refinement was used at the level of the stage. The total length of the refined grid is 1000 ft, which is equal to the longest possible stage with 20 clusters of

50 ft spacing. The grid was refined to 15 ft, which is the smallest possible spacing to ensure that for all cases of sensitivity, one fracture exists per grid [52].

The results were quantified for three and ten year production as follows:

- Offset well by cumulative production normalized by length;
- Primary well cumulative production uplift normalized by length.

The sum of the two quantities represents system production normalized by length. Figure 9 illustrates a simulation case for the sensitivity analysis.

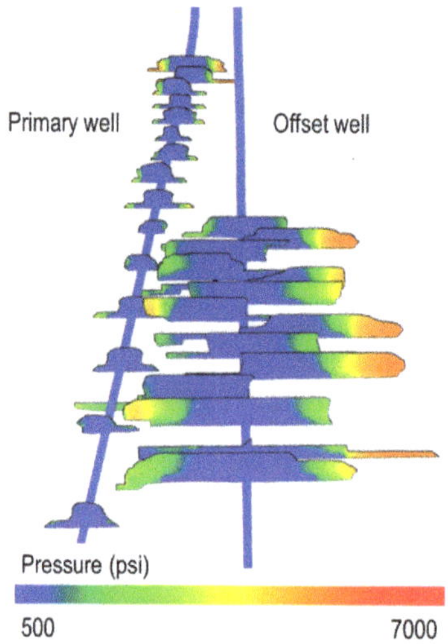

Figure 9. Sensitivity analysis case for one-stage simulation.

6. Results and Discussion

During the sensitivity analysis, there were cases with a tighter well spacing (440 ft) that reached the maximum surface pressure constraint. This is because of high-stress shadow values between clusters. The high-stress shadow is a result of the low stress because of the depletion near the primary well. This causes the net pressure to increase, leading to a higher width, which causes high stress values.

The neural network was trained using 90 data points that were generated using sampling approaches. The model was then incorporated into the NPV function and optimized. Ten simulations were run around the optimum design to increase the accuracy of the results and the model. The neural network training and testing results are illustrated in Figure 10 with a R_score of 0.961 and 0.9678 for the training and the testing, respectively. The RMSE is reported as 4.03 and 3.18 for the training and testing, respectively, as can be seen from Figure 10, which shows the acceptable error between the simulation and the proxy model. The simulations were run on the cloud using a high-performance computer from Microsoft Azure. The minimum simulation time for one case is 20 h, whereas the proxy model can give acceptable results in less than a second.

The NPV function is calculated using the following assumptions:

- Oil price (O_p) USD 80;
- Slurry price (SV_p) USD 100/bbl;
- The cost of adding a new stage (NPT_C) USD 2000.

The neural network was used to run sensitivity analysis and understand the effect of the offset well completion on the performance of the system. Figure 11a illustrates the effect of well spacing and cluster spacing on the NPV. The assumption is that 1000 bbl were injected per cluster, with seven clusters per stage. It can be noted that wider spacing results in better performance. This is attributed to the fact that wider spacing results in a larger drainage volume. The optimum number of clusters is around 22 for all well-spaced cases. Figure 11b illustrates the effect of injected volume and cluster spacing on the NPV. The assumption is that the well has seven stages and 1320 ft spacing between wells. It can be noted that, overall, larger volumes result in a better NPV. However, the optimum cluster spacing was wider for bigger injection volumes. This is because larger injection volumes result in longer fractures, implying larger drainage volumes that cause faster interference. Figure 11c illustrates the effect of cluster number per stage and cluster spacing. The assumption is that the injected volume is 1000 bbl per cluster and the well spacing is 1320 ft. It is observed that fewer clusters per stage result in better performance. This was because the fluid velocity at lower cluster numbers per stage was higher per fracture (limited maximum injection rate). The higher velocity results in better proppant placement in the fracture. The optimum cluster spacing is around 20 ft.

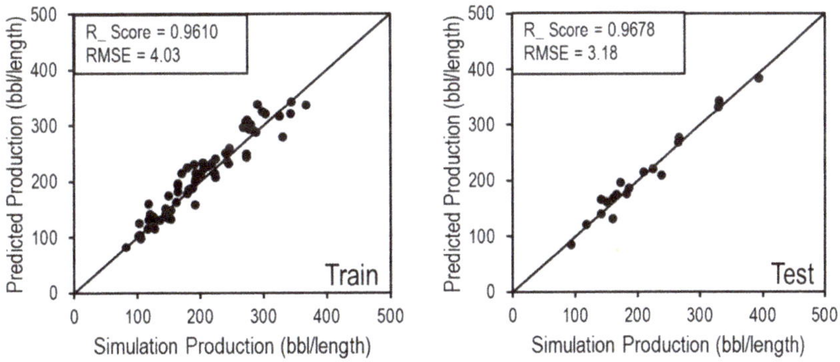

Figure 10. Results of training and testing neural networks.

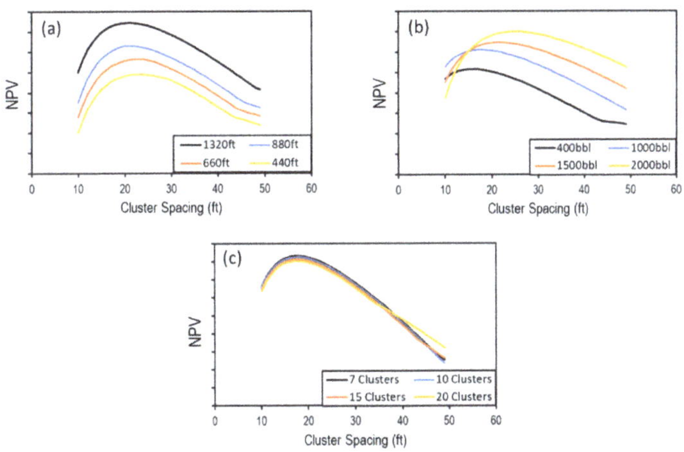

Figure 11. (**a**) The effect of well spacing and cluster spacing on NPV, (**b**) the effect of injected volume per cluster and cluster spacing on NPV, and (**c**) the effect of the number of clusters per stage and the cluster spacing on NPV.

The optimization algorithms were run on the NPV. The algorithms used libraries for the Grey Wolf algorithm [53], and Particle Swarm Optimization [54]. Note that the results reported by these algorithms depend on the economic assumptions of the oil price and the completion cost. Table 5 reports the optimization results. Note that these values assume all clusters contribute equally to production.

Table 5. Optimization results for NPV using different optimization algorithms.

Algorithm	Injected Volume (bbl/Cluster)	Cluster Number	Cluster Spacing (ft)	Well Spacing (ft)	Error from Sim %
GWO	1950	7	25	1320	1%
PSO	1900	7	26	1320	1.2%

7. Conclusions

In this study, we used a surrogate model to understand the effect of stimulation design parameters on the net present value in the Bakken system. We successfully represented the system using a neural network based on a physics-based, fully integrated simulator that accounts for fracture propagation, reservoir fluid flow, and stress changes due to geomechanics. The model was calibrated using different sources of hard data, which gives more confidence in the results. The design was then optimized using metaheuristic algorithms (Grey Wolf Optimization and Particle Swarm Optimization). The numerical simulation was time-consuming, with a total of around 4000 h running on the cloud. The use of a surrogate model reduced the simulation time and provided a proxy for system performance. The model suggests that the optimum design had wider well spacing (1320 ft), tight cluster spacing (22 ft), a low number of clusters per stage (seven clusters), and a large injection volume (1950 STB). The proxy model served the purpose of optimization well and was confirmed by the simulation. Since it has potential, this workflow can be applied to newly drilled wells and for refracturing.

Author Contributions: Conceptualization, A.M.; Methodology, A.M.; Software, A.M.; Validation, A.M.; Formal analysis, A.M.; Writing—original draft, A.M.; Writing—review & editing, V.R.; Supervision, V.R.; Funding acquisition, V.R. All authors have read and agreed to the published version of the manuscript.

Funding: This research was possible through North Dakota Industrial Commission (NDIC) for the financial support contract NDIC G-045-89.

Data Availability Statement: Data available upon request.

Acknowledgments: The authors would like to acknowledge the support from the Hess Corporation for providing the data and validating the results. We appreciate the ResFrac team for their software support and cloud computing power. We would also like to thank Neal Nagel and Marisela Sanchez-Nagel for reviewing the paper.

Conflicts of Interest: The authors declare no conflict of interest.

Appendix A

The mechanical earth model was built using the anisotropic extended Eaton method for transverse isotropic media [55]. The approach assumes that the overburden's stress is applied instantaneously to an elastic rock. The horizontal stresses are a result of tectonic stress and overburden. The approach has several degrees of freedom [56]. In this work, freedom was reduced by using different data sources. The stress was estimated from well logs and calibrated with core data and DFIT values. The used DFIT values are reported in the work of Dohmen et al. [57] and Cipolla et al. [16]. Figure A1 illustrates the calibrated mechanical earth model.

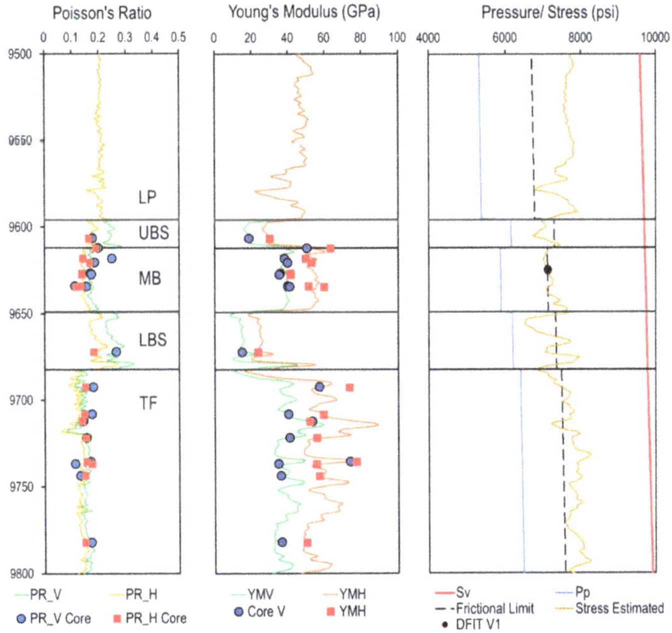

Figure A1. From left to right: Poisson's ratio, Young's modulus, stresses, and pressure. The lines are data predicted from logs and calibrated. The points represent measured core data. V stands for vertical, and H stands for horizontal.

References

1. Alexeyev, A.; Ostadhassan, M.; Bubach, B.; Boualam, A.; Djezzar, S. Integrated Reservoir Characterization of the Middle Bakken in the Blue Buttes Field, Williston Basin, North Dakota. In Proceedings of the SPE Western Regional Meeting, Bakersfield, CA, USA, 23–27 April 2017. [CrossRef]
2. Ouadi, H.; Mishani, S.; Rasouli, V. Applications of Underbalanced Fishbone Drilling for Improved Applications of Under-balanced Fishbone Drilling for Improved Recovery and Reduced Carbon Footprint in Unconventional Plays Recovery and Reduced Carbon Footprint in Unconventional Plays. *Pet. Petrochem. Eng. J.* **2023**, *7*, 000331. [CrossRef]
3. Ajani, A.; Kelkar, M. Interference study in shale plays. In Proceedings of the Society of Petroleum Engineers—SPE Hydraulic Fracturing Technology Conference 2012, The Woodlands, TX, USA, 6–8 February 2012; pp. 36–50. [CrossRef]
4. Daneshy, A.; Au-Yeung, J.; Thompson, T.; Tymko, D. Fracture shadowing: A direct method for determining of the reach and propagation pattern of hydraulic fractures in horizontal wells. *Soc. Pet. Eng.–SPE Hydraul. Fract. Technol. Conf.* **2012**, *2012*, 215–223. [CrossRef]
5. Gupta, I.; Rai, C.; Devegowda, D.; Sondergeld, C.H. Fracture hits in unconventional reservoirs: A critical review. *SPE J.* **2021**, *26*, 412–434. [CrossRef]
6. Cozby, J.; Sharma, M. Parent-Child Well Relationships across US Unconventional Basins: Learnings from a Data Analytics Study. In Proceedings of the SPE Hydraulic Fracturing Technology Conference and Exhibition, The Woodlands, TX, USA, 2 February 2022. [CrossRef]
7. Miller, G.; Lindsay, G.; Baihly, J.; Xu, T. Parent well refracturing: Economic safety nets in an uneconomic market. In Proceedings of the Society of Petroleum Engineers—SPE Low Perm Symposium, Denver, CO, USA, 5–6 May 2016. [CrossRef]
8. Xu, T.; Lindsay, G.; Zheng, W.; Yan, Q.; Patron, K.E.; Alimahomed, F.; Panjaitan, M.L.; Malpani, R. Advanced modeling of production induced pressure depletion and well spacing impact on infill wells in spraberry, Permian basin. In Proceedings of the SPE Annual Technical Conference and Exhibition, Dallas, TX, USA, 24–26 September 2018. [CrossRef]
9. Fowler, G.; McClure, M.; Cipolla, C. Making sense out of a complicated parent/child well dataset: A bakken case study. In Proceedings of the SPE Annual Technical Conference and Exhibition, Virtual, 26–29 October 2020; pp. 1–17. [CrossRef]
10. Gala, D.P.; Manchanda, R.; Sharma, M.M. Modeling of fluid injection in depleted parent wells to minimize damage due to frac-hits. In Proceedings of the SPE/AAPG/SEG Unconventional Resources Technology Conference, Houston, TX, USA, 23–25 July 2018; pp. 1–14. [CrossRef]

11. Manchanda, R.; Bhardwaj, P.; Hwang, J.; Sharma, M.M. Parent-child fracture interference: Explanation and mitigation of child well underperformance. In Proceedings of the Society of Petroleum Engineers—SPE Hydraulic Fracturing Technology Conference and Exhibition 2018, HFTC 2018, The Woodlands, TX, USA, 23–25 January 2018. [CrossRef]
12. Merzoug, A.; Chellal, H.A.K.; Brinkerhoff, R.; Rasouli, V.; Olaoye, O. Parent-Child Well Interaction in Multi-Stage Hydraulic Fracturing: A Bakken Case Study. In Proceedings of the 56th U.S. Rock Mechanics/Geomechanics Symposium, Santa Fe, NM, USA, 26–29 June 2022. [CrossRef]
13. Ratcliff, D.; McClure, M.; Fowler, G.; Elliot, B.; Qualls, A. Modelling of Parent Child Well Interactions. In Proceedings of the SPE Hydraulic Fracturing Technology Conference and Exhibition, The Woodlands, TX, USA, 1–3 February 2022. [CrossRef]
14. Kumar, D.; Ghassemi, A.; Riley, S.; Elliott, B. Geomechanical analysis of frac-hits using a 3D poroelastic hydraulic fracture model. In Proceedings of the SPE Annual Technical Conference and Exhibition, Dallas, TX, USA, 25–26 September 2018. [CrossRef]
15. Rezaei, A.; Dindoruk, B.; Soliman, M.Y. On parameters affecting the propagation of hydraulic fractures from infill wells. *J. Pet. Sci. Eng.* **2019**, *182*, 106255. [CrossRef]
16. Cipolla, C.; Litvak, M.; Prasad, R.S.; McClure, M. Case history of drainage mapping and effective fracture length in the Bakken. In Proceedings of the Society of Petroleum Engineers—SPE Hydraulic Fracturing Technology Conference and Exhibition 2020, HFTC 2020, The Woodlands, TX, USA, 4 February 2020; pp. 1–43. [CrossRef]
17. Morsy, S.; Menconi, M.; Liang, B. Unconventional Optimized Development Strategy Workflow. In Proceedings of the Unconventional Optimized Development Strategy Workflow, Virtual, 21 April 2021; pp. 1–9. [CrossRef]
18. Cai, Y.; Taleghani, A.D. Toward controllable infill completions using frac-driven interactions FDI data. In Proceedings of the SPE Annual Technical Conference and Exhibition, Virtual, 26–29 October 2021. [CrossRef]
19. Daneshy, A. *Fra-Driven Interactions (FDI) Guides for Real-Time Analysis & Execution of Fracturing Treatment*; Daneshy Consultants Int'l: Houston, TX, USA, 2020.
20. Yu, W.; Sepehrnoori, K. Optimization of Multiple Hydraulically Fractured Horizontal Wells in Unconventional Gas Reservoirs. *J. Pet. Eng.* **2013**, *2013*, 151898. [CrossRef]
21. Ma, X.; Gildin, E.; Plaksina, T. Efficient optimization framework for integrated placement of horizontal wells and hydraulic fracture stages in unconventional gas reservoirs. *J. Unconv. Oil Gas Resour.* **2015**, *9*, 1–17. [CrossRef]
22. Plaksina, T.; Gildin, E. Journal of Natural Gas Science and Engineering Practical handling of multiple objectives using evolutionary strategy for optimal placement of hydraulic fracture stages in unconventional gas reservoirs. *J. Nat. Gas Sci. Eng.* **2015**, *27*, 443–451. [CrossRef]
23. Rahmanifard, H.; Plaksina, T. Application of fast analytical approach and AI optimization techniques to hydraulic fracture stage placement in shale gas reservoirs. *J. Nat. Gas Sci. Eng.* **2018**, *52*, 367–378. [CrossRef]
24. Cipolla, C.; Gilbert, C.; Sharma, A.; LeBas, J. Case history of completion optimization in the Utica. In Proceedings of the Society of Petroleum Engineers—SPE Hydraulic Fracturing Technology Conference and Exhibition 2018, HFTC 2018, The Woodlands, TX, USA, 23–25 January 2018. [CrossRef]
25. Kim, K.; Choe, J. Hydraulic Fracture Design with a Proxy Model for Unconventional Shale Gas Reservoir with Considering Feasibility Study. *Energies* **2019**, *12*, 220. [CrossRef]
26. Wang, J.; Olson, J.E. Auto-Optimization of Hydraulic Fracturing Design with Three-Dimensional Fracture Propagation in Naturally Fractured Multi-Layer Formations. In Proceedings of the Unconventional Resources Technology Conference, Virtual, 20–22 July 2020. [CrossRef]
27. Garcia Ferrer, G.; Faskhoodi, M.; Zhmodik, A.; Li, Q.; Mukisa, H. Completion Optimization of Child Wells in a Depleted Environment, a Duvernay Example. In Proceedings of the Completion Optimization of Child Wells in a Depleted Environment, a Duvernay Example, Virtual, 30 September 2020. [CrossRef]
28. Wang, H.; Chen, Z.; Chen, S.; Hui, G.; Kong, B. Production forecast and optimization for parent-child well pattern in unconventional reservoirs. *J. Pet. Sci. Eng.* **2021**, *203*, 108899. [CrossRef]
29. Kang, C.A.; McClure, M.W.; Reddy, S.; Naidenova, M.; Tyankov, Z. Optimizing Shale Economics with an Integrated Hydraulic Fracturing and Reservoir Simulator and a Bayesian Automated History Matching and Optimization Algorithm. In Proceedings of the Optimizing Shale Economics with an Integrated Hydraulic Fracturing and Reservoir Simulator and a Bayesian Automated History Matching and Optimization Algorithm, The Woodlands, TX, USA, 1–3 February 2022. [CrossRef]
30. Pudugramam, S.; Irvin, R.J.; McClure, M.; Fowler, G.; Bessa, F.; Zhao, Y.; Han, J.; Li, H.; Kohli, A.; Zoback, M.D. Optimizing Well Spacing and Completion Design Using Simulation Models Calibrated to the Hydraulic Fracture Test Site 2 (HFTS-2) Dataset. In Proceedings of the 10th Unconventional Resources Technology Conference, Houston, TX, USA, 20–22 June 2022. [CrossRef]
31. Dohmen, T.; Zhang, J.; Li, C.; Blangy, J.P.; Simon, K.M.; Valleau, D.N.; Ewles, J.D.; Morton, S.; Checkles, S. A new surveil-lance method for delineation of depletion using microseismic and its application to development of unconventional reservoirs. In Proceedings of the SPE Annual Technical Conference and Exhibition, New Orleans, LA, USA, 30 September–2 October 2013; Volume 3, pp. 2373–2386. [CrossRef]
32. Boualam, A.; Rasouli, V.; Dalkhaa, C.; Djezzar, S. Advanced petrophysical analysis and water saturation prediction in three forks, williston basin. In Proceedings of the SPWLA 61st Annual Logging Symposium, Virtual Online Webinar, 24 June–29 July 2020. [CrossRef]

33. Sennaoui, B.; Pu, H.; Afari, S.; Malki, M.L.; Kolawole, O. Pore-and Core-Scale Mechanisms Controlling Supercritical Cyclic Gas Utilization for Enhanced Recovery under Immiscible and Miscible Conditions in the Three Forks Formation. *Energy Fuels* **2013**, *37*, 459–476. [CrossRef]
34. Cho, Y.; Uzun, I.; Eker, E.; Yin, X.; Kazemi, H. Water and Oil Relative Permeability of Middle Bakken Formation: Experiments and Numerical Modeling. In Proceedings of the 4th Unconventional Resources Technology Conference, Houston, TX, USA, 1–3 August 2016. [CrossRef]
35. Dontsov, E.V. A Continuous Fracture Front Tracking Algorithm with Multi Layer Tip Elements (MuLTipEl) for a Plane Strain Hydraulic Fracture. *J. Pet. Sci. Eng.* **2021**, *217*, 110841. [CrossRef]
36. Daneshy, A. Use of FDI data for comprehensive evaluation of horizontal well frac treatments. In Proceedings of the SPE Annual Technical Conference and Exhibition, Dubai, United Arab Emirates, 21–23 September 2021; pp. 1–8. [CrossRef]
37. Espinoza, N. Introduction to Energy Geomechanics; DN Espinoza: 2021. Available online: https://dnicolasespinoza.github.io/IPG.html (accessed on 6 November 2022).
38. Cipolla, C.; Wolters, J.; McKimmy, M.; Miranda, C.; Hari-Roy, S.; Kechemir, A.; Gupta, N. Observation Lateral Project: Direct Measurement of Far-Field Drainage. In Proceedings of the Observation Lateral Project: Direct Measurement of Far-Field Drainage, The Woodlands, TX, USA, 25 January 2022. [CrossRef]
39. Naghizadeh, A.; Larestani, A.; Nait Amar, M.; Hemmati-Sarapardeh, A. Predicting viscosity of CO_2–N_2 gaseous mixtures using advanced intelligent schemes. *J. Pet. Sci. Eng.* **2022**, *208*, 109359. [CrossRef]
40. Mirjalili, S.; Mirjalili, S.M.; Lewis, A. Grey Wolf Optimizer. *Adv. Eng. Softw.* **2014**, *69*, 46–61. [CrossRef]
41. Faris, H.; Aljarah, I.; Al-Betar, M.A.; Mirjalili, S. Grey wolf optimizer: A review of recent variants and applications. *Neural Comput. Appl.* **2018**, *30*, 413–435. [CrossRef]
42. Nait Amar, M.; Jahanbani Ghahfarokhi, A.; Ng, C.S.W.; Zeraibi, N. Optimization of WAG in real geological field using rigorous soft computing techniques and nature-inspired algorithms. *J. Pet. Sci. Eng.* **2021**, *206*, 109038. [CrossRef]
43. Xu, C.; Nait Amar, M.; Ghriga, M.A.; Ouaer, H.; Zhang, X.; Hasanipanah, M. Evolving support vector regression using Grey Wolf optimization; forecasting the geomechanical properties of rock. *Eng. Comput.* **2020**, *38*, 1819–1833. [CrossRef]
44. Ouadi, H.; Laalam, A.; Hassan, A.; Chemmakh, A.; Rasouli, V.; Mahmoud, M. Design and Performance Analysis of Dry Gas Fishbone Wells for Lower Carbon Footprint. *Fuels* **2023**, *4*, 92–110. [CrossRef]
45. Kennedy, J.; Eberhart, R. Particle swarm optimization. *Proc. ICNN'95 Int. Conf. Neural Netw.* **1995**, *4*, 1942–1948. [CrossRef]
46. Nait Amar, M.; Zeraibi, N.; Redouane, K. Bottom hole pressure estimation using hybridization neural networks and grey wolves optimization. *Petroleum* **2018**, *4*, 419–429. [CrossRef]
47. Cardoso Braga, D.; Kamyab, M.; Joshi, D.; Harclerode, B.; Cheatham, C. Using Particle Swarm Optimization to Compute Hundreds of Possible Directional Paths to Get Back/Stay in the Drilling Window. In Proceedings of the SPE Annual Technical Conference and Exhibition, Dubai, United Arab Emirates, 23 September 2021. [CrossRef]
48. Gaikwad, G.; Ahire, P. Oil Field Optimization Using Particle Swarm Optimization. In Proceedings of the 2019 5th International Conference on Computing, Communication, Control and Automation (ICCUBEA), Pune, India, 19–21 September 2019; pp. 1–4. [CrossRef]
49. McKay, M.D.; Beckman, R.J.; Conover, W.J. A Comparison of Three Methods for Selecting Values of Input Variables in the Analysis of Output from a Computer Code. *Technometrics* **1979**, *21*, 239. [CrossRef]
50. Sobol', I.M. On the distribution of points in a cube and the approximate evaluation of integrals. *USSR Comput. Math. Math. Phys.* **1967**, *7*, 86–112. [CrossRef]
51. Ng, C.S.W.; Jahanbani Ghahfarokhi, A.; Nait Amar, M. Production optimization under waterflooding with Long Short-Term Memory and metaheuristic algorithm. *Petroleum* **2022**, *9*, 53–60. [CrossRef]
52. McClure, M.; Kang, C.; Medam, S.; Hewson, C. ResFrac Technical Writeup. *arXiv* **2018**, arXiv:1804.02092.
53. Hanisch, L. SwarmLib. Available online: https://github.com/aimacode/aima-python/graphs/contributors (accessed on 8 October 2022).
54. Miranda, L.J.V.; Moser, A.; Cronin, S.K. PySwarms, Version 1.3.0. 2017. Available online: https://github.com/ljvmiranda921/pyswarms (accessed on 8 December 2022).
55. Chellal, H.A.K.; Merzoug, A.; Rasouli, V.; Brinkerhoff, R. Effect of Rock Elastic Anisotropy on Hydraulic Fracture Containment in the Bakken Formation. In Proceedings of the 56th U.S. Rock Mechanics/Geomechanics Symposium, Santa Fe, NM, USA, 26–29 June 2022. [CrossRef]
56. Zoback, M.D. *Reservoir Geomechanics*; Cambridge University Press: Cambridge, UK, 2007; Volume 53. [CrossRef]
57. Dohmen, T.; Zhang, J.; Barker, L.; Blangy, J.P. Microseismic magnitudes and b-values for delineating hydraulic fracturing and depletion. *SPE J.* **2017**, *22*, 1624–1634. [CrossRef]

eng

MDPI

Article

Prediction of Key Parameters in the Design of CO_2 Miscible Injection via the Application of Machine Learning Algorithms

Mohamed Hamadi [1], Tayeb El Mehadji [1], Aimen Laalam [2,*], Noureddine Zeraibi [1], Olusegun Stanley Tomomewo [2], Habib Ouadi [2] and Abdesselem Dehdouh [3]

1 Department of Mining and Petroleum Engineering, University of Boumerdes, Boumerdes 35000, Algeria; mohamed.99.hamadi@gmail.com (M.H.); t.elmehadji@gmail.com (T.E.M.); no_zeraibi@univ-boumerdes.dz (N.Z.)

2 Department of Petroleum Engineering, University of North Dakota, Grand Forks, ND 58202, USA; olusegun.tomomewo@und.edu (O.S.T.); habib.ouadi@und.edu (H.O.)

3 Department of Energy and Petroleum Engineering, University of Wyoming, Laramie, WY 82072, USA; adehdouh@uwyo.edu

* Correspondence: aimen.laalam@und.edu

Abstract: The accurate determination of key parameters, including the CO_2-hydrocarbon solubility ratio (Rs), interfacial tension (IFT), and minimum miscibility pressure (MMP), is vital for the success of CO_2-enhanced oil recovery (CO_2-EOR) projects. This study presents a robust machine learning framework that leverages deep neural networks (MLP-Adam), support vector regression (SVR-RBF) and extreme gradient boosting (XGBoost) algorithms to obtained accurate predictions of these critical parameters. The models are developed and validated using a comprehensive database compiled from previously published studies. Additionally, an in-depth analysis of various factors influencing the Rs, IFT, and MMP is conducted to enhance our understanding of their impacts. Compared to existing correlations and alternative machine learning models, our proposed framework not only exhibits lower calculation errors but also provides enhanced insights into the relationships among the influencing factors. The performance evaluation of the models using statistical indicators revealed impressive coefficients of determination of unseen data (0.9807 for dead oil solubility, 0.9835 for live oil solubility, 0.9931 for CO_2-n-Alkane interfacial tension, and 0.9648 for minimum miscibility pressure). One notable advantage of our models is their ability to predict values while accommodating a wide range of inputs swiftly and accurately beyond the limitations of common correlations. The dataset employed in our study encompasses diverse data, spanning from heptane (C_7) to eicosane (C_{20}) in the IFT dataset, and MMP values ranging from 870 psi to 5500 psi, covering the entire application range of CO_2-EOR. This innovative and robust approach presents a powerful tool for predicting crucial parameters in CO_2-EOR projects, delivering superior accuracy, speed, and data diversity compared to those of the existing methods.

Keywords: CO_2-EOR; solubility; interfacial tension; minimum miscibility pressure; machine learning

check for updates

Citation: Hamadi, M.; El Mehadji, T.; Laalam, A.; Zeraibi, N.; Tomomewo, O.S.; Ouadi, H.; Dehdouh, A. Prediction of Key Parameters in the Design of CO_2 Miscible Injection via the Application of Machine Learning Algorithms. *Eng* **2023**, *4*, 1905–1932. https://doi.org/10.3390/eng4030108

Academic Editor: Reza Rezaee

Received: 2 June 2023
Revised: 1 July 2023
Accepted: 5 July 2023
Published: 7 July 2023

1. Introduction

As our modern society continues to hinge on oil for energy and a wide range of petrochemical products, ranging from everyday household goods to essential medicines, the management of oil resources has become increasingly critical [1]. Of particular concern are the diminishing recovery rates seen in oil fields worldwide, indicating that our current extraction techniques may not be sufficient to satisfy global demand [2]. Estimates suggest that more than half (about 2/3) of the original oil in place (OOIP) remains untapped after primary and secondary recovery methods are applied [3]. For instance, the Rhourde El Baguel (REB) field in Algeria has only managed to recover roughly 21% of the OOIP in over 30 years of production [4]. This points toward an urgent need for enhanced oil recovery (EOR) methods to retrieve substantial quantities of trapped oil [5].

The application of EOR is not just a matter of resource efficiency; it also plays a significant role in environmental preservation. As the oil and gas industry moves towards decarbonization in alignment with global efforts to mitigate climate change, the role of CO_2-EOR becomes even more crucial as part of carbon capture, utilization, and storage (CCUS) strategies [6]. This approach aligns with the industry's goal to remain a leading energy system while addressing environmental concerns. By effectively managing and utilizing CO_2 emissions for oil recovery, the industry not only enhances its resource efficiency but also makes significant strides toward sustainability [7].

Among various EOR techniques, miscible CO_2 gas injection has emerged as the most widely implemented approach in numerous countries, particularly for light oil reservoirs [8]. With nearly 80% of global reservoirs suited for some form of CO_2 injection [9], this method's growing prevalence can be attributed to the economic attractiveness of naturally sourced CO_2, which provides a cost-effective supply [10].

The success of a CO_2-EOR project heavily relies on key parameters such as minimum miscibility pressure (MMP), interfacial tension (IFT), and solubility (Rs) [11]. When CO_2 is injected into oil reservoirs, it dissolves in the oil, causing the oil to swell and reducing its viscosity. This process also lowers the interfacial tension between fluid phases, aiding in the retrieval of trapped oil. Optimal conditions are achieved when the interfacial tension between fluid phases reaches zero, which signifies that CO_2 has become fully miscible with the oil, thereby facilitating the most efficient oil displacement [12].

The oil and gas industry is currently undergoing a significant digital transformation, with advancements in artificial intelligence (AI) and machine learning reshaping traditional practices [13]. Machine learning is being leveraged for tasks such as analysis and modeling, drilling and subsurface characterization, forecasting maintenance requirements, optimizing supply chains, and financial resource management [14]. The integration of these technologies has seen a surge in recent years, and as the industry recognizes the value they add, innovative applications continue to multiply [15].

A substantial number of studies have sought to understand the EOR process via miscible CO_2 injection, employing both experimental and numerical simulation techniques [16]. In recent times, machine learning methods have been increasingly used to gain valuable insights into EOR projects [15]. This study aims to further contribute to this burgeoning field by applying various supervised machine learning techniques to accurately predict key parameters including solubility (Rs), interfacial tension (IFT), and minimum miscibility pressure (MMP) required for effective CO_2-EOR design.

2. Literature Review

The design of a CO_2 miscible injection requires the prediction of key parameters such as the minimum miscibility pressure (MMP), CO_2 solubility, and phase behavior of the CO_2–oil system.

The minimum miscibility pressure (MMP) is a crucial parameter in CO_2 miscible injection, as it indicates the pressure at which the injected CO_2 and the oil become completely miscible [17]. Accurate prediction of the MMP is necessary to optimize the design of the CO_2 injection process and increase oil recovery [18]. Several models and methods have been proposed to predict the MMP in CO_2 miscible injection. These models can be categorized into equation of state (EOS) models and empirical models [19]. EOS models are based on the principle of thermodynamics and can predict the phase behavior of the CO_2–oil system as a function of pressure and temperature. Empirical models, on the other hand, use statistical methods to fit experimental data and predict the MMP [20].

One of the most widely used EOS models for predicting the MMP is the Peng–Robinson (PR) equation of state. This model considers the interactions between the CO_2 and oil molecules and it can predict the phase behavior of the CO_2–oil system [21]. Several modifications have been proposed to improve the accuracy of the PR model for predicting the MMP. For instance, Kiani et al. [22] developed a new PR model that accounts for the impact of asphaltene on MMP prediction. This model was validated using experimental data and demonstrated superior accuracy compared to that of existing models. Additionally, Tahsin Ahmed [23] utilized a modified version of the PR EOS, along with a newly introduced "Miscibility Function", to estimate the injection pressure required for miscible gas injection. Meanwhile, Alshuaibi et al. [24] developed a novel formula for the Abu Dhabi reservoir, which incorporates parameters such as temperature, saturation pressure, and reservoir fluid composition to determine the MMP. Rajak and Ashutosh [25] used multiple EOS models, despite the limited laboratory data, to develop a novel approach for estimating the appropriate MMP value. These methods offer potential ways to optimize the design of CO_2 injection and enhance oil recovery.

Machine learning algorithms are another approach for predicting the MMP. Sinha et al. [26] developed an analytical correlation for calculating the MMP and tuned the correlation coefficients using linear SVM. They also used a hybrid approach that combined random forest (RF) regression and analytical correlation. Shakeel et al. [27] focused on artificial neural network (ANN) and adaptive neuro-fuzzy inference system (ANFIS) techniques to predict MMP for CO_2 miscible flooding. The results showed that the ANN prediction was overall better than the ANFIS technique. Li et al. [28] evaluated the reliability of four machine learning-based prediction models including neural network analysis (NNA), genetic function approximation (GFA), multiple linear regression (MLR), and partial least squares (PLS) using 136 sets of data. Other machine learning models have also been developed for MMP prediction, such as those developed by the authors of [18,29–32].

The prediction of CO_2 solubility in oil is another important parameter that affects the design of CO_2 miscible injection. Various models have been developed to accurately predict CO_2 solubility in crude oil. Zhang et al. [33] developed a novel method using artificial neural networks to predict CO_2 solubility in heavy oil, which was found to be accurate and more efficient than traditional simulation methods. Dadan et al. [34] provided a reliable model to predict CO_2 solubility in formation brines using ion-specific parameters and a binary interaction parameter between ions and CO_2. The solubility of CO_2 in aqueous electrolyte solutions was also described using the electrolyte perturbed hard-sphere chain equation of state (e-PHSC) by Dadan et al. [34]. Zhen et al. [35] employed an artificial neural network (ANN) and support vector machine (SVM) to develop GC models based on 10,116 CO_2 solubility data measured in various ionic liquids (ILs) at different temperatures and pressures. These models can significantly aid in the design of a CO_2 miscible injection.

The phase behavior of the CO_2–oil system is another critical parameter that affects the design of a CO_2 miscible injection. Cheng et al. [36] investigated the effect of phase behavior on the design of a CO_2 miscible injection. The study showed that the CO_2–oil system can exhibit different phase behaviors depending on the pressure and temperature conditions. Therefore, it is important to consider the phase behavior when designing CO_2 miscible injection. Zhao et al. [37] developed a new model to predict the CO_2–oil phase behavior using the Grayson–Streed method. The model was validated using experimental data and was found to be more accurate than existing models.

3. Data Collection

Data collection stands as the cornerstone in resolving any supervised machine learning problem. The efficacy of predictive models hinges largely on the quality of the data they are derived from. As such, meticulous data collection practices have become an indispensable component in crafting highly effective models. The collected data need to be free from errors and brimming with pertinent information directly relevant to the task at hand.

Before embarking on the journey of model development, we must subject our collected data to rigorous statistical analysis. This preliminary step ensures that we gauge the quality of data distribution, isolate and eliminate any outliers, and verify the presence of relationships among our parameters. This data-driven examination lays a solid groundwork for our subsequent machine learning endeavors, facilitating more accurate, reliable, and effective predictive modeling.

3.1. *Solubility (Rs)*

Our dataset for this study was gathered from various published research articles [38–42]. We used laboratory measurements of the solubility of carbon dioxide (CO_2) in oil, taken with the experimental apparatus.

The primary inputs to our dataset were saturation pressure (Ps, MPa), bubble point pressure (Pb, MPa), temperature (T, °C), molecular weight (MW, gr/mol), and specific gravity (γ). We selected these parameters because they are critical to describing CO_2 solubility. Furthermore, these properties are frequently utilized in artificial intelligence projects focusing on solubility.

By focusing on these parameters, we could accurately characterize CO_2 solubility, ensuring that our dataset was relevant and precise. This selection also facilitated the effective development and execution of our machine learning models, allowing a meaningful analysis of the collected data. Table 1 shows a statistical description of the data.

Table 1. A brief description of the experimental data used for the two solubility models (dead oil and live oil).

Oil State	Experimental Data	No. of Samples	Mean	Std	Min	25%	50%	75%	Max
Dead Oil	MW (gr/mole)	105	350.6415	92.0752	196	246	358	424	490
	γ	105	0.9257	0.0481	0.8382	0.8654	0.9452	0.9677	0.9867
	T (°C)	105	53.8450	35.75	18.33	26.17	48.89	69.0275	140
	Ps (MPa)	105	6.9716	4.5963	0.5	3.5475	6.02	9.5725	27.38
	Rs (Mole fraction)	105	0.4575	0.1725	0.1	0.313	0.4789	0.6048	0.847
Live Oil	MW (gr/mole)	74	152.8364	61.9598	80.7	115.7	133.2	173.575	391.6
	γ	74	0.8371	0.0617	0.6748	0.8348	0.8498	0.8789	0.9663
	T (°C)	74	65.9297	19.122	28	59	64.7	67	123.9
	Pb (MPa)	74	8.5052	5.8059	2.15	3.05	6.2	11.91	18.52
	Ps (MPa)	74	13.6241	7.1675	3.23	8.3075	12.33	17.24	32.76
	Rs (Mole fraction)	74	0.4103	0.1677	0.1083	0.2716	0.4182	0.5381	0.7201

A pair plot was executed for both datasets to visually represent the distribution and approximate density of each variable. It also enables us to observe the interrelation between these variables. The variations within each graph can be observed in Figure 1 displayed below.

The graphs are arranged in a matrix format, where the rows represent the *y*-axis and the columns represent the *x*-axis. The diagonal subplots display the individual distributions of each attribute. For instance, when examining the molecular weight distribution in dead oil, it is observed that the values are well-distributed and mostly fall within a range from 200 to approximately 490 gr/mol. The distribution density is higher between 350 and 375 gr/mol. Conversely, in live oil, the molecular weight values are relatively lower compared to those of dead oil, ranging between 13 and about 300 gr/mol. The distribution density is higher between 110 and 170 gr/mol. These molecular weight ranges align with the physical properties of the oils; live oil contains volatile components, resulting in a higher distribution density in the lower molecular weight range. On the other hand, dead oil is a heavier oil or residue that has lost its volatile components, leading to a higher distribution density in the higher molecular weight range.

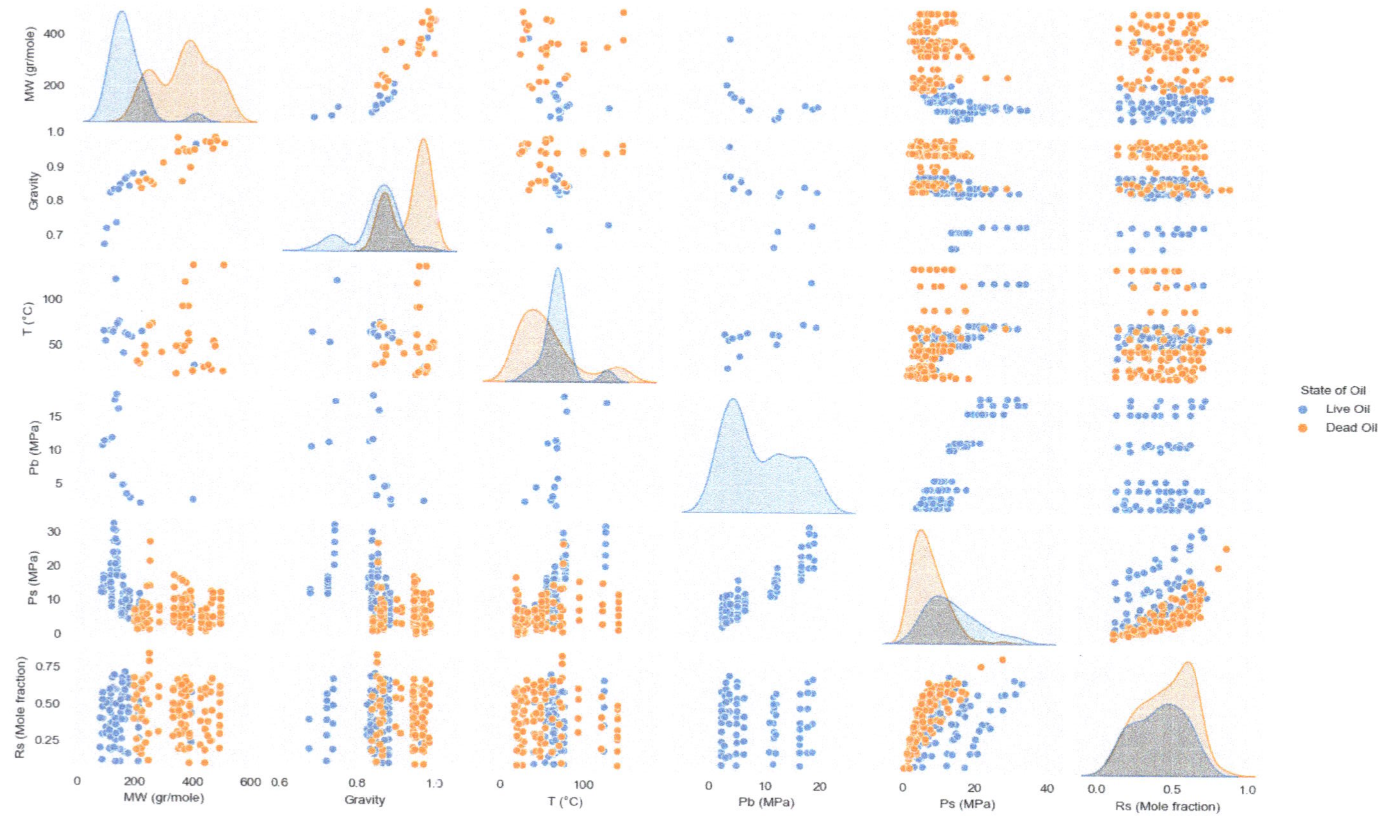

Figure 1. Pair plot of CO_2 solubility data for dead oil and live oil.

Furthermore, the graphs reveal a significant correlation between saturation pressure (Ps) and the solubility of CO_2 (Rs) in both models. As the saturation pressure increases, the solubility also increases.

Figure 2 depicts a graph with linear curves, providing a clearer illustration of the strong relationship between these variables.

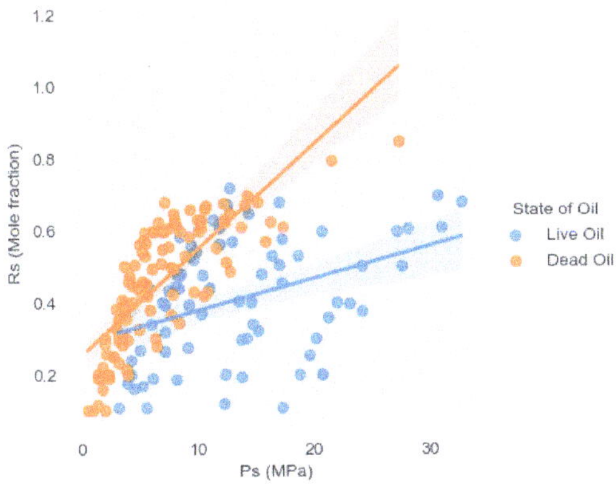

Figure 2. Rs versus Ps for both models with their linear curves.

The Pearson correlation coefficient was employed to quantify the degree of association between the input variables and solubility, further validating the aforementioned observations. Table 2 and the heat maps in Figure 3 presented below depict the correlation coefficients for each parameter.

Table 2. Correlation coefficients between solubility and other parameters.

Oil State	Experimental Data	MW (gr/mole)	γ	T (°C)	Pb (MPa)	Ps (MPa)
Dead Oil	Rs (Mole fraction)	−0.0713	−0.0934	−0.1696	-	0.7813
Live Oil	Rs (Mole fraction)	0.0231	0.0181	0.0774	−0.0132	0.3844

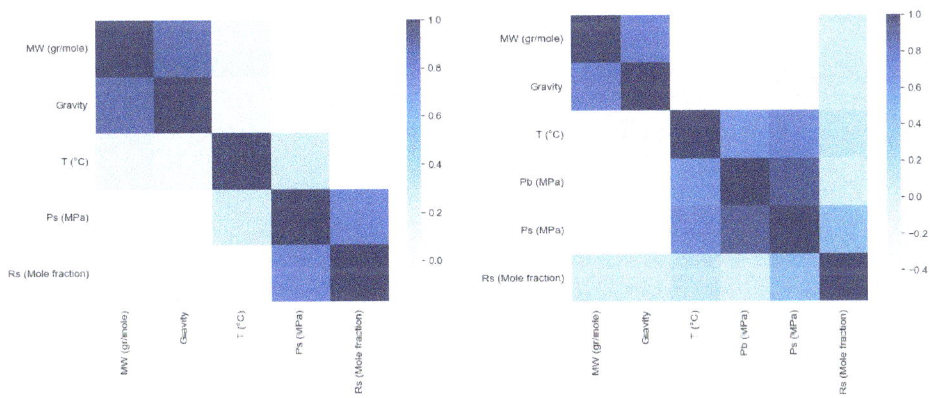

Figure 3. Heatmaps of correlation coefficients. (**a**) dead oil; (**b**) live oil.

The heatmaps clearly indicate that certain input variables exhibit a weak linear relationship with solubility. This implies that a linear model may not be suitable for capturing these relationships effectively. Consequently, a nonlinear implementation is required to accurately identify and model these relationships.

3.2. *Interfacial Tension (IFT)*

Data regarding CO_2–n-alkane interfacial tension (IFT, mN/m) were gathered from various research sources, including works by Zolghadr et al. [43], Philip T. Jaeger [44], and Georgiadis et al. [45]. It is important to note that the sessile drop technique at high pressures was the primary method used for experimentally determining the interfacial tension in most of these sources. The histogram displayed below (Figure 4) illustrates the data distribution for each component.

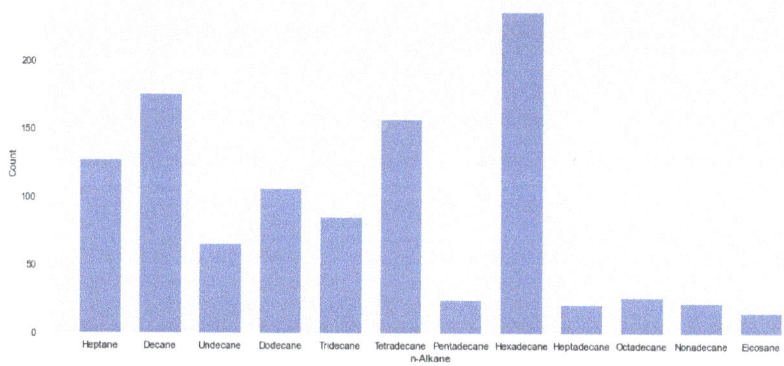

Figure 4. Data distribution for each component.

The parameters that characterize the interfacial tension include pressure (P, MPa), temperature (T, K), molecular weight (MW, g/mol), critical temperature (Tc, K), critical pressure (Pc, MPa), and the acentric factor (ω) of the n-alkane. Table 3 provides a statistical description of the dataset. These properties were chosen because of their significant impact on interfacial tension, making them crucial inputs to our dataset. This careful selection of features ensured that our machine learning models were informed by relevant and precise data, leading to accurate and meaningful results.

Table 3. A brief description of the data used for the interfacial tension model.

Experimental Data	No. Of Samples	Mean	Std	Min	25%	50%	75%	Max
MW (g/mol)	1071	175.6069	64.6520	96	134	175	222	275
P (MPa)	1071	6.3848	4.1064	0.097	3.025	6	9.085	17.1
T (K)	1071	350.6999	31.6949	297.85	323.175	344.3	373.1	443.05
IFT (mN/m)	1071	9.8366	5.8556	0.001	5.225	9.37	14.15	27.05

By examining the histograms presented in Figure 5, it is evident that the data for each parameter are distributed effectively within their minimum and maximum ranges. Taking molecular weight and pressure as examples, we observe a high-density distribution between 210 and 230 g/mol, particularly peaking at 222 g/mol, which corresponds to hexadecane. As for pressure, there is a notable concentration of values below 10 MPa. This is of particular interest because, for economic reasons, it is desirable to achieve low interfacial tension (IFT) values at the lowest possible pressure.

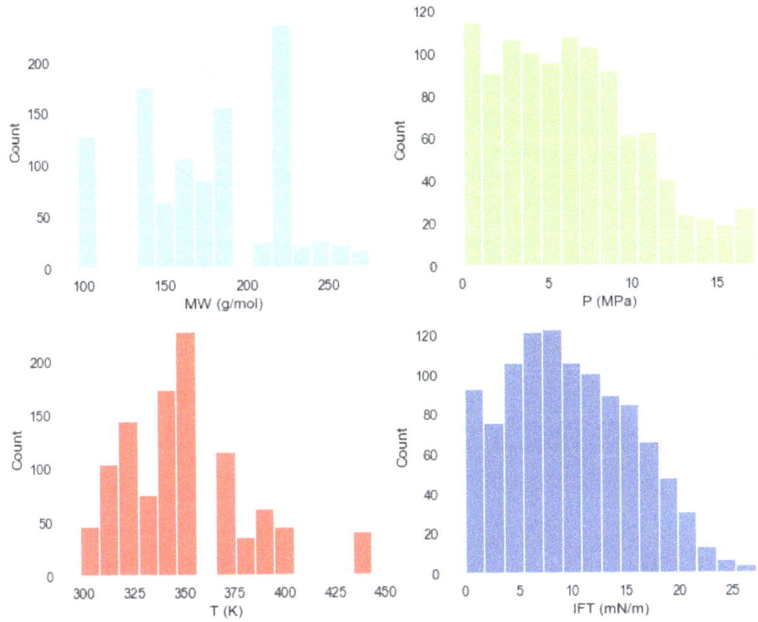

Figure 5. Data distribution for each parameter.

To gain deeper insights into the impact of pressure on interfacial tension, a scatter plot (Figure 6) was created to visualize the relationship between pressure and interfacial tension at different temperature values. The first graph demonstrates a uniform distribution of pressure for each temperature, reflecting the experimental principle outlined by the authors in the literature. The experimental method, known as the sessile drop technique, involves gradually increasing pressure to observe the behavior of interfacial tension across multiple temperature values (in this case, 11 values). The experiment was repeated with diverse compositions, and the results were recorded. In the second graph, a prominent association between interfacial tension (IFT) and pressure is evident. As pressure rises, there is a noticeable reduction in interfacial tension. This correlation holds true for all compositions tested, indicating the consistent influence of pressure on interfacial tension.

Figure 6. Pressure–interfacial tension relationship.

The correlation coefficients displayed in Table 4 and Figure 7 below reveal notable relationships between the variables. Pressure exhibits a strong negative correlation with interfacial tension, indicated by a coefficient of −0.8577. Similarly, temperature shows a negative correlation, albeit weaker, with a coefficient of −0.2042. Conversely, molecular weight displays a positive linear relationship with interfacial tension, reflected by a coefficient of 0.2918.

Table 4. Correlation coefficients between interfacial tension and the other parameters.

Experimental Data	MW (gr/mole)	P (MPa)	T (K)
IFT (mN/m)	0.2918	−0.8577	−0.2042

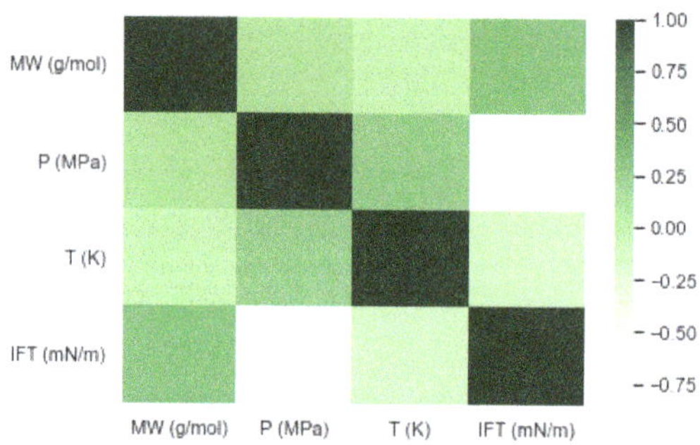

Figure 7. The heatmap of correlation coefficients between interfacial tension and the other parameters.

3.3. Minimum Miscibility Pressure (MMP)

The data utilized for the model's development were obtained from various literature sources, notably Cronquist [46], Metcalfe [47], Alston et al. [48], Yuan et al. [49], and Zhang et al. [50]. Multiple slim tube tests were conducted under varying conditions, and the minimum miscibility pressure (MMP, MPa) values were recorded in each instance.

The key factors that influence the MMP are reservoir temperature, oil composition, and the components of the injected gas. Accordingly, the inputs chosen for our model included reservoir temperature (TR, K), the critical temperature of the injected gas (TC, K), an oil composition represented by a molecular weight of C5 and heavier (MW_{C5+}, g/mol), and the ratio of volatile to intermediate components ($_{xvol}/_{xint}$). This selection of inputs ensured that our model was guided by factors directly influencing the MMP, providing a reliable basis for accurate predictions.

The histograms displayed in Figure 8 effectively visualize the distribution of the data, and Table 5 provides a statistical description of the MMP dataset.

The histograms provide visual evidence that although the dataset covers a wide range of values, there are certain variables that are not well-distributed and may not be statistically significant. Taking MMP (minimum miscibility pressure) values as an example, we observe that the 75th percentile of the data is 19.12 MPa, while the maximum value reaches 38.52 MPa. Upon closer examination of the MMP histogram, it becomes apparent that only a small number of samples (six samples) fall above the 30 MPa threshold. To further validate and identify these values as outliers, boxplots serve as excellent visualization tools. They enable the identification of abnormal and outlier data points, which can aid in making informed decisions about their inclusion or exclusion from the dataset.

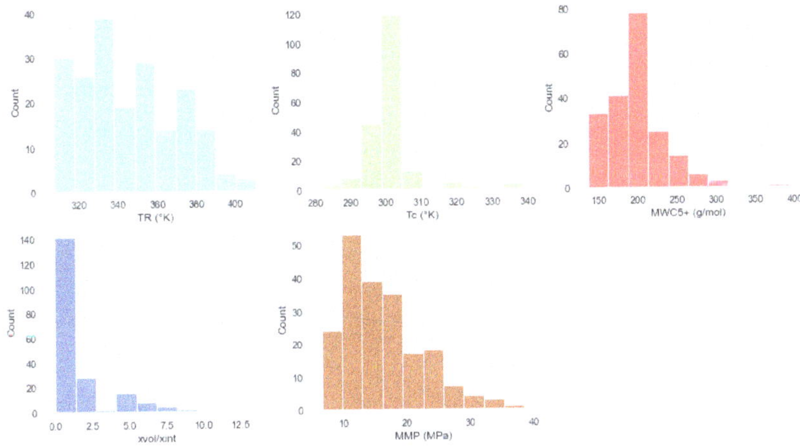

Figure 8. The distribution of data for each parameter.

Table 5. A brief description of the data used for the minimum miscibility pressure model.

Experimental Data	No. of Samples	Mean	Std	Min	25%	50%	75%	Max
TR (K)	201	345.4395	24.3101	307.55	327.59	338.71	362.040	410.37
Tc (K)	201	302.7178	8.3058	281.45	295.29	304.19	304.190	338.77
MW_{C5+} (g/mol)	201	194.6348	40.1033	136.26	171.1	187.80	211.213	391
$^{xvol}/_{xint}$	201	1.5955	2.0928	0	0.51	0.74	1.5	13.6067
MMP (MPa)	201	16.0235	6.1184	6.50	11.138	14.80	19.12	38.52

The box plot operates by identifying outliers as values that fall below the limit on the left (Q1 − 1.5 ∗ IQR) and above the limit on the right (Q3 + 1.5 ∗ IQR), where Q1 represents the first quartile (25th percentile), Q3 denotes the third quartile (75th percentile), and IQR corresponds to the interquartile range (the width of the box being from the 25th to 75th percentile). In Figure 9, the box plot reveals the presence of six outliers (represented by diamonds) that surpass the 30 MPa threshold, indicating the need for their removal from the dataset.

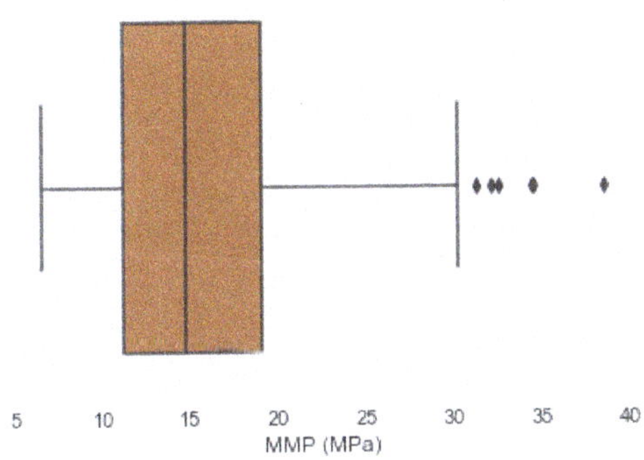

Figure 9. Boxplot of MMP data. The presence of six outliers (indicated by the six black diamonds) exceeds the right threshold (30 MPa).

It is crucial to perform this step prior to model development to ensure optimal results, as retaining these outliers would likely lead to higher error values and a lower correlation coefficient. Attempting to train the model effectively with only six values above 30 MPa would be challenging. Figure 10 demonstrates the updated box plot visualizations and data distribution histogram after removing the outliers, enabling a more accurate representation of the dataset.

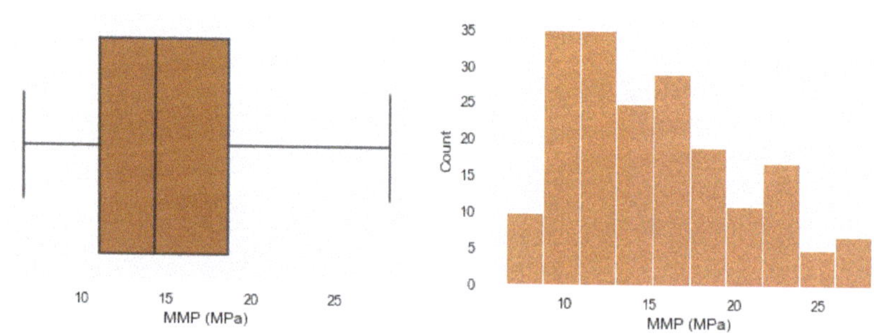

Figure 10. Boxplot and histogram of MMP after outliers' removal.

As depicted in Table 6 and Figure 11 below, a clear pattern emerges regarding the influence of various parameters on MMP variation. Reservoir temperature stands out as the most influential factor, displaying a strong positive correlation with a coefficient of 0.68. This indicates that as the temperature rises, the MMP tends to increase as well. Additionally, the molecular weight exhibits a moderate positive relationship with MMP, evident from its correlation coefficient of 0.47. Similarly, volatile to intermediate components show a modest positive correlation with a coefficient of 0.31. On the other hand, the critical temperature demonstrates a small negative linear relationship with the other parameters. Although this negative correlation is relatively weak, it still provides valuable insights and adds value to our predictive model.

Table 6. Correlation coefficients between MMP and the other parameters.

Experimental Data	T_R (K)	Tc (K)	MW_{C5+} (g/mol)	x_{vol}/x_{int}
MMP (MPa)	0.6845	−0.1829	0.4657	0.3133

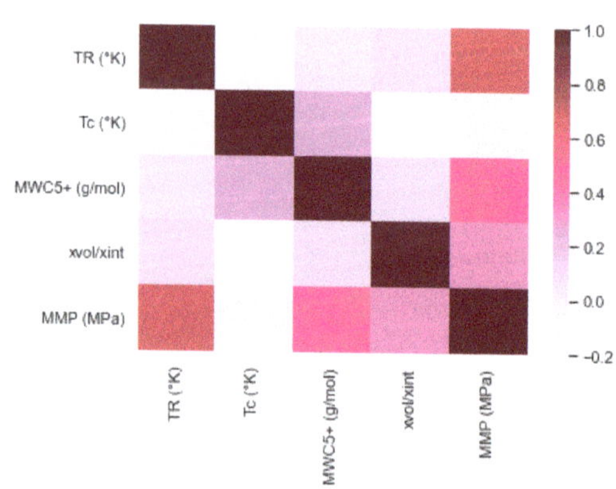

Figure 11. Heatmap of correlation coefficients between MMP and the other parameters.

4. Model Implementation

In the process of training machine learning models, it is often observed that the models might start to overfit or memorize the training data. While this might lead to good performance on the training set, it could also result in poor predictive accuracy for unseen data. To counteract overfitting and ensure the model's generalization, the dataset is commonly partitioned. Thus, the datasets in this study were randomly divided into distinct subsets:

- Dead oil solubility model: the training and validation set comprised 85% of the dataset (90 samples), and a test set formed 15% of the dataset (15 samples).
- Live oil solubility model: the training set contained 80% of the dataset (60 samples), and a test set held 20% of the dataset (14 samples).
- Interfacial tension model: the training set included 80% of the dataset (856 samples), a cross-validation set made up 1/8 of the training set (107 samples), and a test set represented 20% of the dataset (215 samples).
- Minimum miscibility pressure model: the training set consisted of 84% of the dataset (162 samples), and a test set incorporated 16% of the dataset (31 samples).

In addition, normalization of data was conducted before inputting them into machine learning models to ensure consistent ranges. For example, in the case of solubility, the molecular weight values extended up to 490 gr/mol, while specific gravity values remained as less than 1. To balance this, z-score normalization was applied.

It should also be noted that Python 3.8 and its associated libraries were utilized for the development of all models.

4.1. Dead Oil Solubility

Initially, a multilayer perceptron (MLP) model was constructed, owing to its robust nonlinear representation capability, and its foundational unit being a neuron. The configuration of varying numbers of neurons and layers enables the characterization of mapping relationships of differing complexity levels. The inputs for this model were saturation pressure (Ps, MPa), temperature (T, °C), molecular weight (MW, gr/mol), and density (γ). As a result, a four-layer structure was established, with the input layer, two hidden layers, and an output layer, respectively containing 4, 12, and 1 neuron, as illustrated in Figure 12.

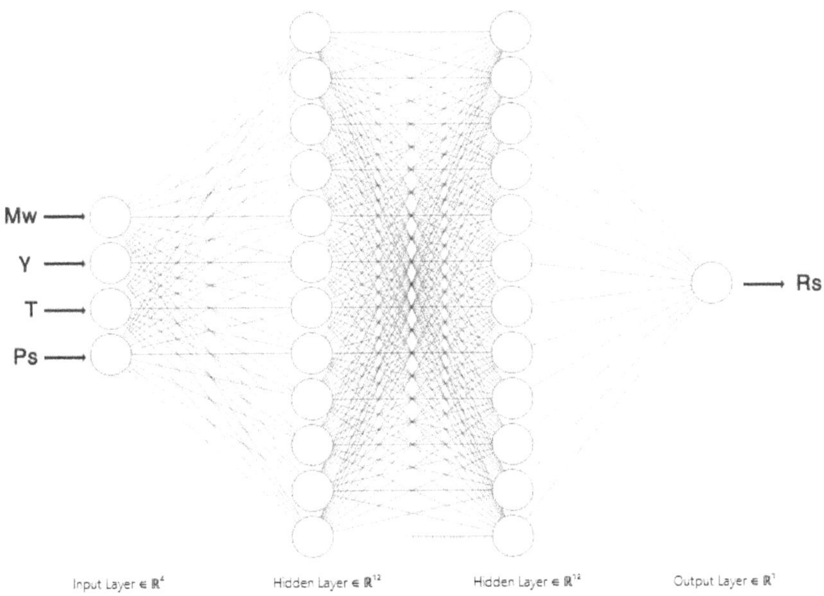

Figure 12. Architecture of the MLP-Adam solubility model for dead oil.

The flowchart presented in Figure 13 outlines the primary steps involved in constructing the MLP-Adam model and determining the optimal parameters that yield the lowest possible error. Appendix B provides a comprehensive overview of the feed forward equation in its general form, along with the corresponding weight and bias values. Additionally, it includes a detailed example illustrating the calculations using these specific weight and bias values. Table 7 provides details on the structure of the MLP-Adam model.

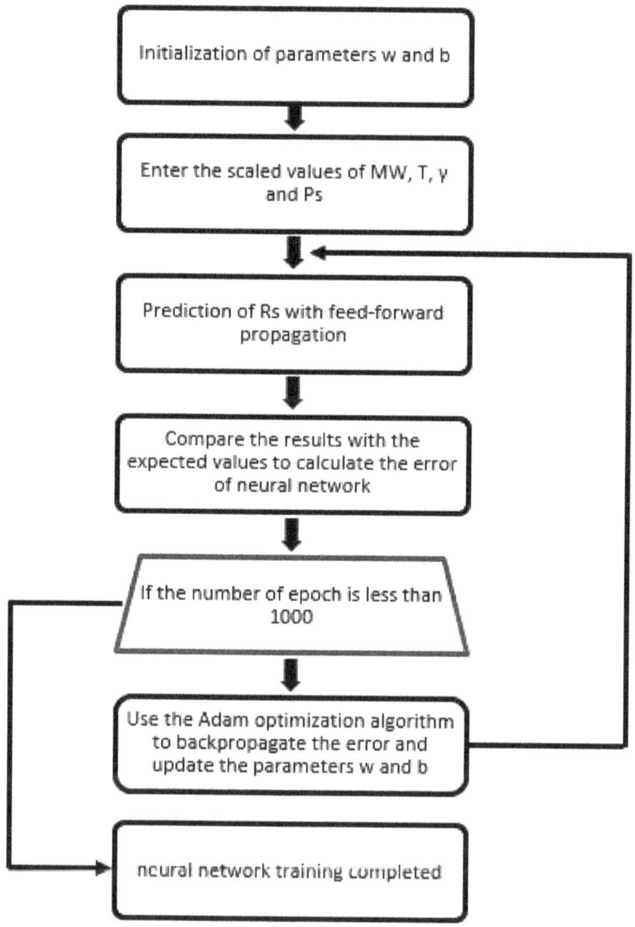

Figure 13. Flowchart of the multilayer perceptron using the Adam optimization algorithm for the proposed model.

Table 7. Structure of the proposed MLP-Adam model.

Number of hidden layers	2
Number of neurons in the hidden layers	12
Number of epochs	1000
Optimization algorithm	Adam
Activation function	Relu
Performance Indicator	MSE, MAE
Validation dataset	16 Samples

In order to evaluate the accuracy and predictive ability of the MLP-Adam model for Rs in dead oil, the average absolute relative deviation (AARD (%)), root mean square error (RMSE), and coefficient of determination (R^2) were computed (please refer to Appendix A for the definition and mathematical formulation of these metrics). The outcomes of these calculations are presented in Table 8. For visual validation, the predicted values versus the actual values for both the training and test data are depicted in Figure 14.

Table 8. Statistical analysis of MLP-Adam performance.

Model	Training Data			Test Data			All Data		
	AARD (%)	RMSE	R^2	AARD (%)	RMSE	R^2	AARD (%)	RMSE	R^2
MLP-Adam	2.0161	0.0123	0.9948	3.9629	0.0234	0.9807	2.3099	0.0145	0.9928

Figure 14. Comparative plot of predicted and experimental dead solubility values: an analysis of training data, test data, and the complete dataset.

The efficacy of the model was eventually benchmarked against some of the most commonly employed correlations in the field. The selected models from the literature include the Chung et al. [51] correlation, the Rostami et al. [52] correlation, and the genetic algorithm-based correlations of Emera and Sarma [53]. The comparative analysis was conducted utilizing the statistical parameters AARD (%), RMSE, and R^2 (check Table 9)and supplemented with an error histogram plot of the different correlations as depicted in Figure 15 below. Upon examination, the histogram of Chung et al. showcases a significant error in comparison to the other models. While the model by Emera and Sarma holds a considerable number of zero-error values, its distribution is skewed to the right with a somewhat wide error range. The model from Rostami et al. [52] presents a favorable error distribution with minimal values; nevertheless, the MLP-Adam model is still considered superior in comparison to those outlined in the literature.

Table 9. The comparison between the statistical parameters of MLP-Adam and the different correlations found in the literature.

Model	AARD (%)	RMSE	R^2
MLP-Adam	2.3099	0.0145	0.9928
Chung et al., 1988 [51]	99.4213	0.5138	0.0083
GA—Emera and Sarma, 2011 [53]	6.1521	0.0546	0.8987
Rostami et al., 2017 [52]	3.8709	0.02045	0.9858

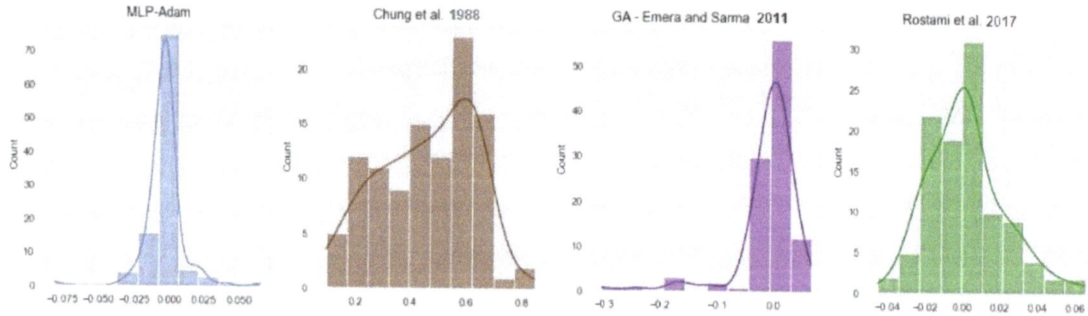

Figure 15. The error histogram of the different correlations [51–53].

4.2. Live Oil Solubility

In the instance of live oil, a support vector regression (SVR) model was constructed, with the radial basis function (RBF) being selected as the kernel function in the SVR configuration. The selection of RBF over other kernel functions can be attributed to its lower number of parameters requiring optimization and reduced computational cost [54]. Of the 74 available samples, 60 were utilized for model construction, while the remaining data served to assess model performance. In this section, an additional input, bubble point pressure (Pb, MPa), was included alongside those employed in the dead oil model. To produce a model of high accuracy, it is crucial to ascertain the optimal values of the SVR-RBF hyperparameters. In this study, the grid search method was employed to identify these optimal values in a comprehensive manner. The search range for epsilon, gamma, and C, along with the corresponding optimal values yielded via the global search, are detailed in Table 10. In total, 30 support vectors were used to construct the decision function.

Table 10. Search interval and optimal values of the SVR-RBF parameters.

Hyperparameter	C	Epsilon	Gamma
Range	0.1–50,000	0.0001–0.1	0.001–10
Optimal value	950	0.039	0.01035

The solubility values forecasted by the SVR-RBF model are plotted with the empirically determined solubility values, encompassing the training data, test data, and the complete dataset, in Figure 16. Subsequently, the statistical parameters AARD (%), RMSE, and R^2 were computed, with the corresponding results presented in Table 11.

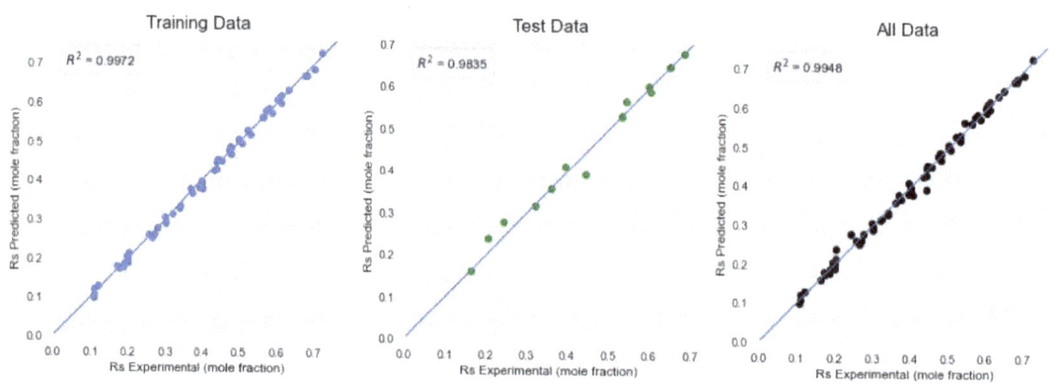

Figure 16. Plot of predicted versus experimental values of solubility in live oil.

Table 11. Statistical analysis of SVR-RBF performance.

Model	Training Data			Test Data			All Data		
	AARD (%)	RMSE	R^2	AARD (%)	RMSE	R^2	AARD (%)	RMSE	R^2
SVR-RBF	2.4618	0.0088	0.9972	4.2742	0.0209	0.9835	2.8047	0.0120	0.9948

Finally, the process implemented for the dead oil model was replicated. The performance of the SVR-RBF model was benchmarked against the most prevalent correlations in the literature, with the comparison based on previously described statistical parameters, as shown in Table 12. To bolster this comparison, an error histogram was produced, visualizing the different correlations, as depicted in Figure 17.

Table 12. The comparison between the statistical parameters of SVR-RBF and the different correlations found in the literature.

Model	AARD (%)	RMSE	R^2
SVR-RBF	2.8047	0.0120	0.9948
Chung et al. [51]	99.9250	0.4425	0.0097
GA—Emera and Sarma [53]	4.9734	0.0295	0.9686
Rostami et al. [52]	3.7642	0.0203	0.9851

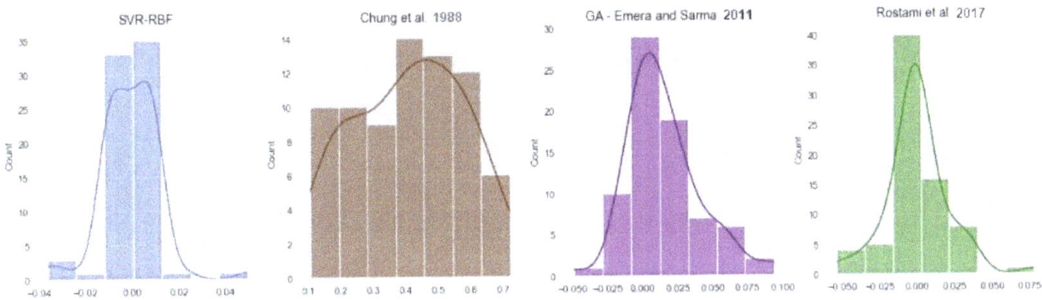

Figure 17. Error histogram of the different correlations [51–53].

Upon scrutinizing the table along with the distributions and ranges of the histograms, it becomes apparent that the SVR-RBF model outperformed the processed correlations, considering its error range and the amount of values with exceedingly low error. Nonetheless, the model proposed by Rostami et al. [52] demonstrated satisfactory accuracy when compared to the models of Chung et al. [51] and Emera and Sarma [53].

4.3. *Interfacial Tension*

To construct a robust model adept at handling extensive datasets, an XGBoost model was employed, based on the decision tree approach. An 8-fold cross-validation scheme was utilized on the input set to evade the selection bias associated with training and testing data. The hyperparameters of XGBoost that delivered optimal performance are listed in Table 13. The main procedures in the construction of the model are outlined in the accompanying flowchart of Figure 18.

Table 13. Selection of hyperparameters for the proposed XGBoost model of IFT.

Model	Hyperparameter	Range	Optimal Value
XGBoost	Number of trees	100, 200, 400, 800, 1000, 2000	1000
	Regularization parameter λ	0.0001, 0.001, 0.1, 0.3, 10, 100	0.001
	Regularization parameter α	0.01, 0.04, 0.09, 0.1	0.09
	Gamma γ	0, 0,1, 1, 10	0
	Max. depth	2, 4, 6, 8	4
	Learning rate	0.001, 0.01, 0.1	0.1

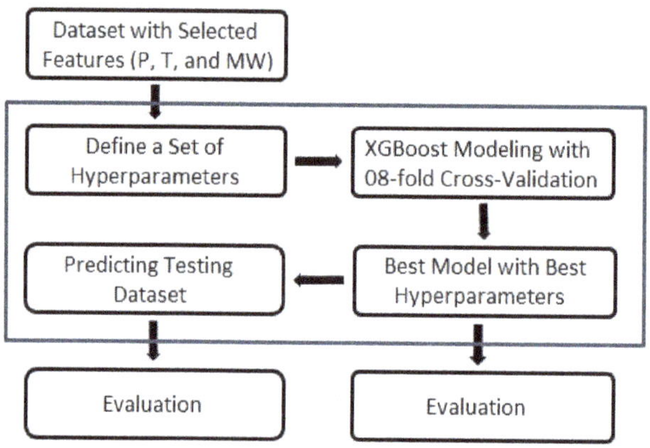

Figure 18. Flowchart of the proposed XGBoost model.

To obtain an understanding of the model's decision-making process, and to discern which parameters held the most and least significance during prediction, the XGBoost model offers a remarkable feature that enables the visualization of parameter importance. This feature is demonstrated in Figure 19 below.

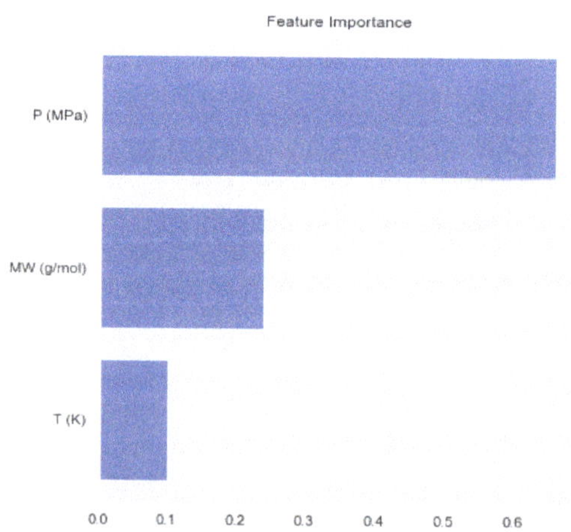

Figure 19. Importance of inputs in the prediction of IFT.

The interfacial tension values, predicted by the XGBoost model, are plotted against the corresponding experimentally measured values for the training set, the test set, and the entire dataset in Figure 20. The associated average absolute relative deviation (AARD (%)), root mean square error (RMSE), and coefficient of determination (R^2) were computed and the resulting performance are provided in Table 14.

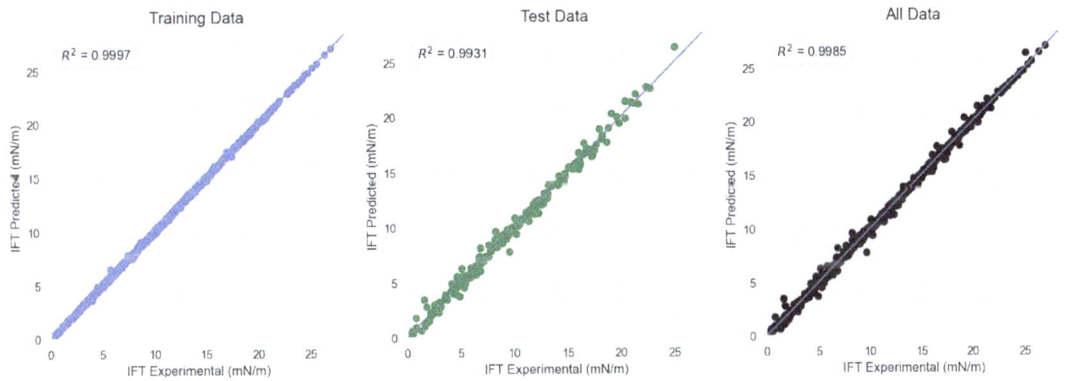

Figure 20. Plot of predicted versus experimental values of interfacial tension.

Table 14. Statistical analysis of XGBoost performance on IFT data.

Model	Training Data			Test Data			All Data		
	AARD (%)	RMSE	R^2	AARD (%)	RMSE	R^2	AARD (%)	RMSE	R^2
XGBoost	1.9386	0.0952	0.9997	8.6422	0.4698	0.9931	3.2844	0.2271	0.9985

Ultimately, the reliability of the model was evaluated through a comparison of its predictive accuracy with the Peng–Robinson equation of state (PR EOS) and the GEP model put forward by Mirzaie et al. [55]. This comparative analysis was performed using the statistical parameters AARD (%), RMSE, and R^2 (refer to Table 15), as well as through the construction of scatter plots that juxtapose the experimental IFT values with the respective predictions made by each model (see Figure 21).

Table 15. Statistical comparison of XGBoost and literature-based correlations for the IFT dataset.

Model	AARD (%)	RMSE	R^2
XGBoost	3.2844	0.2271	0.9984
PR EOS	60.5471	2.6261	0.7949
GEP	219.1053	1.4437	0.9391

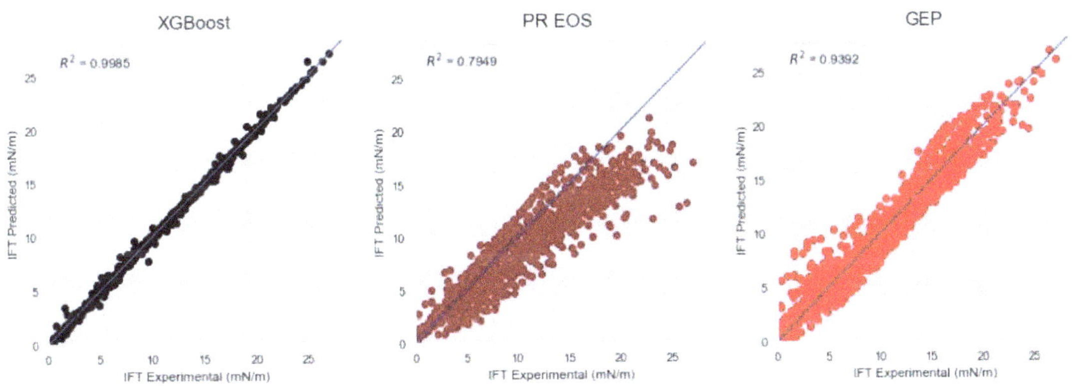

Figure 21. Scatter plots of the experimental IFT versus the values predicted by each model.

The equation-of-state model delivered satisfactory results for IFT < 15 mN/m, and the GEP model demonstrated its predictive efficacy across all data with an accuracy of

94%. However, the XGBoost model ultimately emerged as superior, boasting outstanding statistical parameters in comparison to the models currently available in the literature. Figure 22 depicts the absolute discrepancy between the experimental and predicted IFT values for the XGBoost, GEP, and EOS models. It's evident that the XGBoost model displays the most minimal error values among the three, ranging from −2 to 2 with most values hovering around zero. On the other hand, the other models exhibit error values reaching up to 12.5 and lack a normal distribution of errors centered around zero.

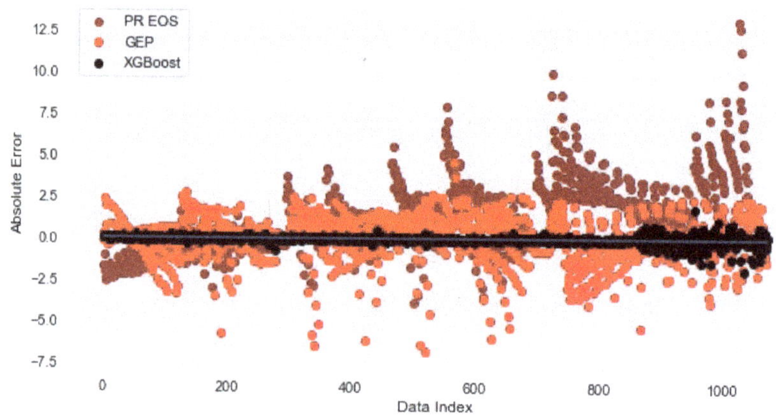

Figure 22. Scatter plot of the absolute error between predicted and experimental values.

4.4. Minimum Miscibility Pressure

XGBoost is used again on the MMP data, and it gave excellent prediction performance. The hyperparameters that fit the model best are shown in Table 16.

Table 16. Selection of hyperparameters for our XGBoost model of MMP.

Model	Hyperparameters	Range	Optimal Value
XGBoost	Number of trees	100, 1000, 4000, 5000, 8000	8000
	Regularization parameter λ	0.0001, 0.001, 0.1, 0.3, 15, 100	15
	Regularization parameter α	0.01, 0.02, 0.09, 0.1	0.02
	Gamma γ	0, 0,1, 01, 10	0
	Maximum depth	2, 4, 6, 8	2
	Learning rate	0.001, 0.01, 0.1	0.1

The XGBoost model's predicted minimum miscibility pressure values are graphed against the experimentally determined values for the training data, the test data, and the complete dataset in Figure 24. Statistical metrics-average absolute relative deviation (AARD (%)), root mean square error (RMSE), and coefficient of determination (R^2)-were computed, and the results are presented in Table 17.

Following the approach adopted for the preceding IFT model, the significance of parameters for the MMP model is assessed, pinpointing those of utmost and least importance, as depicted in Figure 23. It's readily apparent that the molecular weight of C5 plus stands out as the most significant variable, contributing 37.76%, followed by reservoir temperature at 32.93%, the ratio of intermediate to volatile components at 16.36%, and finally the critical temperature with 12.95%.

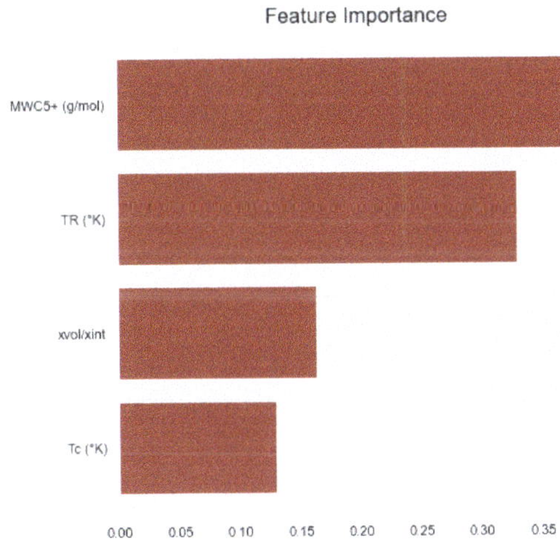

Figure 23. Importance of inputs in MMP prediction.

Figure 24. Plot of predicted versus experimental values of minimum miscibility pressure.

Table 17. Statistical analysis of XGBoost performance on MMP data.

Model	Training Data			Test Data			All Data		
	AARD (%)	RMSE	R^2	AARD (%)	RMSE	R^2	AARD (%)	RMSE	R^2
XGBoost	0.9326	0.1893	0.9986	4.0043	0.941	0.9648	1.4262	0.4151	0.9934

Upon completion of the evaluation process, the proposed model was compared to the most prevalent correlations in existing literature. Given the existence of specific correlations for pure CO_2 (100% CO_2) and others for impure CO_2 (CO_2 containing percentages of C_1, N_2, H_2S, etc.), the data was bifurcated into 'pure' and 'impure' based on the critical temperature. For pure CO_2, the correlations of Alston et al. (pure) [48], Lee [56], and Emera-Sarma [57] were used, while for impure CO_2, the correlations of Alston et al. (impure) [48] and Fathinasab-Ayatollahi [58] were utilized. Table 18 summarizes the results of the comparison.

Table 18. Statistical comparison of XGBoost and literature-based correlations for the MMP dataset.

	Model	AARD (%)	RMSE	R^2
Pure CO_2	XGBoost (Pure)	0.9161	0.1936	0.9988
	Lee [56]	18.781	5.1538	0.5146
	Alston et al. (Pure) [48]	18.177	5.5472	0.7063
	Emera-Sarma [57]	13.2203	3.7385	0.6161
Impure CO_2	XGBoost (Impure)	1.9525	0.558	0.9856
	Alston et al. (Impure) [48]	34.5324	6.4668	0.5967
	Fathinasab-Ayatollahi [58]	15.0134	2.702	0.7019

As can be seen in the aforementioned table, both the pure and impure XGBoost models exhibit the lowest AARD (%) and RMSE values, along with the highest coefficient of determination in comparison to the other models. Upon scrutinizing the error histograms for the pure CO_2 case (Figure 25), it becomes apparent that while all correlations reasonably predict an acceptable quantity of values (roughly 20), they are subject to extensive error ranges and less satisfactory distributions when compared to the XGBoost model. The XGBoost model stands out with more than 50 values concentrated around 0, and an error range restricted to -1 to 0.5. This stark contrast emphasizes the superior performance and reliability of the XGBoost model when handling pure data.

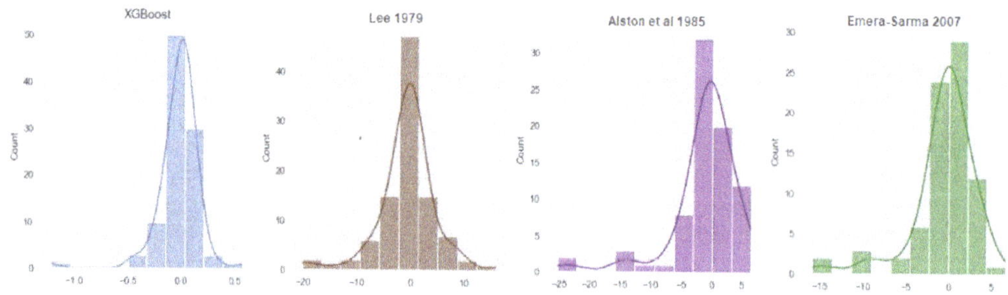

Figure 25. Error histogram of pure XGBoost and the different correlations [48,56,57].

In the scenario involving impure CO_2 (Figure 26), the Fathinasab-Ayatollahi [58] correlation delivered a relatively low error margin and a fairly decent distribution compared to that of Alston et al. [48]. However, it still could not rival the predictive efficiency of the XGBoost model, which exhibited a minimal error margin ranging from -2 to 2 and recorded over 60 values clustered around 0. This further emphasizes the robustness and precision of the XGBoost model in estimating impure CO_2 data.

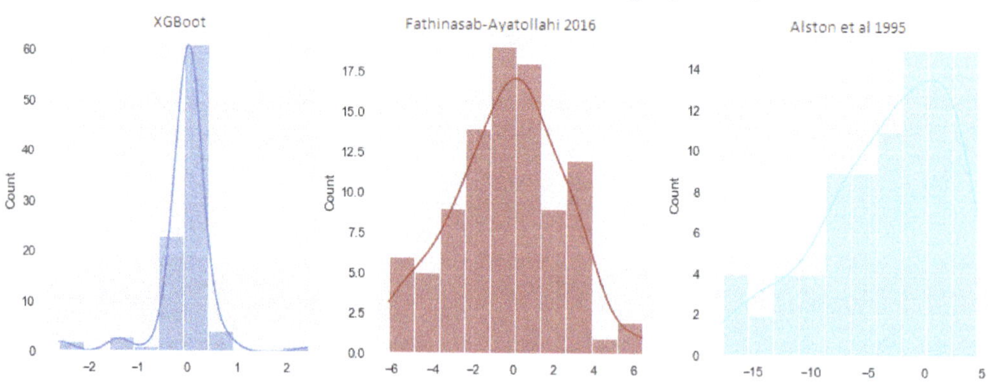

Figure 26. Error histogram of impure XGBoost and the different correlations [38,58].

5. Conclusions

This study introduces efficient and reliable models for estimating key parameters in CO_2-enhanced oil recovery (CO_2-EOR) operations: the solubility of CO_2 in both dead and live oil, the interfacial tension, and the minimum miscibility pressure. These parameters are critical as they play a significant role in the planning and implementation of CO_2-EOR projects. For instance, accurate estimation of the CO_2 solubility in oil can inform on oil displacement efficiency, while a precise calculation of interfacial tension aids in assessing the mobility of the injected CO_2, and understanding the minimum miscibility pressure is essential for the economic feasibility of the operation.

Our models, based on advanced machine learning algorithms—MLP, SVR, and XGBoost—and Adam's optimization algorithm, present an innovative approach to estimate these parameters. They not only offer a high degree of precision and reliability but also showed a promising improvement over the existing correlations in the tests conducted.

However, it is worth mentioning that the real-world validation of these models in CO_2-EOR projects remains an area for future exploration. Potential variability in the underlying data is another factor that could influence the models' performance.

We recommend future work to focus on validating these models under diverse real-world conditions, and to explore emerging machine learning algorithms and optimization techniques for potential improvements. Such research directions can further enhance the planning and implementation of CO_2-EOR projects, contributing to the advancements in the field of petroleum reservoir studies.

Author Contributions: Methodology, M.H., T.E.M. and N.Z.; Validation, A.L. and N.Z.; Investigation, M.H. and T.E.M.; Data curation, M.H. and T.E.M.; Writing–original draft, M.H., T.E.M. and A.D.; Writing–review & editing, A.L., O.S.T. and H.O. All authors have read and agreed to the published version of the manuscript.

Funding: This research received no external funding.

Data Availability Statement: The data presented in this study are available in the different articles declared in each section of the manuscript.

Conflicts of Interest: The authors declare no conflict of interest.

Appendix A

In this section, we present the definitions and mathematical formulas of the three metrics used to evaluate the models in this work.

Appendix A.1. Average Absolute Relative Deviation (AARD (%))

This is a measure of prediction accuracy in statistical modeling and forecasting. The AARD is expressed as a percentage, and lower values generally indicate better predictive accuracy. It is calculated as the average of absolute errors relative to the actual values.

The formula to calculate AARD is as follows:

$$AARD\ (\%) = \left(\frac{1}{n}\right) \sum \frac{(|Actual - Predicted|)}{Actual} * 100 \tag{A1}$$

where

- n is the total number of observations;
- *Actual* refers to the actual value;
- *Predicted* refers to the predicted value.

Appendix A.2. Root Mean Square Error (RMSE)

This is a standard way to measure the error of a model in predicting quantitative data. *RMSE* is essentially the standard deviation of the residuals (prediction errors). Lower values of *RMSE* indicate a better fit of the data. The formula for calculating *RMSE* is as follows:

$$RMSE = \sqrt{\left[\left(\frac{1}{n}\right)\sum(Actual - Predicted)^2\right]} \quad \text{(A2)}$$

where:

- n is the total number of observations;
- *Actual* refers to the actual value;
- *Predicted* refers to the predicted value.

Appendix A.3. Coefficient of Determination (R_2)

This is a statistical measure that represents the proportion of the variance for a dependent variable that is explained by an independent variable or variables in a regression model. So, if the R^2 of a model is 0.50, then approximately half of the observed variation can be explained by the model's inputs.

The formula for calculating R^2 is as follows:

$$R^2 = 1 - \left(\frac{SSres}{SStot}\right) \quad \text{(A3)}$$

where:

- *SSres* is the sum of squares of the residual errors.
- *SStot* is the total sum of squares.

Appendix B

In this section, we present the derivation of the feed forward equation for our proposed MLP-Adam model. The feed forward equation describes the mathematical relationship between the input features, hidden layers, and output prediction. Additionally, we provide tables of the weight and bias values for each layer, as well as an example calculation for a specific set of input features.

Appendix B.1. Feed Forward Equation of our MLP-Adam Model

Below is the step-by-step process of forwarding the input data through the layers of the network to generate the final output.

1. Initialize the input data. Let us denote the input vector as X.
2. Calculate the activations of the neurons in the first hidden layer by applying the *ReLU* activation function (this function computes the maximum value between 0 and the input x. If x is positive, the output is equal to x, and if x is negative, the output is set to 0) to the resulting sum to introduce non-linearity. This is carried out using the following equation:

$$a1_j = f(z1_j) = ReLU\left(\sum_{i=1}^{n} w_{ji} \cdot X_i + b_j\right) \quad \text{(A4)}$$

   where w_{ji} is the interconnection weight between the input vector X_i and the hidden layer neurons, j, $z1_j$ is the sum of the weighted inputs and the bias, b_j, and n is the number of neurons in the input layer. $a1_j$ represents the resulting activation value.

3. The same process is repeated for the second hidden layer. The output of the second hidden layer is denoted as $a2$.

$$a2_k = f(z2_k) = ReLU\left(\sum_{k=1}^{p} w_{jk} \cdot a1_j + b_k\right) \quad \text{(A5)}$$

where w_{jk} represents the weights connecting the first hidden layer neurons j to the second hidden layer neurons k, $z2_k$ Represents the weighted sum of inputs for the neurons in the second hidden layer, and b_k is the bias term. $a2_k$ represents the activation values for this second hidden layer.

4. Finally, the output of our MLP-Adam model can be calculated by applying the purelin function to the output of the *ReLU* function as shown below:

$$Y_P = \sum_{k=1}^{p} w_{kl} \cdot a2_k + b_l \tag{A6}$$

where Y_P is the predicted output value, w_{kl} represents the weights connecting the second hidden layer neurons, k, to the output layer neurons, l, b_l is the bias term, and p is the number of neurons in the hidden layer.

The combination of Equations (A4)–(A6) yields the following general form of the proposed neural network model:

$$R_S = \sum_{k=1}^{p} w_{kl} \cdot ReLU\left(\sum_{k=1}^{p} w_{jk}\ ReLU\left(w_{j,1} \cdot MW + w_{j,2} \cdot \gamma + w_{j,3} \cdot T + w_{j,4} \cdot Ps + b_j\right) + b_k \right) + b_l \tag{A7}$$

The values of the weights and biases are listed in Tables A1 and A2 below.

Appendix B.2. Example Calculations using MLP-Adam Model

The example calculation uses the following values for the four input variables: MW = 490 gr/mol, γ = 0.967786, T = 140 °C, and Ps = 10.48 MPa. These values are utilized in the MLP-Adam model to derive the corresponding output prediction. The values were normalized using z-score normalization, which involved applying the following formula to each value:

$$X_{istd} = \frac{X_i - \mu}{\sigma} \tag{A8}$$

where X_{istd} represents the standardized value of a specific data point, X_i denotes the original value of that data point, μ is the mean of the input data points, and σ is the standard deviation of the input data points. The normalized values of MW, γ, T, and Ps are 1.54606763, 0.90499909, 2.66878506, and 0.76176893, respectively.

By applying Equations (A4)–(A6), the predicted output is computed. The step-by-step calculations are outlined in Table A3, providing a comprehensive overview of the process.

Table A1. Weights and biases of the first hidden layer of the proposed MLP-Adam model.

$w_{j,MW}$	$w_{j,\gamma}$	$w_{j,T}$	$w_{j,Ps}$	b_j
0.423479229	−0.518270671	0.088841140	0.164605036	−0.182387754
−0.316850155	0.578180193	−0.627018213	−0.370452255	−0.037461437
−0.260930061	0.121095933	0.302609562	0.177341118	0.035739433
0.153113961	−0.273656278	0.023316100	−0.014185284	−0.152793422
0.326721847	0.163599714	0.017112899	0.437370806	−0.370236605
0.467836350	−0.183758318	−0.116376496	0.173847764	0.190825283
0.207402825	−0.402902960	0.277075022	0.077882327	−0.256408870
0.430666834	0.488847017	0.382416307	0.316209614	−0.437328159
−0.378489106	−0.191637143	−0.586777627	0.073175244	−0.207403078
−0.280519455	−0.169934719	−0.038683220	0.464787781	0.129119664
−0.012112551	−0.279909700	0.314301490	−0.553606331	0.127572730
0.203990727	0.348036944	0.120888933	−0.571946859	−0.362548828

Table A2. Weights and biases of the second hidden layer and the output layer of the proposed MLP-Adam model.

$w_{1,k}$	$w_{2,k}$	$w_{3,k}$	$w_{4,k}$	$w_{5,k}$	$w_{6,k}$	$w_{7,k}$	$w_{8,k}$
0.144444540	0.227294683	−0.281868785	−0.386379957	−0.244969561	0.250844776	−0.042056944	0.090741582
−1.246394872	−0.613879323	−0.806254267	0.332979083	0.174128487	−0.160888448	−0.905039012	0.223389938
0.158317938	0.136602625	0.250266492	−0.048559281	−0.043032091	−0.009495512	0.364784896	−0.316569924
−0.292102873	0.049241617	0.113946393	0.185241475	−0.189562544	0.473260581	0.171075671	−0.035240747
−0.311310201	−1.128083109	−0.132358402	−0.147601380	0.150322437	−0.051223963	−0.059710107	0.302232533
−0.527317762	0.004510418	−0.090777598	0.033773034	0.003524607	0.325446367	−0.200799241	−1.144739747
0.641047120	−0.064388409	0.391169577	−0.684768438	−0.434764891	0.371954649	−0.063837923	−0.090706437
−0.190623462	0.257651656	0.394092589	0.200460493	−0.200868785	0.064583137	0.155178993	0.315470844
0.193483933	−0.301786810	0.255001187	−0.513664782	−0.427212923	−0.234824061	−0.042243052	0.111917041
−1.080619454	0.096860095	0.129510939	0.049882758	0.238265812	−1.272954463	0.236488863	−0.735467910
−0.364739000	−0.515439033	−0.178362324	−0.179078683	−0.595661461	−0.054487861	−0.096768409	−0.003158351
−0.499953687	0.379382699	−0.177857115	−0.423149019	−0.938039004	0.343048214	−0.956486344	0.245499372

$w_{9,k}$	$w_{10,k}$	$w_{11,k}$	$w_{12,k}$	b_k	$w_{k,l}$	b_l
−0.792608916	−0.343328714	−0.205415770	−0.539200484	0.158580690	0.079246789	0.300516456
0.365020424	−0.149115592	−0.426100313	0.130489438	0.123922713	0.187488675	
0.137588575	0.520926713	−0.278029352	−0.333180844	−0.322128087	−0.186551764	
0.191870614	0.492062687	−0.308154106	−0.205118045	0.259233176	0.440697550	
−0.230481609	−0.726262688	0.058385573	−0.124779440	−0.023145271	0.217374727	
0.379300296	0.162133157	0.567164421	0.756009399	−0.201348185	−0.275265455	
−0.633766531	0.062475737	0.018612951	−0.710203170	0.197099491	0.088092155	
0.158039510	−0.123929366	0.011550034	0.471806019	−0.221232160	−0.171224877	
0.196399033	−0.388778716	−0.568655312	0.230788096	−0.103322580	−0.453178435	
0.274261921	−0.640708744	0.155315384	0.250834226	0.017402615	−0.166323795	
0.512196242	−0.019978577	−0.330687165	0.177631750	0.079844228	0.371093213	
−0.256439089	0.436899453	−0.405297756	0.383212924	−0.086818188	0.109248526	

Table A3. Example calculation using the proposed MLP-Adam Model.

MW	γ	T	Ps	$z1$	$a1$	$z2$	$a2$	Rs-Pred	Rs-Exp
1.54606763	0.90499909	2.66878506	0.76176893	0.3657942	0.3657942	0.01417779	0.01417779	0.4256788	0.42
				−1.9596565	0	−0.9757543	0		
				0.68460845	0.68460845	−0.6840449	0		
				−0.1123078	0	0.2759657	0.2759657		
				0.6618012	0.6618012	0.39354511	0.39354511		
				0.56967878	0.56967878	−2.3092059	0		
				0.49840513	0.49840513	0.31163391	0.31163391		
				1.93238358	1.93238358	0.64580446	0.64580446		
				−2.4762450	0	−0.1985719	0		
				−0.2075494	0	−2.0788913	0		
				0.2726109	0.2726109	−0.7179282	0		
				0.15474298	0.15474298	−0.8703367	0		

References

1. Holdren, J.P. Population and the energy problem. *Popul. Env.* **1991**, *12*, 231–255. [CrossRef]
2. Laherrere, J.; Hall, C.B.; Bentley, R. How much oil remains for the world to produce? Comparing assessment methods, and separating fact from fiction. *Curr. Res. Environ. Sustain.* **2022**, *4*, 100174. [CrossRef]
3. Ozotta, O.; Ostadhassan, M.; Lee, H.; Pu, H.; Kolawole, O.; Malki, M.L. Time-dependent Impact of CO_2-shale Interaction on CO_2 Storage Potential. In Proceedings of the 15th Greenhouse Gas Control Technologies Conference, Abu Dhabi, United Arab Emirates, 18 March 2021; pp. 15–18.
4. Clonts, M.; Mazighi, M.; Touami, M. Reservoir simulation of the planned miscible gas injection project at Rhourde El Baguel, Algeria. In Proceedings of the European Petroleum Conference, Milan, Italy, 22–24 October 1996; OnePetro: Richardson, TX, USA, 1996.

5. Malki, M.L.; Rasouli, V.; Saberi, M.R.; Sennaoui, B.; Ozotta, O.; Chellal, H.A. Effect of CO_2 on Mineralogy, Fluid, and Elastic Properties in Middle Bakken Formation Using Rock Physics Modeling. In Proceedings of the ARMA US Rock Mechanics/Geomechanics Symposium, Santa Fe, NM, USA, 26–29 June 2022. [CrossRef]
6. Hasan, M.M.F.; First, E.L.; Boukouvala, F.; Floudas, C.A. A multi-scale framework for CO_2 capture, utilization, and sequestration: CCUS and CCU. *Comput. Chem. Eng.* **2015**, *81*, 2–21. [CrossRef]
7. Merzoug, A.; Mouedden, N.; Rasouli, V.; Damjanac, B. Simulation of Proppant Placement Efficiency at the Intersection of Induced and Natural Fractures. In Proceedings of the ARMA US Rock Mechanics/Geomechanics Symposium, Santa Fe, NM, USA, 26–29 June 2022. [CrossRef]
8. Afari, S.; Ling, K.; Sennaoui, B.; Maxey, D.; Oguntade, T.; Porlles, J. Optimization of CO_2 huff-n-puff EOR in the Bakken Formation using numerical simulation and response surface methodology. *J. Pet. Sci. Eng.* **2022**, *215 Pt A*, 110552. [CrossRef]
9. Taber, J.J.; Martin, F.D.; Seright, R.S. EOR screening criteria revisited -Part 1: Introduction to screening criteria and enhanced recovery field projects. *SPE Reserv. Eng.* **1997**, *12*, 189–198. [CrossRef]
10. Sennaoui, B.; Pu, H.; Afari, S.; Malki, M.L.; Kolawole, O. Pore- and Core-Scale Mechanisms Controlling Supercritical Cyclic Gas Utilization for Enhanced Recovery under Immiscible and Miscible Conditions in the Three Forks Formation. *Energy Fuels* **2023**, *37*, 459–476. [CrossRef]
11. Almobarak, M.; Wu, Z.; Daiyu, Z.; Fan, K.; Liu, Y.; Xie, Q. A review of chemical-assisted minimum miscibility pressure reduction in CO_2 injection for enhanced oil recovery. *Petroleum* **2021**, *7*, 245–253. [CrossRef]
12. El-Hoshoudy, A.; Desouky, S. CO_2 Miscible Flooding for Enhanced Oil Recovery. In *Carbon Capture, Utilization and Sequestration*; InTech eBooks: London, UK, 2018. [CrossRef]
13. Mouedden, N.; Laalam, A.; Mahmoud, M.; Rabiei, M.; Merzoug, A.; Ouadi, H.; Boualam, A.; Djezzar, S. A Screening Methodology Using Fuzzy Logic to Improve the Well Stimulation Candidate Selection. In *All Days*; OnePetro: Richardson, TX, USA, 2022. [CrossRef]
14. Boualam, A.; Rasouli, V.; Dalkhaa, C.; Djezzar, S. Stress-Dependent Permeability and Porosity in Three Forks Carbonate Reservoir, Williston Basin. In Proceedings of the 54th U.S. Rock Mechanics/Geomechanics Symposium, Physical Event Cancelled, Golden, CO, USA, 28 June–1 July 2020.
15. Boualam, A.; Rasouli, V.; Dalkhaa, C.; Djezzar, S. Advanced Petrophysical Analysis and Water Saturation Prediction in Three Forks, Williston Basin. In Proceedings of the SPWLA Annual Logging Symposium, Online, 24 June–29 July 2020. [CrossRef]
16. Koroteev, D.; Tekic, Z. Artificial intelligence in oil and gas upstream: Trends, challenges, and scenarios for the future. *Energy AI* **2021**, *3*, 100041. [CrossRef]
17. Dargahi-Zarandi, A.; Hemmati-Sarapardeh, A.; Shateri, M.; Menad, N.A.; Ahmadi, M. Modeling minimum miscibility pressure of pure/impure CO_2-crude oil systems using adaptive boosting support vector regression: Application to gas injection processes. *J. Pet. Sci. Eng.* **2020**, *184*, 106499. [CrossRef]
18. Sambo, C.; Liu, N.; Shaibu, R.; Ahmed, A.A.; Hashish, R.G. A Technical Review of CO_2 for Enhanced Oil Recovery in Unconventional Oil Reservoirs. *Geoenergy Sci. Eng.* **2022**, *221*, 111185. [CrossRef]
19. Fath, A.H.; Pouranfard, A.-R. Evaluation of miscible and immiscible CO_2 injection in one of the Iranian oil fields. *Egypt. J. Pet.* **2014**, *23*, 255–270. [CrossRef]
20. Lv, Q.; Zheng, R.; Guo, X.; Larestani, A.; Hadavimoghaddam, F.; Riazi, M.; Hemmati-Sarapardeh, A.; Wang, K.; Li, J. Modelling minimum miscibility pressure of CO_2-crude oil systems using deep learning, tree-based, and thermodynamic models: Application to CO_2 sequestration and enhanced oil recovery. *Sep. Purif. Technol.* **2023**, *310*, 123086. [CrossRef]
21. Yang, G.; Li, X. Modified Peng-Robinson equation of state for CO_2/hydrocarbon systems within nanopores. *J. Nat. Gas Sci. Eng.* **2020**, *84*, 103700. [CrossRef]
22. Kiani, S.; Saeedi, M.; Nikoo, M.R.; Mohammadi, A.H. New model for prediction of minimum miscibility pressure and CO_2 solubility in crude oil. *J. Nat. Gas Sci. Eng.* **2020**, *80*, 103431. [CrossRef]
23. Ahmed, T. Minimum Miscibility Pressure from EOS. In Proceedings of the Canadian International Petroleum Conference, Calgary, AB, Canada, 4–8 June 2000. [CrossRef]
24. Alshuaibi, M.; Farzaneh, S.A.; Sohrabi, M.; Mogensen, K. An Accurate and Reliable Correlation to Determine CO_2/Crude Oil MMP for High-Temperature Reservoirs in Abu Dhabi. In Proceedings of the Abu Dhabi International Petroleum Exhibition and Conference, Abu Dhabi, United Arab Emirates, 11–14 November 2019. [CrossRef]
25. Jhalendra, R.K.; Kumar, A. Reliable estimate of minimum miscibility pressure from multiple possible EOS models for a reservoir oil under data constraint. *Pet. Sci. Technol.* **2022**, *40*, 1898–1913. [CrossRef]
26. Sinha, U.; Dindoruk, B.; Soliman, M. Prediction of CO_2 Minimum Miscibility Pressure MMP Using Machine Learning Techniques. In Proceedings of the SPE Improved Oil Recovery Conference, Virtual, 31 August–4 September 2020. [CrossRef]
27. Shakeel, M.; Khan, M.R.; Kalam, S.; Khan, R.A.; Patil, S.; Dar, U.A. Machine Learning for Prediction of CO_2 Minimum Miscibility Pressure. In *Proceedings of the Society of Petroleum Engineers—Middle East Oil, Gas and Geosciences Show, MEOS, Manama, Bahrain, 19–21 February 2023*; SPE Middle East Oil and Gas Show and Conference, MEOS, Proceedings; Society of Petroleum Engineers (SPE): Richardson, TX, USA, 2023. [CrossRef]
28. Li, D.; Li, X.; Zhang, Y.; Sun, L.; Yuan, S. Four Methods to Estimate Minimum Miscibility Pressure of CO_2-Oil Based on Machine Learning. *Chin. J. Chem.* **2019**, *37*, 1271–1278. [CrossRef]

29. Ekechukwu, G.K.; Falode, O.; Orodu, O.D. Improved Method for the Estimation of Minimum Miscibility Pressure for Pure and Impure CO_2–Crude Oil Systems Using Gaussian Process Machine Learning Approach. *ASME J. Energy Resour. Technol.* **2020**, *142*, 123003. [CrossRef]
30. Dong, P.; Liao, X.; Chen, Z.; Chu, H. An improved method for predicting CO_2 minimum miscibility pressure based on artificial neural network. *Adv. Geo-Energy Res.* **2019**, *3*, 355–364. [CrossRef]
31. Huang, C.; Tian, L.; Zhang, T.; Chen, J.; Wu, J.; Wang, H.; Wang, J.; Jiang, L.; Zhang, K. Globally optimized machine-learning framework for CO_2 hydrocarbon minimum miscibility pressure calculations. *Fuel* **2022**, *329*, 125312. [CrossRef]
32. Ge, D.; Cheng, H.; Cai, M.; Zhang, Y.; Dong, P. A New Predictive Method for CO_2-Oil Minimum Miscibility Pressure. *Geofluids* **2021**, *2021*, 8868592. [CrossRef]
33. Chemmakh, A.; Merzoug, A.; Ouadi, H.; Ladmia, A.; Rasouli, V. Machine Learning Predictive Models to Estimate the Minimum Miscibility Pressure of CO_2-Oil System. In Proceedings of the Abu Dhabi International Petroleum Exhibition & Conference, Abu Dhabi, United Arab Emirates, 15–18 November 2021. [CrossRef]
34. Ramdan, D.; Najmi, M.; Rajabzadeh, H.; Elveny, M.; Alizadeh, S.M.S.; Shahriari, R. Prediction of CO_2 solubility in electrolyte solutions using the e-PHSC equation of state. *J. Supercrit. Fluids* **2022**, *180*, 105454. [CrossRef]
35. Song, Z.; Shi, H.; Zhang, X.; Zhou, T. Prediction of CO_2 solubility in ionic liquids using machine learning methods. *Chem. Eng. Sci.* **2020**, *223*, 115752. [CrossRef]
36. Cheng, Y.; Zhang, X.; Lu, Z.; Pan, Z.J.; Zeng, M.; Du, X.; Xiao, S. The effect of subcritical and supercritical CO_2 on the pore structure of bituminous coals. *J. Nat. Gas Sci. Eng.* **2021**, *94*, 104132. [CrossRef]
37. Zhao, W.; Zhang, T.; Jia, C.; Li, X.; Wu, K.; He, M. Numerical simulation on natural gas migration and accumulation in sweet spots of tight reservoir. *J. Nat. Gas Sci. Eng.* **2020**, *81*, 103454. [CrossRef]
38. Srivastava, R.K.; Huang, S.S.; Dyer, S.B. *Measurement and Prediction of PVT Properties of Heavy and Medium Oils with Carbon Dioxide*; No. CONF-9502114-Vol. 1; UNITAR: New York, NY, USA, 1995.
39. Kokal, S.L.; Sayegh, S.G. Phase behavior and physical properties of CO-saturated heavy oil and its constitutive fractions. In Proceedings of the Annual Technical Meeting, Calgary, AB, Canada, 9–12 June 1990; OnePetro: Richardson, TX, USA, 1990.
40. Simon, R.; Graue, D.J. Generalized correlations for predicting solubility, swelling and viscosity behavior of CO_2-crude oil systems. *J. Pet. Technol.* **1965**, *17*, 102–106. [CrossRef]
41. Simon, R.; Rosman, A.; Zana, E. Phase-behavior properties of CO_2-reservoir oil systems. *Soc. Pet. Eng. J.* **1978**, *18*, 20–26. [CrossRef]
42. Sim, S.S.K.; Udegbuanam, E.; Haggerty, D.J.; Baroni, J.; Baroni, M. Laboratory experiments and reservoir simulation studies in support of CO_2 injection project in Mattoon field, Illinois, USA. In Proceedings of the Annual Technical Meeting, New Orleans, LA, USA, 25–28 September 1994; OnePetro: Richardson, TX, USA, 1994.
43. Zolghadr, A.; Escrochi, M.; Ayatollahi, S. Temperature and Composition Effect on CO_2 Miscibility by Interfacial Tension Measurement. *J. Chem. Eng. Data* **2013**, *58*, 1168–1175. [CrossRef]
44. Jaeger, P.T.; Alotaibi, M.B.; Nasr-El-Din, H.A. Influence of Compressed Carbon Dioxide on the Capillarity of the Gas−Crude Oil−Reservoir Water System. *J. Chem. Eng. Data* **2010**, *55*, 5246–5251. [CrossRef]
45. Georgiadis, A.; Llovell, F.; Bismarck, A.; Blas, F.J.; Galindo, A.; Maitland, G.C.; Trusler, J.P.M.; Jackson, G. Interfacial tension measurements and modelling of (carbon dioxide + n-alkane) and (carbon dioxide + water) binary mixtures at elevated pressures and temperatures. *J. Supercrit. Fluids* **2010**, *55*, 743–754. [CrossRef]
46. Cronquist, C. Carbon dioxide dynamic miscibility with light reservoir oils. In Proceedings of the Fourth Annual US DOE Symposium, Tulsa, OK, USA; 1978.
47. Yellig, W.; Metcalfe, R. Determination and Prediction of CO_2 Minimum Miscibility Pressures (includes associated paper 8876). *J. Pet. Technol.* **1980**, *32*, 160–168. [CrossRef]
48. Alston, R.; Kokolis, G.; James, C. CO_2 minimum miscibility pressure: A correlation for impure CO_2 streams and live oil systems. *Soc. Pet. Eng. J.* **1985**, *25*, 268–274. [CrossRef]
49. Yuan, H.; Johns, R.T.; Egwuenu, A.M.; Dindoruk, B. Improved MMP correlations for CO_2 floods using analytical gas flooding theory. In *Proceedings of the Society of Petroleum Engineers—SPE/DOE Symposium on Improved Oil Recovery, IOR, Tulsa, OK, USA, 17–21 April 2004*; (Proceedings—SPE Symposium on Improved Oil Recovery; Vol. 2004-April); Society of Petroleum Engineers (SPE): Richardson, TX, USA, 2004.
50. Chen, B.L.; Huang, H.D.; Zhang, Y. An Improved Predicting Model for Minimum Miscibility Pressure (MMP) of CO_2 and Crude Oil. *J. Oil Gas Technol.* **2013**, *35*, 126–130.
51. Chung, F.T.H.; Jones, R.A.; Burchfield, T.E. Recovery of Viscous Oil Under High Pressure by CO_2 Displacement: A Laboratory Study. In Proceedings of the International Meeting on Petroleum Engineering, Tianjin, China, 1–4 November 1988. [CrossRef]
52. Rostami, A.; Arabloo, M.; Kamari, A.; Mohammadi, A.H. Modeling of CO_2 solubility in crude oil during carbon dioxide enhanced oil recovery using gene expression programming. *Fuel* **2017**, *210*, 768–782. [CrossRef]
53. Emera, M.K.; Sarma, H.K. Prediction of CO_2 Solubility in Oil and the Effects on the Oil Physical Properties. *Energy Sources Part A Recovery Util. Environ. Eff.* **2007**, *29*, 1233–1242. [CrossRef]
54. Yu, H.; Xie, T.; Paszczynski, S.; Wilamowski, B.M. Advantages of Radial Basis Function Networks for Dynamic System Design. *IEEE Trans. Ind. Electron.* **2011**, *58*, 5438–5450. [CrossRef]

55. Mirzaie, M.; Tatar, A. Modeling of interfacial tension in binary mixtures of CH_4, CO_2, and N_2 -alkanes using gene expression programming and equation of state. *J. Mol. Liq.* **2020**, *320 Pt B*, 114454. [CrossRef]
56. Lee, I. *Effectiveness of Carbon Dioxide Displacement under Miscible and Immiscible Conditions*; U.S. Department of Energy Office of Scientific and Technical Information: Oak Ridge, TN, USA, 1979.
57. Emera, M.K.; Javadpour, F.; Sarma, H.K. Genetic algorithm (GA)-based correlations offer more reliable prediction of minimum miscibility pressures (MMP) between reservoir oil and CO_2 or flue gas. *J. Can. Pet. Technol.* **2007**, *46*, 19–25. [CrossRef]
58. Fathinasab, M.; Ayatollahi, S. On the determination of CO_2–crude oil minimum miscibility pressure using genetic programming combined with constrained multivariable search methods. *Fuel* **2016**, *173*, 180–188. [CrossRef]

eng

Review

Carbon Capture Utilisation and Storage in Extractive Industries for Methanol Production

Antonis Peppas *, Sotiris Kottaridis, Chrysa Politi and Panagiotis M. Angelopoulos

School of Mining and Metallurgical Engineering, National Technical University of Athens (NTUA), 15780 Athens, Greece
* Correspondence: peppas@metal.ntua.gr

Abstract: The elevated increase of CO_2 emissions related to activities of the extractive industry is becoming a challenging issue gradually affecting climate change and global warming. In this frame, the effective utilisation of CO_2 through the techniques of Carbon Capture and Storage (CCS) as well as Carbon Capture and Utilisation (CCU) can alleviate the greenhouse effect. Converting CO_2 into a value-added chemical or liquid fuel (e.g., methanol, hydrocarbons, propylene, dimethyl ether, ethylene, etc.) is a promising approach in this regard. Methanol (MeOH) synthesis offers a key feedstock for industries, being both an industrial commodity for several chemical products and an efficient transportation fuel. This article presents a review of the CCS and CCU technologies for the production of MeOH in extractive industries. The CCS technologies investigated in this framework are the amine-based absorption and the WGS-enhanced CCS. The CCU technologies are CO_2 hydrogenation and enhanced CO_2 transformation by the Fischer-Tropsch reaction. Incorporating these systems for the processing of the flue-gases of the extractive industries significantly reduces the CO_2 emissions, while creating new revenues by the production of valuable MeOH.

Keywords: extractive industry; methanol; Carbon Capture and Utilisation; Carbon Capture and Storage; amine-based carbon capture; sorption-enhanced carbon capture and storage; CO_2 hydrogenation; Fischer-Tropsch

Citation: Peppas, A.; Kottaridis, S.; Politi, C.; Angelopoulos, P.M. Carbon Capture Utilisation and Storage in Extractive Industries for Methanol Production. *Eng* **2023**, *4*, 480–506. https://doi.org/10.3390/eng4010029

Academic Editors: Antonio Gil Bravo, Reza Rezaee and Yujie Yuan

Received: 24 November 2022
Revised: 10 January 2023
Accepted: 29 January 2023
Published: 1 February 2023

1. Introduction

In recent decades, the increasing research interest in carbon dioxide (CO_2) was driven by its immensely high effect on the global warming challenge. In 2019, extraction and primary processing of metals and minerals were responsible for 26% of Global Greenhouse Gas (GHG) emissions, as well as 20% of all health impacts [1,2]. In 2021, the metals and mining sector accounted for approximately 4.5 Gt of CO_2 equivalent [3]. Many studies and research projects have focused on the mitigation of GHG emissions, especially CO_2, which accounts for more than 80% of these emissions [4].

Over the years, various technologies have been investigated, focusing mostly on energy efficiency, fuel switching and innovative use of existing knowledge. However, in steel industry, they can only account for a 15% reduction in CO_2 emissions [5].

The concept of Carbon Capture Utilisation and Storage (CCUS) has received significant attention in recent years, as the means to further decarbonize the extractive industries. CCUS comprised methods and technologies to collect CO_2 from the flue-gas and from the atmosphere and provide safe and permanent storage options, i.e., Carbon Capture and Storage (CCS), followed by recycling the CO_2 for the production of valuable products, such as methanol (MeOH), i.e., Carbon Capture and Utilisation (CCU) [4].

CCS technologies can be classified into three different categories, depending on the suitability of each approach to industrial processes. These categories include pre-combustion, post-combustion and oxy-combustion technologies. Pre-combustion capture refers to the removal of CO_2 from fossil fuels before the combustion's completion [6]. In these applications, a feedstock, such as coal, is partially oxidized in steam and oxygen/air at high temperature and pressure by a gasification process to produce synthesis gas, also known as syngas. The syngas, containing mostly carbon monoxide (CO), CO_2 and hydrogen (H_2), typically undergoes the water gas shift (WGS) reaction, for the conversion of unexploitable CO to CO_2, which can be captured and later stored or transported. Post-combustion capture refers to the capturing of CO_2 contained in flue-gas streams. These technologies typically operate at low pressures, and are most suitable for the processing of gases with low CO_2 concentrations, 5–15%. Post-combustion capture is an already mature technology, finding application in various fields such as oil refineries, petrochemical plants and other industries [7]. Of all the available techniques, absorption-based capture is widely researched and utilised, with proven high efficiencies from laboratory to commercial scale. Specifically, amine-based CO_2 capture is the most commonly used approach, reaching capture efficiencies up to 95% and CO_2 purities higher than 99%. Oxy-combustion capture refers to the capturing of CO_2 deriving from the combustion of fuels with pure or nearly pure oxygen instead of air. In oxy-combustion, nitrogen is separated from the air, resulting in almost zero nitrogen oxides (NO_x) formation. As a result, the flue-gases of the combustion have CO_2 content as much as 90% [8]. These high contents enable larger amounts of CO_2 to be captured and utilised in carbon transformation units.

In 2021, 27 CCS units were already operating globally, with 4 more being under construction [9]. On the global scale, these facilities account for a 36.6 Mtpa CO_2 captured and stored. The field of activity and capture capacities of these units are summarized in Table 1. Concerning extractive industries, the Ultra-Low CO_2 Steelmaking (ULCOS) research project proposed various novel CCS routes aiming for the reduction of CO_2 emissions by 50% in the steelmaking industry [10]. It was estimated that, while process modifications, such as the utilisation of top gas recycle blast furnaces (TGR-BF) for replacing hot blast with pure oxygen, allow for 20–25% reduction, only CCS technologies are capable of 55–60% overall reduction.

Table 1. Main characteristics of operating CCS plants globally in 2021 [9].

CCS Plant	Location	Starting Date	Main Processes	Capacity (Mtpa CO_2)
Terrell NG Processing Plant	United States	1972	NG Processing	0.50
Enid Fertilizer	United States	1982	Fertilizer Production	0.20
Shute Creek Gas Processing Plant	United States	1986	NG Processing	7.00
MOL Szank field CO_2 EOR	Hungary	1992	NG Processing	0.16
Sleipner CO_2 Storage	Norway	1996	NG Processing	1.00
Great Plains Synfuels Plant & Weyburn-Midale	United States	2000	Synthetic NG	3.00
Core Energy CO_2-EOR	United States	2003	NG Processing	0.35
Sinopec Zhongyuan CCUS	China	2006	Chemical Production	0.12
Snøhvit CO_2 Storage	Norway	2008	NG Processing	0.70
Arkalon CO_2 Compression Facility	United States	2009	Ethanol Production	0.29
Century Plant	United States	2010	NG Processing	5.00
Petrobras Santos Basin Pre-Salt Oil Field CCS	Brazil	2011	NG Processing	4.60
Bonanza BioEnergy CCUS EOR	United States	2012	Ethanol Production	0.10
Coffeyville Gasification Plant	United States	2013	Fertilizer Production	0.90
Air Products Steam Methane Reformer	United States	2013	Hydrogen Production	1.00
PCS Nitrogen	United States	2013	Fertilizer Production	0.30
Boundary Dam 3 CCUS Facility	Canada	2014	Power Generation	1.00
Quest	Canada	2015	Hydrogen Production	1.20
Uthmaniyah CO_2-EOR Demonstration	Saudi Arabia	2015	NG Processing	0.80
Karamay Dunhua Oil Technology CCUS EOR	China	2015	Methanol Production	0.10
Abu Dhabi CCS	United Arab Emirates	2016	Iron And Steel Production	0.80
Illinois Industrial CCS	United States	2017	Ethanol Production	1.00
CNPC Jilin Oil Field CO_2 EOR	China	2018	NG Processing	0.60
Gorgon Carbon Dioxide Injection	Australia	2019	NG Processing	4.00
Qatar LNG CCS	Qatar	2019	NG Processing	2.20
Alberta Carbon Trunk Line—North West Redwater Sturgeon Refinery CO_2 Stream	Canada	2020	Hydrogen Production	1.60
Alberta Carbon Trunk Line with Nutrien CO_2 Stream	Canada	2020	Fertilizer Production	0.30

CCU technologies refer to the utilisation of CO_2 for the production of valuable products including urea, MeOH, formaldehyde, formic acid, carbamates, polymer-building blocks and fine chemicals. CCU can be divided into five main utilisation pathways including CO_2 through chemical conversion, mineral carbonation, enhanced oil recovery, biological conversion and direct utilisation [11]. Among them, the leading category in the sector is chemical conversion which refers to the conversion of carbon into urea, formic acid, organic carbonates, and fine chemicals such as biotin, etc. In addition, CO_2 can be converted to fuels such as methane, MeOH, and syngas through CO_2 hydrogenation and the Fischer–Tropsch reaction. The CO_2 to fuels and chemicals pathway offers several advantages for the extractive industry, contributing to the decarbonisation goals as well as the additional sustainable production of fuels and chemicals, and the reduction of fossil-fuel consumption.

Among CO_2 by-products, urea and MeOH are the leading consumers of CO_2 in industry with an annual consumption of CO_2 of more than 110 Mt/year [12]. Respectively, in 2021, global MeOH demand reached more than 164 million tonnes, seeing constant rise [13]. MeOH sees use in multiple sectors, from fuel for the automotive and marine sectors, to feedstock for plastics manufacturing, to pharmaceutical applications. For the automotive and marine sector specifically, production of MeOH-based fuels accounted for 31% of the global MeOH consumption, with 11.7 million tonnes being used for gasoline blending and combustion in 2021 [14,15]. The growing popularity of renewable MeOH as a fuel is due to its many environmental benefits, accounting for reductions of up to 95% CO_2, 80% NO_x and almost 100% in sulphur oxides and particulate matter [16].

This study focuses on CCUS technologies oriented to the production of MeOH on industrial scale. The benefits for extractive industries comprise the significant reduction of the GHG emissions that end up in the atmosphere and increased economic viability, not only by reducing emissions related costs (i.e., carbon taxation), but also creating revenues from the production of valuable products. The study is divided in two sections analysing the CCS technologies and the CCU.

2. Carbon Capture and Storage Technologies for Extractive Industries

The suitability of the CCS technologies for the different industrial applications is highly associated with the related composition of the flue-gases. Considering that typical flue-gases produced by extractive industry activities have low CO_2 content, amine-based absorption is considered the most suitable for CO_2 capture. Despite this, the flue-gases of extractive industries typically have high concentrations of CO, which cannot be captured and is more toxic than CO_2. For this reason, many applications utilise the WGS-enhanced CCS technologies, where CO is transformed to CO_2 by the WGS reaction, to increase the overall carbon capture efficiency of the CCS. In this line, the CCS technologies presented in this section are the amine-based absorption for post-combustion capture and storage and the WGS-enhanced CCS, focusing on relative applications in the extractive industry.

2.1. *Amine-Based Absorption for Post-Combustion Capture and Storage*

Amine-based absorption for post-combustion capture and storage is one of the most commonly utilised technologies for CO_2 capture from flue-gases of energy intensive industries, such as extractive industries. These technologies typically comprise of the following main parts: (1) a cooling system for lowering the temperature of flue-gases, (2) an absorption column, where CO_2 is captured by the use of amines, (3) a regeneration column, for the removal of CO_2 from the amine stream and (4) a multistage compression system, raising the pressure of CO_2 to be stored or transported [1]. A conventional amine-based CCS system is shown in Figure 1.

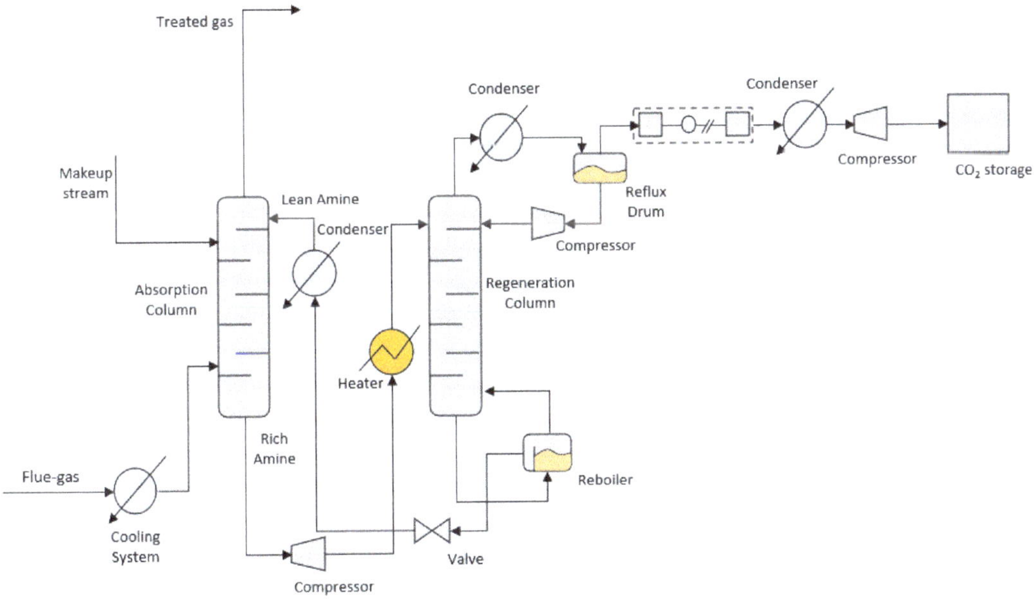

Figure 1. Schematics of a conventional amine-based CCS system.

Flue-gases are introduced to the cooling system where the temperature is decreased to the absorption column operating conditions. The cooled gases are then fed to the absorption column's bottom where, moving upwards, make contact with a liquid amine stream moving downwards from the column's top. CO_2 absorption by the amine takes place on the column's trays, designed to increase the contact area.

Liquid amine containing the absorbed CO_2 (rich amine) is then compressed, preheated and fed to the regeneration column, where separation of CO_2 from the amine occurs. The amine stream enters the regeneration column and concentrates at its bottom. The bottom is continuously heated for the volatile CO_2 to be vaporized and exit the column from the top. After initial separation, the liquid amine in the column's bottom is fed to a reboiler, for further CO_2 vaporization and separation. The low CO_2 containing amine stream (lean amine) is then cooled and fed to the top of the absorption column, closing the amine loop.

The CO_2 stream exiting from the regeneration column's top is condensed and fed to a flash separator where impurities such as amine and water are removed and sent back to the column. The purified CO_2 stream then passes through a multistage compression system and is afterwards condensed and repressurized to be stored in tanks.

2.2. Amine-Based Absorbents

2.2.1. Primary Amines

Monoethanolamine (MEA) is one of the most widely used primary amines in amine-based CCS technologies, achieving removal capacities from 87.1 to 100% and producing CO_2 of more than 99% purity [17]. Other primary amines used are 2-(2-aminoethoxy) ethanol (DGA), 1-amino-2-propanol (1A2P) and 2 amino-1-butanol (2A1B). Primary amines are very commonly used due to their fast kinetics, high water solubility, as well as their low price. They are typically used in aqueous form rather than in their pure state.

CO_2 capture by amines is a four-step reaction process, including the ionization of water, the CO_2 hydrolysis and ionization, the protonation of alkanolamine and lastly the formation of carbamate [17]. The reactions taking place are shown below:

$$H_2O(aq) \rightarrow H^+(aq) + OH^-(aq) \tag{1}$$

$$CO_2(g) + H_2O(aq) \rightarrow HCO_3^-(aq) + H^+(aq) \quad (2)$$

$$RNH_2(aq) + H^+(aq) \rightarrow RNH_3^+(aq) \quad (3)$$

$$RNH_2(aq) + CO_2(aq) \rightarrow RNHCOO^-(aq) + H^+(aq) \quad (4)$$

Due to its commercial popularity, MEA has drawn the most attention. In 2021, Akram et al. used a typical 30 wt%. MEA system as a benchmark for comparing with a 40 wt% MEA [18]. The study considered a 6.6 vol% CO_2 gas stream. It was demonstrated that the second system was able to achieve CO_2 capture efficiency of 89.6%, practically equal to the 90% of the benchmark, with the main advantage being the 25.1% reduction in regeneration energy consumption. In addition, it was concluded that increasing the regeneration temperature from 125.7 to 126.8 °C increased the capture efficiency and capacity to 91% and 74.6 g CO_2/kg MEA, respectively. On the other hand, increase in the regeneration temperature accounted for a 12.3% increase in overall energy consumption, as well as faster thermal degradation of the solvent.

An amine-based CCS system using the conventional 30 wt% MEA was also studied in the pilot plant of Niederaussem, from 2017 to 2019 [19]. The flue-gas of the plant had a CO_2 content of 14.2%. The system achieved carbon capture efficiency of 90%, with a capacity of 0.25 tonnes CO_2 daily. The focus of the particular study was the degradation of MEA, defined by the formation of oxidative degradation products, such as acetate, formate and oxalate, as well as the increase in trace elements, such as chloride, sulphur, sulphate, nitrate and iron. The results showed a 0.47 wt% degradation after 100 h, accounting for MEA consumption of 0.21–0.35 kg/t CO_2.

Pellegrini et al. investigated the absorption capacity of DGA as an alternative to the conventional MEA [20]. In this study, the efficiency of a 13–27 wt%. MEA system was compared to an 8–27 wt% DGA., on a flue-gas of 7 wt% CO_2. The results showed that the MEA system achieved 95% capture efficiency in comparison with the 88% of the DGA system, both operating at 20 °C. Despite this, it was concluded that the temperature in which the DGA system achieved efficiency of 90% was lower than what the MEA system required to achieve the same results, 63 °C compared to 67.5 °C. In addition, the reported regeneration efficiency of DGA was higher. For a reboiler operating at 110 °C, DGA regeneration efficiency was 90%, in comparison with the 66% of MEA. The studies conducted for MEA and DGA systems prove there are advantages to both approaches, depending on the application. The carbon efficiencies of various primary amine solvents in CCS systems are summarized in Table 2.

Table 2. Capture efficiencies of various primary amine solvents in CCS systems.

Solvent	CO_2 Content (%)	Capture Efficiency (%)	Reference
30 wt% MEA 40 wt% MEA	6.6 vol	90.0 86.9	[18]
30 wt% MEA	14.2	90.0	[19]
13–27 wt% MEA 8–27 wt% DGA	7 wt	95.0 88.0	[20]

2.2.2. Non-Primary Amines

In addition to primary amines, various secondary and tertiary amines also see use in numerous applications across the industry. Typical secondary amines for CCS systems include 2-(methylamino) ethanol (MMEA), N-ethylmonoethanolamine (EMEA) and the most commonly used diethanolamine (DEA) [6,7,21,22]. These amines display similar behaviour with primary amines, presenting high capture efficiencies of up to 90% as well as high regeneration energy demand, due to carbamate formation. Typical regeneration temperature for such amines is 70–200 °C [6,23]. The reaction mechanisms are also similar and can be expressed simply by substituting primary amines (RNH_2) with secondary amines (R_2NH) [24].

Barzagli et al. compared the performance of MMEA and EMEA and other amines, such as MEA, for a gas mixture of 15% CO_2 [25]. All experiments used 0.12 mol of amine. MMEA achieved capture efficiency of 92%, at 21 °C and 1.0145 bar. EMEA achieved 92.4%, at 24 °C and 1.0013 bar. In comparison, the 30% MEA solvent achieved 82.2% capture efficiency despite operating at slightly higher temperature and pressure than EMEA, 25 °C and 1.0067 bar.

Wang et al. conducted a study in 2021 over a lab-scale CO_2 capture system including DEA [26]. The study proposed an integrated absorption and mineralization process where, after the absorption, CO_2-rich solvents are regenerated by mineralization with semidry desulphurization slag. The experiment utilised a 15 vol% CO_2 and 85 vol% N_2 gas. After absorption, the CO_2-rich solvent was mixed with Ca-rich compound for the formation of calcium carbonate and the regeneration of the solvent. The results showed CO_2 loading of 0.73 and 0.18 mol CO_2/mol DEA for the rich and lean DEA, respectively. The desorption process achieved efficiency up to 85%. The main advantage of this approach was the lack of need for a compressor and reboiler in the regeneration system, thus reducing the energy consumption by as much as 93.7% in relation to conventional systems.

Tertiary amines for CCS systems include 2-(dimethylamino) ethanol (DMAE), N-methyldiethanolamine (MDEA) and 2-diethylaminoethanol (DEAE). In recent years, such amines have drawn much attention due to their high CO_2 absorption capabilities of 1 mol CO_2/mol solvent, almost two times higher than primary and secondary amines [26]. In addition, the inability of these amines to react with CO_2 accounts for no carbamate's formation [24]. Instead, bicarbonate ions are formatted, which require far less regeneration energy, approximately 58.5 kJ/mol compared to 86.9 and 68.9 kJ/mol for primary and secondary amines, respectively [6,23,27].

The main disadvantage of tertiary amines is the low CO_2 absorption rate of approximately 2.13 mol CO_2/hr [26]. To counter this issue, these amines are usually mixed with promoters such as piperazine (PZ) or carbonic anhydrase (CA) to increase absorption [22,28]. The reaction taking place is the following [23]:

$$R_3N(aq) + CO_2(g) + H_2O(aq) \leftrightarrow R_3NH^+(aq) + HCO_3\text{-}(aq) \quad (5)$$

Antonini et al. examine the use of an MDEA carbon capture system for producing low-carbon H_2 [29]. The system utilised a high-pressure and a low-pressure flash separator for the recycling of CO_2 not absorbed and semi-lean amine solvent that can further contribute to absorption in the absorption column. The gas considered in this study contained 16.27 mol% CO_2. The solvent was a conventional 40 wt% MDEA. The results of the study showed that utilising these recycling streams in the system accounted for a 54.5% reduction in regeneration energy demand, and capture efficiency of 90%.

Zhao et al. also examined post-combustion CCS using an MDEA+PZ solvent for a 650 MW power plant [30]. The flue-gases treated had a CO_2 content of 12%. The study examined the effect of the MDEA/PZ ratio as well as the targeted carbon capture capacity on the overall performance of the system and energy demand. The results showed that 30% and 20% of MDEA and PZ, respectively, allowed for high carbon capture efficiencies of 90%, without significantly increasing the reboiler heat duty for the amine regeneration.

Another study of a MDEA CCS system was conducted by Jaffary et al. in 2021 [31]. In this study, polyamines were used as activators for the absorption of 100% CO_2 gas. The polyamines examined were polyethylenimine (PEI-B), tetraethylenepentamine (TEPA) and PZ. The results of the study showed that MDEA enhanced with PEI-B had an increased CO_2 loading and specific desorption rate of 1.08 mol CO_2/mol solvent and 0.0385 mol CO_2/mol amine/min, respectively. Despite these benefits, incorporating PEI-B in the solvent accounted for increased viscosity, which results in decreased CO_2 absorption rates. PZ on the other hand accounted for a moderate increase in both the CO_2 loading and desorption rate, 0.67 mol CO_2/solvent and 0.0352 mol CO_2/mol amine/min, respectively, without increasing viscosity. The performance of various secondary and tertiary amine solvents is summarized in Table 3.

Table 3. Capture efficiencies of various secondary and tertiary amine solvents in CCS systems.

Solvent	CO_2 Content (%)	Capture Efficiency (%)	Reference
DEA	15 vol	85.0	[26]
MMEA EMEA	15	92.0 92.4	[25]
MDEA	16.27	90.0	[29]
MDEA/PZ	12 vol	90.0	[30]

CCS systems in extractive industries aim for the exploitation of CO_2 after the separation from the amine solvents, therefore, the most efficient amines are those having high capture efficiencies but also low regeneration energy, to allow for higher separation and reduced energy consumption.

Primary amines offer high capture efficiencies, up to 95% for the commercial MEA. Other primary amines, such as DGA, require significantly less regeneration energy while also having high capture efficiencies and thus can be considered another viable option. Secondary amines exhibit similar capture efficiencies to primary amines. Studies comparing MEA with EMEA specifically, have shown that EMEA can achieve carbon efficiency as high as 92.4% in comparison to MEA's 82.2%, with a slight increase in temperature and pressure. Tertiary amines, however, specifically MDEA, provide similarly high carbon capture efficiencies, while also requiring significantly less regeneration energy, 1.34 MJ/kg CO_2, than other amines [32]. For these reasons, MDEA is considered one of the most efficient options, used in various applications in CCS systems in the industry. While these amines have lower absorption rates on their own, these disadvantages are easily dealt with when mixed with promoters, especially PZ. In that sense, the use of MDEA+PZ solvents in novel systems utilizing recycle streams can be considered one of the most optimal approaches for CCS systems.

2.3. Innovative Amino Acids

Amino acids have seen significant research over the years as an energy-efficient replacement of conventional amines. The use of these solvents allows for the mitigation of CO_2 emissions in plants while also reducing the economic impact of these technologies.

In 2019, Moioli et al. examined a CCS system capturing CO_2 by precipitating potassium taurate solvent [33]. The unit's layout is nearly identical to the conventional amine-based, with the main differences lying in the optional components related to the processing of formed solids. Specifically, the reactions taking place result in the formation of solid taurine. Precipitation of taurine shifts the equilibrium toward the formation of more products, resulting in increased CO_2 capture. The production of solid taurine results in a slurry, comprising of solid taurine and a mixture of ionic species and water, obtained at the absorption's column bottom, which can optionally be fed to a solid-liquid separator, where supernatant liquid is returned to the absorption column, while the solids are led to the regeneration column, after being dissolved.

The advantage of this approach lies in the lower regeneration energy required. Specifically, the amino acid rich slurry stream entering the regeneration column has a lower pH which favours the desorption of CO_2. The case studies examined both the inclusion and not inclusion of a solid-liquid separator in the CCS unit for a 500 MW output power plant. The CO_2 removal efficiency was fixed at 90%. The analysis highlighted the benefits of incorporating the optional solid-liquid separator to allow for a fraction of liquid to be recycled to the absorption column without the need of regeneration, thus reducing energy requirements. For a recycle split fraction (liquid recycled/total liquid fed to the separator) of 0.2, corresponding to 0.255 lean loading, the reboiler duty was at its minimum, lower than 2.9 MJ/kg CO_2. For comparison, conventional amines such as MEA, have shown reboiler duties as high as 3.8 MJ/kg CO_2 for 0.23 lean loading.

Garrabrant et al. also examined the use of amino acid salts for energy-efficient CCS systems [34]. Specifically, the study examined CO_2 absorption by a hybrid potassium glycinate and sarcosinate followed by the amino acid regeneration and bicarbonate removal crystallization with a simple bis-iminoguanidine base (GBIG), and examined the regeneration energy with that of industrial benchmarks such as 30% aqueous MEA and 30% aqueous sodium glycinate (SG). The results showed that, while the required regeneration energy for SG was higher than MEA, approximately 5.7 MJ/kg CO_2 in comparison to 4.5 MJ/kg CO_2, the required regeneration energy for the potassium glycinate/sarcosinate + GBIG case was 3.4 kJ/kg CO_2, a reduction of 24% in comparison with MEA. This reduction in the regeneration energy lies in the ability of the GBIG process to regenerate the solvent in ambient temperatures.

In 2021, Xu et al. also examined the use of amino acid salts for CO_2 capture from a flue-gas containing 15% CO_2, proposing the encapsulation of amino acid salts into solid particle matrices for the production of liquid amino acid salts hydrogel (LAHP) [35]. The proposed approach allowed for bigger contact areas and more intense interactions with CO_2 uptakes increased as much as 40%, compared to stirred aqueous amino acid salts solutions. The study concluded that the heat of absorption for potassium sarcosinate LAHP was 35% lower than MEA's, 53.3 kJ/mol CO_2 in comparison to 81,77 kJ/mol CO_2., suggesting lower regeneration energy requirements. In addition, the study proposed the replacement of water with solvents of higher boiling points, i.e., ethylene glycol, to further reduce the required energy. With a boiling point of 60 °C, ethylene glycol opens up the approach of using waste industrial steam to further reduce operational costs. The regeneration energy for the different solvents is shown in Table 4.

Table 4. Regeneration energy for innovative solvents.

Solvent	CO_2 Content (%)	Capture Efficiency (%)	Regeneration Energy (MJ/kg CO_2)	Reference
MEA	-	90	3.8	[33]
Potassium taurate		90	2.9	
30% MEA	12.8	-	4.5	[34]
30% SG			5.7	
Potassium Glycinate/Sarcosinate + GBIG			3.4	

2.4. WGS-Enhanced Carbon Capture and Storage

Many industries, including extractive industries, produce flue-gases of very small CO_2 concentrations and high concentrations of CO, which is far more toxic and cannot be captured. To tackle this issue, many CCS technologies utilise the WGS-enhanced CCS, which is the incorporation of a CO to CO_2 transformation system before capture, in the form of a WGS reactor. Water reacts with CO for the production of CO_2 and H_2. The produced H_2 is separated and can be used in the carbon transformation technologies. The reaction taking place is the following [36]:

$$CO(g) + H_2O(g) \leftrightarrow CO_2(g) + H_2(g),\ \Delta H = -41\ kJ/mol \tag{6}$$

CO conversion to CO_2 is highly exothermic and thus is favoured in low temperatures. Specifically, studies have shown than a reduction of the reaction temperature from 800 to 200 °C can increase the equilibrium constant up to 80 times [37]. In addition, the reaction's equilibrium is proven to be unaffected by pressure increase, but CO conversion is favoured due to higher reaction rates.

WGS reaction commercially takes place in two adiabatic catalytic reactors arranged in series for maximum CO transformation [38]. The first reactor is a high-temperature WGS (HT-WGS) operating with a Fe-based catalyst at 350–500 °C and 10–60 bar, while the second reactor is a low-temperature WGS (LT-WGS) operating with a Cu-based catalyst at approximately 200 °C and 10–40 bar [39]. HT-WGS are characterized by fast kinetics, but equilibrium-limited conversion. On the other hand, LT-WGS have slower kinetics, but are much less limited by the thermodynamic equilibrium, allowing for higher CO conversion. The combination of the two allows for fewer thermodynamic limitations and the ability to adjust the H_2:CO ratio in the final gas product.

Despite this, most CCS units using WGS reactors for enhanced decarbonization incorporate a single HT-WGS reactor. The Fe-based catalysts used in such reactors can tolerate small quantities of particles commonly found in the flue-gases of extractive industry's processes, such as sulphur-based compounds, in contrast with Cu-based catalysts used in LT-WGS reactors [40]. In addition, any traces of hydrogen sulphides (H_2S) in the flue-gas can be separated along with CO_2 in the absorption process. For these reasons, WGS-enhanced CCS is considered one of the most efficient approaches for CCS in extractive industries. A conventional WGS-enhanced CCS system is shown in Figure 2.

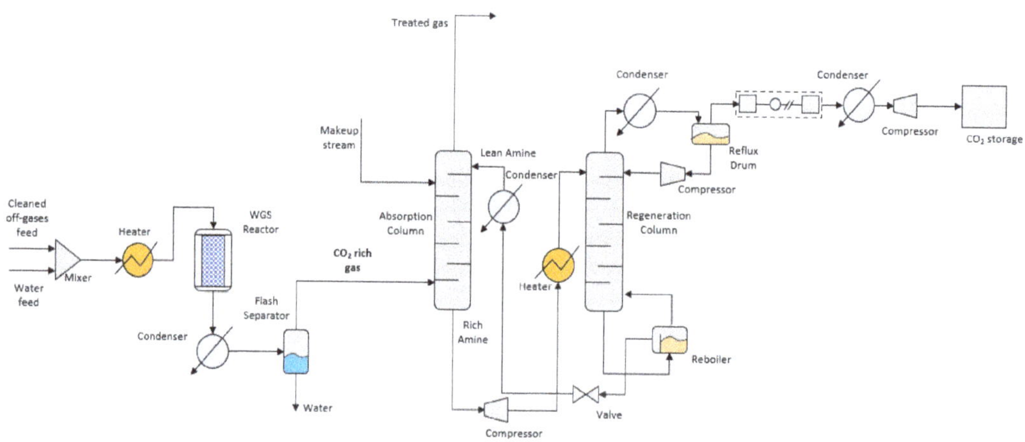

Figure 2. Schematics of a conventional WGS-enhanced CCS system.

Catalytic Membrane Reactors for WGS Reaction

Commonly used reactors for WGS reaction are catalytic membrane reactors (CMR). Such technologies allow for the immediate separation of either CO_2 or H_2 from the product stream. This continuous separation allows for higher CO conversion than a conventional reactor limited by equilibrium. Numerous studies have been conducted for high-grade H_2 recovery using dense Pd-based, composite Pd-based or silica and zeolite membranes. A conventional CMR for WGS reaction is shown in Figure 3.

For Pd-based CMR, several studies indicate CO conversion up to 98% and H_2 recovery up to 94%, but for small-scale reactors [41]. In large-scale applications, multi-tubular systems with membranes of 50–60 μm thickness or composite planar membranes based on foils of 20–25 μm are the most commonly used. Of the two approaches, composite Pd membranes generally achieve higher H_2 production rates, due to higher permeances of the thinner membrane.

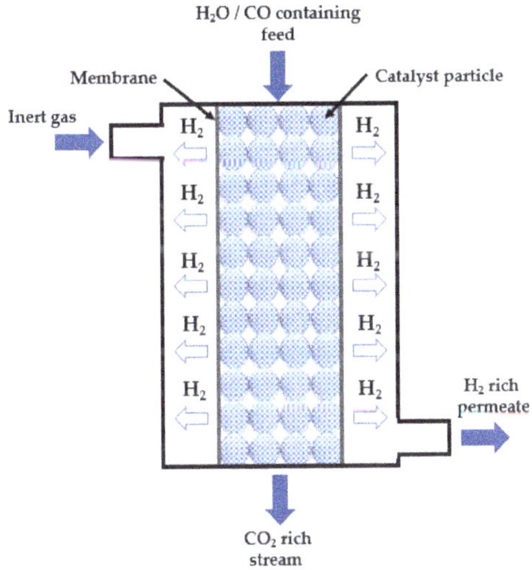

Figure 3. Schematics of a conventional CMR reactor for WGS reaction.

Catalano et al. investigated the performance of <10 μm Pd membranes over a porous (316 L) stainless steel (PSS) tubular support [41]. The tube-and-shell reactor used was filled with a commercial Fe/Cr catalyst for high temperature WGS reaction. The gas used was decided to simulate actual gasifier syngas, with a composition of 40% H_2, 42.2% CO and 17.8% CO_2. The H_2O:CO ratio ranged from 2.5 to 3.5 and the total feed flow was up to 1.5 Nm^3h^{-1}. The temperature raged from 420–440 °C and the pressure was up to 20 bar. The study concluded that for an 8.3 μm Pd membrane and operating temperature and pressure of 440 °C and 20 bar, respectively, CO conversion reached 84.9%, H_2 recovery 42.8% and H_2 purity 99.5%.

Augustine et al. also investigated the performance of a CMR using Pd-membranes, over different H_2O:CO ratios, temperatures, gas compositions, and gas hourly space velocity (GHSV), at a fixed 14.4 bar pressure [42]. The reactor was using an iron-chrome oxide catalyst. The experiments conducted examined both the use of a CO feed as well as a syngas feed with a composition of 22.7% CO, 45.4% H_2O, 22.0% H_2 and 9.9% CO_2, a H_2O:CO ratio of 2. For the syngas, results showed that a H_2O:CO ratio of 1.3:1 was the optimal, reaching CO conversion and H_2 recovery of more than 95% and 85%, respectively, at 450 °C. Increase of the ratio beyond this point had a minor effect on CO conversion, but a negative effect on H_2 recovery. Indicatively, at a H_2O:CO ratio of 2.5:1, CO conversion and H_2 recovery were 94% and 78%, respectively.

Nishida et al. investigated the performance of a silica membrane developed with the counter-diffusion chemical vapor deposition (CVD) method [43]. The silica membrane had a H_2 permeance of 1.29×10^{-6} mol $m^{-2}s^{-1}Pa^{-1}$. The reactor tube was filled with 1.05 g of a commercial $Cu/ZnO/Al_2O_3$ catalyst. The test was conducted with a blast furnace gas (BFG) feed with a composition of 52% N_2, 22% CO, 22% CO_2 and 4% H_2, at 300 °C and 3.05 bar. The study showed that operating the reactor with a space velocity of 7000 h^{-1} accounted for a CO conversion of 95–97%, H_2 recovery of approximately 60% and an H_2 purity around 94%. The study also examined the effect of space velocity in the reaction, in the range of 3500 to 10,500 h^{-1} and concluded that the increase in space velocity resulted in decreased CO conversion and H_2 recovery but increased H_2 purity.

Battersby et al. also examined the effect of H_2O:CO molar ratio in a silica membrane reactor using $Cu/ZnO/A_2O_3$ catalyst, prepared by $NaCO_3$ co-precipitation of precursor salts and followed by consequent aging, filtration, washing drying and calcination [44]. The tests were conducted with a fixed reactants' flow rate of 80 mL/min and varying H_2O:CO ratios. It was concluded, that increasing the H_2O:CO molar ratio from 1 to 4 increased the CO conversion from 20% to 29%, at 250 °C. The study also examined the effect of temperature on CO conversion for a H_2O:CO molar ratio of 1. The results showed that increasing the temperature resulted in increased CO conversion, reaching almost 80% at 300 °C. The performances of the various CMR reactors are summarized in Table 5.

Table 5. Performance of CMR reactors for WGS reaction.

Membrane	H_2O:CO Ratio	Temp. (°C)	Pressure (bar)	CO Conversion (%)	H_2 Recovery (%)	H_2 Purity (%)	Reference
Pd-based	2.5:1	440	20	84.9	42.8	99.5	[41]
Pd-based	1.3:1 2:1	450	14.4	96.0 94.0	85.0 78.0	-	[42]
Silica	-	300	3.05	95–97.0	60.0	94	[43]
Silica	1:1 4:1	300 250	-	80.0 29.0	-	-	[44]

The studies conducted over the CMR reactors using Pd-based membranes show the highest CO conversion and H_2 recovery at a temperature of approximately 450 °C. In terms of the H_2O:CO ratio, near stoichiometric conditions are generally preferred, as higher ratios negatively affect both the CO conversion and H_2 recovery. Considering the operating pressure, the reaction is generally unaffected.

On the other hand, CMR reactors using silica membranes exhibit high CO conversion at relatively moderate temperatures of 300 °C, as well as significantly lower operating pressures. For these reasons, CMR reactors with silica-based membranes can be considered one of the most efficient options. Incorporating such reactors before amine-based CCS allows for increased CO_2 content in the stream and thus maximum carbon capture.

2.5. CCS Applications in Extractive Industries

The transformation of an energy-intensive industrial sector towards the transition to a low carbon economy, without reducing the production volume, requires joint, collaborative efforts to develop and invest in novel process designs and solutions. In this context, CCS applications in extractive industries are highlighted with main focus on the capture, transport and storage of CO_2 that would otherwise have been emitted to the atmosphere.

The ULCOS project investigated various CCS technologies for implementation in extractive industries. These technologies were pressure swing adsorption (PSA), vacuum pressure swing adsorption (VPSA), VPSA and cryogenic separation and compression, amine-based capture and compression and PSA and cryogenic distillation and compression [10]. The process gas composition was almost identical for all technologies, with compositions of 2.7–3 vol% CO_2, 67.8–71.4 vol% CO, 12.1–13 vol% H_2, 13.5–15.7 vol% N_2 and 0–2.1 vol% H_2O. The results showed that of all the aforementioned technologies, amine-based capture and compression and PSA and Cryogenic distillation are the most efficient approaches accounting for almost 100% CO yield in the recycled gas and 100% captured CO_2 purity. For comparison, PSA accounted for 88% CO yield and 79.7% captured CO_2 purity. In addition, these technologies were concluded to be the only two 100% suitable for CO_2 transport and storage.

WGS-enhanced CCS has received significant attention for utilisation in extractive industries over the last years. European project C4U, started in 2020, examined the utilisation of a WGS-enhanced CCS system in iron and steel industries [5]. The performance of the system was compared with a base case scenario of a conventional amine-based CCS without a WGS reactor. The flue-gas examined was a BFG with a composition of 21.2 mol_{dry}% CO_2, 22.7 mol_{dry}% CO, 0.2 mol_{dry}% C_2H_4, 2.4 mol_{dry}% H_2, 53.5 mol_{dry}% N_2 and a total flow rate of 125 kg/s. The amine-based CCS systems utilised a 25 wt% MDEA solvent. In the base case, the CCS system delivered 36.8 kg/s of 98.2% pure CO_2. The system also accounted for approximately 0.92 kmol/s of CO and 0.03 kmol/s of uncaptured CO_2. The WGS-enhanced

CCS system delivered 66.3 kg/s of 98.1% pure CO_2, an increase of approximately 80%. The system also accounted for approximately 0.01 kmol/s CO and 0.0036 kmol/s uncaptured CO_2, a decrease of almost 99% and 88%, respectively. The results of this examination clearly displayed the advantages of the WGS-enhanced CCS technologies over the conventional amine-based.

The STEPWISE project executed within the European H2020 LCE program also aimed for the demonstration of an advanced pre-combustion CO_2 removal technology for iron and steel industries [45]. The technology was based on sorption enhanced water gas shift (SEWGS) using regenerative solid adsorbents at elevated temperatures. The solid adsorbent was regenerated by pressure swing, resulting in CO_2-rich streams suitable for storage and transport. The H_2 produced by the WGS reaction was considered suitable for power generation. In the project, the process was demonstrated at a scale of 14 t/day CO_2 removal. For simple integration of the technology, carbon intensity was reduced to 1.26 t CO_2/t steel, with an energy consumption of 1.95 MJ/kgCO_2 and CO_2 avoidance costs of 32 €/t CO_2. For intensified implementation, focusing on internal uses and electricity production, carbon intensity was reduced to 0.60 t CO_2/t steel.

The technology developed utilised a WGS active sorbent, allowing for the in-situ removal of CO_2, to account for limitations set by equilibrium. Thus, there is no need of two or more WGS reactors [46]. The dry gas considered had a conventional composition of 19% CO, 25% CO_2, 3% H_2, 15–20 ppm H_2S + carbonyl sulphide (COS) and 53% N_2. Implementation of the WGS reactor allowed for the reduction of CO content to 5%. The SEWGS unit utilised a catalytically active potassium-promoted hydrotalcite-based sorbent for both CO conversion completion and separation of CO_2 from H_2.

In 2020, Manzolini et al. performed a techno-economic assessment of SEWGS technology for integration in a steel-plant, as a follow-up study of the STEPWISE project [47]. The study compared the performance of a conventional amine wet scrubbing technology and the solid adsorption technology of the STEPWISE project. Simulation results showed that the SEWGS using solid absorbents accounted for a specific energy consumption for CO_2 avoided (SPECCA) of around 2.5 MJ/kg CO_2, in comparison to the 4.8 MJ/kg CO_2 of the conventional amine scrubbing.

3. Carbon Capture and Utilisation Technologies for Extractive Industries

CCU solutions typically opt for the transformation of CO_2 to valuable products, such as MeOH, using an external H_2 feed. Commonly used approaches comprise the CO_2 hydrogenation, for the immediate utilisation of the captured CO_2, and the enhanced CO_2 transformation by the Fischer-Tropsch reaction, where CO_2 is transformed into CO by the reverse water gas shift (RWGS) reaction to allow for higher carbon conversions and MeOH production. In this frame, CO_2 hydrogenation and the Fischer–Tropsch reaction process will be reviewed, as the one of the most commonly used and promising approaches for MeOH production in extractive industries.

3.1. CO_2 Hydrogenation

CO_2 hydrogenation utilises reactors where CO_2 reacts with H_2 for the production of methane or MeOH. A typical CO_2 hydrogenation system is comprised of the following main components: (1) a CO_2 hydrogenation reactor and (2) a distillation column for the separation and collection of produced MeOH. A conventional system using CO_2 hydrogenation for MeOH production is shown in Figure 4.

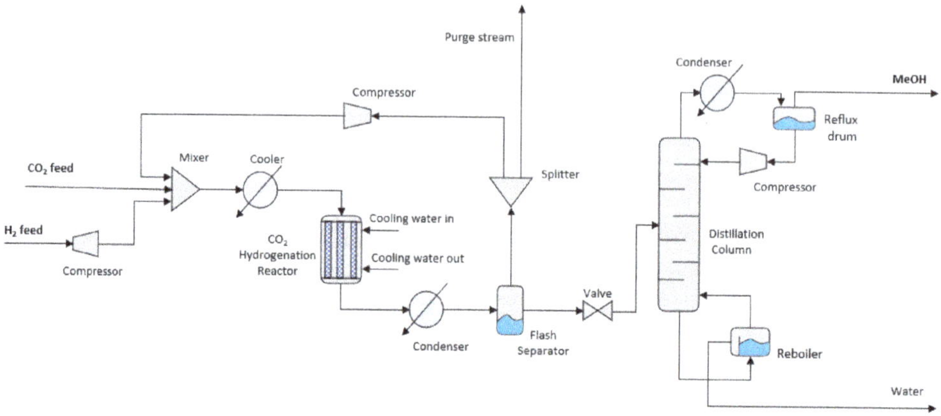

Figure 4. Schematics of a conventional CO_2 hydrogenation system for MeOH production.

CO_2 feed is mixed with an H_2 stream and preheated before entering the reactor. Inside the reactor, the CO_2 hydrogenation occurs, and MeOH is produced, as well as by-product water and CO. CO specifically results from the RWGS reaction, naturally occurring in the reactor. CO produced reacts with the H_2 in the Fischer-Tropsch reaction for enhanced production of MeOH. The reactions taking place in the reactor are the following [48]:

$$CO_2 + 3H_2 \leftrightarrow CH_3OH + H_2O,\ \Delta H_{298K} = -49.16\ \text{kJ/mol} \quad (CO_2\ \text{hydrogenation}) \tag{7}$$

$$CO_2 + H_2 \leftrightarrow CO + H_2O,\ \Delta H_{298K} = +41.21\ \text{kJ/mol} \quad (\text{RWGS}) \tag{8}$$

$$CO + 2H_2 \leftrightarrow CH_3OH,\ \Delta H_{298K} = -90.77\ \text{kJ/mol} \quad (\text{Fischer-Tropsch}) \tag{9}$$

The stream exiting the reactor consists of MeOH, water and unreacted CO_2, CO and H_2, and therefore, requires separation. The stream is condensed and passes through a flash separator where the unreacted gas is separated from liquid MeOH and water. A significant part of this unreacted gas is repressurized and recycled back to the reactor to enhance the final CO_2 conversion.

The liquid MeOH and water are separated in the distillation column of the system. The stream introduced at the middle trays concentrates at the bottom of the column, which is constantly heated for the vaporization of volatile MeOH and separation from water. After initial vaporization, the liquid in the bottom of the column is fed to a reboiler which facilitates further separation. The gaseous MeOH exits the distillation column from the top, and the stream is condensed and passes through a flash separator for the separation of any water vaporized along, and the production of the final MeOH product.

3.1.1. CO_2 Hydrogenation Reactors

The main concern when developing a CO_2 hydrogenation reactor is the efficient removal of the heat produced by the highly exothermic reaction. Efficient heat removal facilitates avoiding the formation of by-products, achieving high CO_2 conversion by low outlet temperatures and good energy efficiency by internal heat recovery [49]. Current designs have shifted from the previously used quench reactors to the quasi-isothermal steam-raising fixed bed reactors, or steam-raising converters (SRC), which allow for more efficient heat recovery and temperature control [49].

One of the most commonly used SRC reactors is the multi-tubular packed bed catalytic reactor, also known as the Lurgi reactor. Lurgi reactor comprises multiple tubes filled with catalyst particles to enhance the reaction rate and overall CO_2 conversion. Such designs incorporate an inner cooling system, in the form of a coolant, typically water, flowing between the space of the tubes. The feed gas feed flows through the tubes in an axial direction. Approximately 80% of the reaction heat is converted to medium pressure steam [50]. A conventional Lurgi reactor for CO_2 hydrogenation is shown in Figure 5.

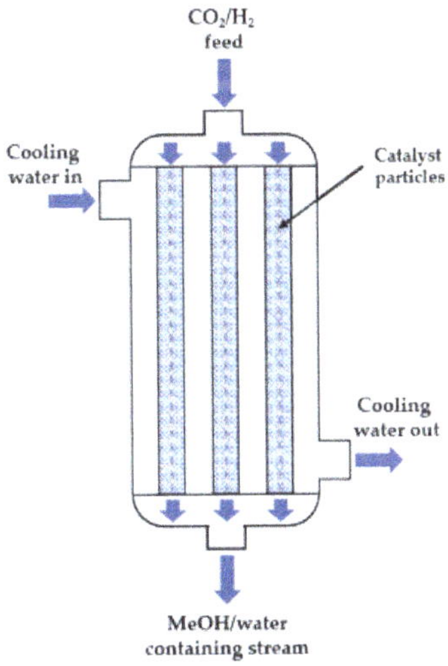

Figure 5. Schematics of a conventional Lurgi reactor for CO_2 hydrogenation.

3.1.2. CO_2 Hydrogenation Catalysts

Currently, the majority of commercial low-pressure catalysts are CuO and ZnO based. In most cases, Al_2O_3 carriers are utilised, along with various stabilizing additives and promoters such as Zr, Cr, Mg and rare earth metals [49]. Several studies showed space time yields (STY) for CO gases being 0.7–2.3 kg of MeOH per litre of catalyst. For CO_2 gases the STY was significantly lower, 0.4–0.8 $kgl_{cat}{}^{-1}h^{-1}$. The studies considered operational pressures 40–100 bar and GHSV about 10 000 h^{-1} [45,50,51]. Modern copper catalysts achieve selectivity of more than 99% for conventional gases [52]. Impurities in the gas feed, high pressure and temperature, high $CO:H_2$ and $CO:CO_2$ ratios and low space velocities are some of the factors reducing MeOH selectivity.

The typical lifetime of industrial catalysts is 4–6 years [49]. Deactivation as a result of poisoning or thermal sintering is the main limitation of the catalysts' lifetime. For copper-based catalysts, sulphur compounds and chlorides are the major causes of poisoning, resulting in blocking of active sites and acceleration of sintering, respectively. Arsenic and carbonyls are other poisons reported, promoting the Fischer-Tropsch reaction and thus decreasing selectivity [53]. To prevent such issues, a prior gas cleaning step is usually implemented, as well as guard beds, such as ZnO targeting sulphur, for the catalyst's protection [52]. For commercial Cu/ZnO catalysts, typical gas requirements are: 0.05–0.5 ppm H_2S, 1 ppb HCI, few ppb metal carbonyl, less than 0.1 mg N m^{-3} particles, less than 1 mg N m^{-3} tar and less than 0.25 mg N m^{-3} alkalis [49].

Liao et al. examined the effect of the $H_2:CO_2$ ratio and temperature of the reaction over a Cu/rod ZnO/Al_2O_3 and Cu/plate ZnO/Al2O$_3$ catalyst at 45 bar [54]. Results showed that increasing the $H_2:CO_2$ ratio from 2.2:1 to 2.5:1 had a positive effect on the CO_2 conversion, but decreased MeOH selectivity for both catalysts at 270 °C, as well as 280 °C. Increase in temperature also resulted in increased CO_2 conversion but decreased MeOH selectivity for both catalysts and the different $H_2:CO_2$ ratios. The results also suggested that the Cu/rod ZnO/Al_2O_3 catalysts exhibited slightly higher CO_2 conversion but significantly lower MeOH selectivity.

Sahki et al. examined the effect of pressure in the reaction over a $CuO/ZnO/Al_2O_3$ catalyst [55]. The reaction took place at 230 °C at varying pressures of 1–75 bar, with a flow rate of 2 L h^{-1} and a stoichiometric H_2:CO_2 ration of 3:1. The results showed that increasing pressure from 1 to 75 bar increased the CO_2 conversion from 3.1 to 11.9%, and MeOH selectivity from 5.3 to 47.1%. CO selectivity decreased from 94.66 to 52.9%.

To highlight the effect of the H_2:CO_2 ratio and pressure in the reaction, Bansode et al. examined the performance of a $Cu/ZnO/Al_2O_3$ catalyst for a 10:1 ratio and a high pressure of 360 bar [56]. The results showed a CO_2 conversion of 65.8% on a single pass, almost six times higher than other applications opting for near stoichiometric ratios and moderate pressures. The MeOH selectivity was 77.3%.

Table 6 summarizes the performance of the various Cu-based catalysts for CO_2 hydrogenation found in literature.

Table 6. Performance of Cu-based catalysts for CO_2 hydrogenation.

Catalysts	H_2:CO_2 Ratio	Temperature (°C)	Pressure (bar)	CO_2 Conversion (%)	MeOH Selectivity (%)	Catalyst Cost (€/kg cat.)	Reference
Cu/plate ZnO/Al_2O_3	2.2:1	270	45	10.9	72.7	~18.1	[54]
		280		12.0	71.6		
	2.5:1	270		15.5	64.5		
		280		14.7	63.3		
$CuO/ZnO/Al_2O_3$	3:1	230	1	3.1	5.3		[55]
			75	11.9	47.1		
$Cu/ZnO/Al_2O_3$	10:1	260	360	65.8	77.3		[56,57]

The studies on the performance of these catalysts show the importance of the operational parameters in the CO_2 hydrogenation reaction. In terms of H_2 supply, high H_2:CO_2 ratios tend to benefit the overall CO_2 conversion. Despite this, raising the H_2:CO_2 ratios results in decreased MeOH selectivity, as well as increased energy demand and cost for the production of H_2. For these reasons, the majority of applications opt for a stoichiometric H_2:CO_2 ratio of 3:1.

Operational temperature also significantly affects the reaction. Higher temperatures result in increased overall CO_2 conversion but decreased MeOH selectivity. The most influential parameter is pressure. While an increase in pressure results in significantly higher conversion, applications need to account for the safety of the systems. Thus, reactors typically operate at pressures up to 100 bar.

3.2. *Enhanced CO_2 Transformation by Fischer-Tropsch Reaction*

CCU commonly utilise RWGS reactors for the conversion of CO to CO_2 to increase carbon conversion efficiency. This approach aims for optimising the CO_2 hydrogenation reaction with the Fischer-Tropsch reaction. As the Fischer-Tropsch reaction for MeOH production is almost twice as much exothermic, $\Delta H_{298K} = -90.77$ kJ/mol compared to $\Delta H_{298K} = -49.16$ kJ/mol, the required activation energy is lower, resulting in increased reaction rate and conversion to account for the decreased selectivity. A conventional system for MeOH production from CO_2 utilising a RWGS reactor is shown in Figure 6.

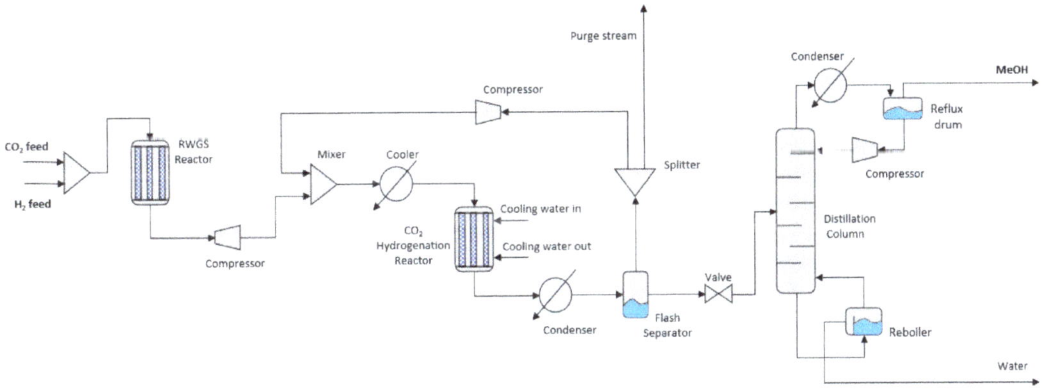

Figure 6. Schematics of a conventional CO_2 to MeOH production system utilizing a RWGS reactor.

The RWGS reaction is endothermic, thus high reaction temperatures are required, increasing the energy consumption of such systems. Specifically, CO becomes the main product of the reaction at temperatures above 700 °C [58]. The H_2:CO_2 ratio for the RWGS reaction ranges from 1:1 to 4:1. In general, the chemical equilibrium of the reaction is pressure independent.

For the RWGS reaction, numerous reactor designs have been studied such as the packed-bed, the fluidized-bed and the moving-bed reactors. A number of catalysts have been investigated for the RWGS reaction, with precious metals such as Pt, Pd, Ru and Au finding successful application as the active sites of the catalysts [59]. Recent studies have also focused on Cu, Ni and Fe catalysts, as the high prices of noble metals such as Pt limit their availability. The results show that, while Pt catalysts exhibit the highest CO_2 conversions, Cu and Fe display the highest CO selectivity [60].

Zhuang et al. specifically examined the performance of a 0.5 wt% Ru-promoted Cu/ZnO/Al_2O_3 catalyst compared to a conventional Cu/ZnO/Al_2O_3 catalyst, under identical operational parameters. The results showed that under 500 °C, CO_2 conversion of the Ru-promoted catalyst was 46%, in comparison to 17%. Metal catalysts are typically coupled with reductive oxides such as CeO_2, MN_2O_3, CrO_2, FeO_X or TiO_2 [61]. In addition, many catalysts incorporate transitional metals or rare earths such as Zr, Fe, La and Y into the reducible oxide lattice [62].

To demonstrate the efficiency of high temperatures in the reaction, various studies examined the performance of catalyst under different reaction temperatures. Zhuang et al. also examined a similar Pt/TiO_2 at a slightly higher H_2:CO_2 ratio, 3:1, and higher temperature, 600 °C. The study showed that when operating under these conditions, CO_2 conversion reached 56% [63]. Kim et al. in their study examined the performance of a Pt/Al_2O_3 and a Pt/TiO_2 catalyst under different temperatures, keeping other operational parameters such as H_2:CO_2 (3:2.1) ratio and space velocity the same [64]. The results showed that, at 500 °C, CO_2 conversion was approximately 34% and 39%, while at 300 °C it was only 6% and 15%, for the Pt/Al_2O_3 and Pt/TiO_2 catalyst, respectively.

For even higher temperatures, Zonetti et al. examined the performance of a Ni/$Ce_{0.75}Zr_{0.25}O_2$ catalyst at 700 °C, at a H_2:CO_2 ratio of 3:1. The study showed that when operating at these conditions, CO_2 conversion reached 62.5%. The performance of the various catalysts is summarized in Table 7.

Table 7. Performance of catalysts for RWGS reaction.

Catalyst	H_2:CO_2 Ratio	Temperature (°C)	CO_2 Conversion (%)	CO Selectivity (%)	Catalyst Cost (€/kg cat.)	Reference
$Cu/ZnO/Al_2O_3$	4:1	500	17.0	100	~18.1	[57,63]
$Ru\text{-}Cu/ZnO/Al_2O_3$			46.0			
Pt/Al_2O_3	3:2.1	500	34.0	100	6520	[64,65]
		300	6.0			
Pt/TiO_2		500	39.0		-	
		300	15.0			
Pt/TiO_2	3:1	600	56.0	100	-	[66]
$Ni/Ce_{0.75}Zr_{0.25}O_2$	3:1	700	62.5	99.5	-	[67]

Studies conducted on the effect of CO_2:CO ratio in MeOH production have shown that the MeOH yield for CO_2-based gases is 18–58% at 200–250 °C and 50–100 bar, significantly lower than for CO-based gases, which accounted for 55–89% [49].

Klier et al. examined the effect of CO_2:CO ratio in the production of MeOH over a Cu/ZnO-based catalyst [68]. The results showed that the complete absence of CO_2 from the syngas accounted for reduced carbon conversion, as a small amount is required as a promoter of the reaction. Despite this, at higher concentrations, carbon conversion was significantly hindered. At 250 °C, carbon conversion to MeOH for a CO_2:H_2 gas of zero CO content was only 9.7%. For a syngas of 2:28 CO_2:CO ratio, at the same temperature and H_2 supply, carbon conversion was up to 69.7%.

Ng et al. also examined the influence of the CO_2:CO ratio in the feed composition for the production of MeOH and dimethyl ether (DME) over the commercial $CuO/ZnO/Al_2O_3$ catalyst [69]. At 250 °C, 50 bar and a GHSV of 27,500 h^{-1}, the MeOH yield was the highest, more than 50% at less than 10% CO_2:(CO_2+CO) ratio. In the absence of CO from the gas, MeOH yield was lower than 20%. The effect of the CO_2:CO ratio in the production of MeOH for the various catalysts is summarized in Table 8.

Table 8. Effect of CO_2:CO ratio in the production of MeOH for various catalysts.

Catalyst	CO_2:CO Ratio	Temperature (°C)	Carbon Conversion to MeOH (%)	Catalyst Cost (€/kg cat.)	Reference
Cu/ZnO-based	2:28	250	69.7	-	[68]
	No CO		9.7	-	
$CuO/ZnO/Al_2O_3$	1:9	250	>50	18.1	[57,69]
	No CO		<20		

Enhanced CO_2 transformation by the Fischer-Tropsch reaction is a widely used approach for the production of MeOH. For the transformation of CO to CO_2, the commercial $Cu/ZnO/Al_2O_3$ catalyst is used, mixed with precious metals such as Ru to enhance the conversion. As the RWGS reaction is highly endothermic, high temperatures of 500 °C are generally preferable. In terms of H_2:CO_2 ratio, near stoichiometric ratios are generally preferred.

Incorporating a RWGS reactor prior to the CO_2 transformation process significantly increases the overall carbon conversion and MeOH production. Studies have shown that the production of MeOH from CO gases can be up to 6–7 times higher than from CO_2 gases. Despite this, such systems account for higher energy consumption, thus incorporating them or not depends on the techno-economical specifications and the desired MeOH production capacity of each system.

3.3. Applications of CO_2 Transformation to MeOH in Extractive Industries

MeOH production from the flue-gases of extractive industries, such as steel mills, is an already commercial technology. For instance, in 2011, 11% of the total MeOH production of China, which is currently the largest market, derived from such processes [70]. However, most of these processes were not focused on exploiting renewable electricity for the production of the H_2 required for the MeOH synthesis, often increasing the total CO_2 emissions [71]. To tackle such issues, numerous studies and projects have focused on modernizing the process to achieve the industries' decarbonization.

Carbon2Chem is one of the projects focusing on the power to MeOH concept, promoting numerous studies on the subject [72]. Bender et al. examined the coupling of steel and chemical production, proposing different scenarios, one of them being the direct reduced iron (DRI) process coupled with MeOH production by green H_2, using the flue-gases of a steel mill [73]. The study was conducted for the implementation of the scenario in the Duisburg site, with a nominal production rate of 12 MMTA of iron products. The flue-gases derived by a blast furnace by 85.5%, a coke-oven by 9.45% and a basic oxygen furnace by 5.05%, and comprised mostly of CO, CO_2, N_2 and H_2. The results of the study suggested that the proposed scenario can reduce the total CO_2 emissions from 21.9 to 0.5 MMTA, a reduction of almost 98%, and is dimed the most efficient approach for the decarbonization of the site's activities.

Schlüter et al. developed a model for the catalytic conversion of steel mill gases for MeOH synthesis [70]. Three different cases were examined, with different rates of coke oven gas (COG), BFG and basic oxygen furnace gas (BOFG) in the flue-gas. The study examined the commercial $Cu/ZnO/Al_2O_3$ catalyst with a bed porosity of 0.38. The reactor was 7.5 m in length, with tubes of 44 mm diameter. The number of tubes varied for every examined case. The reaction took place at 240 °C and 85 bar. The steel mill had a production capacity of 16.4 kt_{steel}/day, with a CO_2 production of 2.74 kt/day (0.34 $ktCO_2$/day by COG, 1.38 $ktCO_2$/day in BFG and 1.02 $ktCO_2$/day in BOFG). Case 1 examined the MeOH production without additional H_2 demand, by limiting the BOFG content in the flue-gas. Case 2 examined the use of additional H_2 for maximum utilisation of BOFG. Both these cases excluded BFG, accounting for MeOH productions of 155.81 and 370.32 kt/a, respectively. Case 3 examined the utilisation of the BFG gases, accounting for a MeOH production of 669.05 kt/a. The carbon efficiency of cases 1,2 and 3 were 90.6%, 92.8% and 82.5%, respectively.

The FReSMe project, which concluded in 2021, also aimed for the production of MeOH from CO_2 deriving from the residual steel gases recovered from an industrial blast furnace [74]. The produced MeOH was to be used as fuel in ship transportation. The pilot plant was designed for a nominal production rate of 50 kg/hr with an input of 800 m^3/hr CO_2. The pilot plant constructed utilised a SEWGS CCS system for the production of a CO_2 rich stream and a H_2/N_2 stream from the BFG [75]. The CO_2 rich stream was cooled, buffered and compressed before entering the desulphurization step. After desulphurization, the pure CO_2 stream was introduced to the reactor system for the production of MeOH to be stored in tanks. The H_2/N_2 stream passed through a membrane skid for the separation of H_2 and N_2. The produced H_2 was used as a feed for the CO_2 hydrogenation reaction, along with the feed produced by an electrolyser. The FReSMe project is the integration of the CCS technology developed in the aforementioned STEPWISE project and the improved technology for MeOH production by CO_2 developed in the MefCO_2 project [74,76].

MefCO_2 project designed a Lurgi reactor utilising a commercial $CuO/ZnO/Al_2O_3$ catalyst for CO_2 hydrogenation [77]. The experiments showed that at 20 bar, 280 °C and 12,030 h^{-1} GHSV, CO_2 conversion reached 25% in a single pass, with MeOH selectivity being 10%. At 200 °C, keeping the other operational parameters constant, CO_2 conversion was less than 5%, while MeOH selectivity reached as high as 80%. The results proved that higher temperatures result in higher conversion while lower temperatures result in higher selectivity, thus moderate temperatures are considered optimal. The project's pilot plant at Niederaussem was designed for 1 tonne of pure MeOH production daily. The reactor developed was a 7.022 m Lurgi reactor of 2962 tubes of 0.053 m diameter each. The catalyst used was a conventional $Cu/ZnO/Al_2O_3$ with a bed porosity of 0.39. The reactor operated at 80 bar, with a feed of 10.2 kg/hr of H_2 and 66.1 kg/hr of CO_2, a conventional H_2:CO_2 molar ratio of 3:1 [77]. The system also incorporated a recycling stream, with a CO_2 recycle ratio (fresh CO_2 feed/CO_2 feed of reactor) of 0.25, thus the system accounted for 4.89 kmol/h and 1.63 kmol/h of fresh H_2 and CO_2, respectively, and a MeOH production of 1.2 ton/day.

4. Sustainability Analysis of MeOH Production from CO_2

MeOH production through CO_2 is an approach seen constant rise in popularity and application. Various studies over the recent years have focused on the environmental impact and the sustainability of these systems. Key points comprise the comparison of MeOH production from CO_2 with the conventional processes, and the energy efficiency of the process.

In 2021, Ryoo et al. conducted a Life Cycle Assessment (LCA) for the CO_2 to MeOH conversion process and its energy optimization [78]. The study examined three different scenarios, MeOH production from the CO_2 deriving from coal gasification, coal coking and production by photocatalytic conversion. All systems utilised an amine-based CCS system using MEA. For CO_2 hydrogenation, a $La/Cu/ZrO_2$ catalyst was examined. The study considered various energy sources for the supply of the required heat and electricity in the system. Solar thermal, natural gas (NG), oil, coke (coal) and steam were examined as heat sources. Nuclear, wind, photovoltaic, hydro and grid electricity were examined for electricity. The results showed that, when using grid electricity and steam, the photocatalytic conversion had the least GHG emissions, 2.28 kg CO_2 eq/kg MeOH, followed by the coke coking conversion process, with 2.9 kg CO_2 eq/kg MeOH. Coal gasification conversion on the other hand was concluded to have the greatest environmental impact, with 17.7 kg CO_2 eq/kg MeOH. CO_2 hydrogenation accounted for 10.7 kg CO_2 eq/kg MeOH. The results also highlighted the importance of energy optimization in the processes, especially for CO_2 hydrogenation. With the use of nuclear and solar thermal energy, the process accounted for 1.67 kg CO_2 eq/kg MeOH, a reduction of more than 84%.

Rosental et al. conducted an LCA on CCU for the production of large-volume organic chemicals [79]. The study examined various CO_2 sources, and the corresponding carbon capture technologies. One of these was the amine-base carbon capture from flue-gases. The amine used by the system was MEA. The MeOH production took place over a $CuO/ZnO/MgO/Al_2O_3$ catalyst. The thermal energy for the carbon capture was supplied by the excess heat of the chemical reactions with a 100% efficiency. The required electricity was provided by offshore wind turbines. The results showed that MeOH produced by CCU had significantly less impact on GHG, approximately 1.5 kg CO_2 eq/kg MeOH, in comparison to the production from syngas, which accounts for 2.002 kg CO_2 eq/kg MeOH.

Hoppe et al. also performed an LCA for the production of MeOH from CO_2 [80]. The study compared the production of MeOH from CO_2, captured by various gases, and H_2, produced by electrolysis from wind power electricity, with the conventional production method from Natural Gas (NG). The results showed that production of MeOH from CO_2 captured besides from air (direct air capture), resulted in <0 kg CO_2 eq/kg MeOH, as a result of utilizing the CO_2 otherwise released into the atmosphere. On the other hand, production from NG and CO_2 from direct air capture accounts for approximately 0.9 kg CO_2 eq/kg MeOH and 0.5 kg CO_2 eq/kg MeOH, respectively.

The environmental benefits of MeOH production from CCUS lie in both the CO_2 emissions reduction from the process, as well the use of the eco-friendlier MeOH, compared to conventional fuels. Rigamonti et al. performed an LCA for MeOH production from the CO_2 of steel mill gases, to be used in ship transportation [81]. The gases examined were a mix of BOFG and COG. The system was a conventional CCU, utilizing the SEWGS CCS technology and CO_2 hydrogenation technology. The CO_2 hydrogenation reactor utilised the commercial $Cu/ZnO/Al_2O_3$ catalyst. The required electricity was provided from a power plant utilizing the remaining flue-gases of the steel mill, to allow for their maximum exploitation. In this frame, Table 9 depicts the environmental impact of CO_2 to MeOH conversion processes in relation to the Global warming potential indicator.

Table 9. Global warming potential of CO_2 to MeOH conversion processes.

CO_2 to MeOH Conversion Processes	Global Warming Potential [kg CO_2 eq/kg MeOH]	Life Cycle Identification	Reference
Coal gasification conversion	17.7	1 kg MeOH production ReCiPe midpoint impact assessment	[78]
CO_2 hydrogenation	10.7		
Coke coking conversion	2.90		
Photocatalytic conversion	2.28		
CO_2 hydrogenation (nuclear and solar thermal energy)	1.67		
Amine-based CCS + CO_2 hydrogenation	1.05	1000 kg CCU products IPCC, 2013	[79]
Production from syngas	2.002		
Production from NG	~0.9	1 kg MeOH production ecoinvent 3.1	[80]
Direct air capture + CO_2 hydrogenation	~0.5		
No direct air capture + CO_2 hydrogenation	<0		

The studies conducted on the sustainability of MeOH production from CO_2 show the environmental benefits of this approach, in comparison with conventional production methods. The studies indicate that CO_2 hydrogenation is significantly less GHG emissive than other conventional MeOH production methods. While other processes, such as photocatalytic conversion also allow for a very high reduction in GHG emissions, CO_2 hydrogenation, as a mature and relatively simple approach, can have even greater mitigation capabilities when coupled with renewable energy sources. Compared to coal gasification, CO_2 hydrogenation using green energy produces approximately 10 times less kg CO_2 eq/kg MeOH. Amine-based CCS coupled with CO_2 hydrogenation processes accounts for as much as 47% reduction in the overall CO_2 emissions of MeOH production, compared to the production from NG. In addition, non-direct air capture, such as amine-based capture, can account for almost 100% reduction in GHG emissions, compared to direct air capture. The use of CCUS allows for significant mitigation of GHG emissions thanks to the avoidance of MeOH production by highly emissive conventional processes.

5. Cost Comparison of CCUS for MeOH Production

A short literature review of studies focused on the techno-economic analysis of the CCS and CCU systems for MeOH production is presented. The evaluation of the economic sustainability and the cost comparison of the different practises is key for the optimisation and consequently the large-scale commercialisation of the different solutions.

Roussanaly et al. performed a techno-economic analysis over a CCS system using MEA for the CO_2 capture from a flue-gas of approximately 20 vol% CO_2 [82]. The study concluded that, with a capture efficiency of 90%, the cost of CO_2 capture was 63.2 €/ton CO_2. The study also concluded that the CO_2 avoided cost was 83.2 €/ton CO_2, highlighting the economic benefits of incorporating a CCS unit. Panja et.al also performed a techno-economic analysis for an amine-based CCS unit, using the PacifiCorp's Hunter Plant as the case study [83]. The study examined the use of MEA for different capture capacities. For a 65% capture efficiency, the estimated carbon capture cost was approximately 52 €/ton CO_2 [84]. Increasing capture efficiency to a conventional 90% resulted in decreased costs, to approximately 43 €/ton CO_2.

Moioli et al. performed a cost comparison of the conventionally used MEA scrubbing to potassium taurate absorption [85]. The study examined a potassium taurate system with an almost identical layout to the amine-based systems, without the use of the optional solid-liquid separator for separating solid taurine. The study assessed the costs for the two practices, in terms of total capital and operating costs and the capital costs of the absorption and regeneration processes. For MEA, the analysis concluded capital and operational costs of approximately 180 million €/year and 43 million €/year, respectively. For potassium taurate the corresponding costs were approximately 150 million €/year and 38 million €/year. In terms of the absorption and regeneration processes, the capital costs for MEA were approximately 17 million €/year and 26.4 million €/year, respectively. For potassium taurate, these costs were 13.6 million €/year and 6.4 million €/year. The aforementioned costs were estimated for a 90% capture efficiency, for a flue-gas of 13 mol% CO_2 content, a flowrate of 19.6 kmol/s, and an operating load of 8000 h/year. Thus, the capital and operational costs for MEA can be expressed as 62 €/ton CO_2 and 14.7 €/ton CO_2, respectively. For potassium taurate, the corresponding figures were 51.6 €/ton CO_2 and 13.1 €/ton CO_2, highlighting the potential of using energy-efficient solvents such as amino acids.

Khallaghi et al. performed a techno-economic analysis comparing a base case of a conventional amine-based system to a WGS-enhanced system, used for carbon capture from the BFG of a steel mill [40]. The results of the study showed that the base case accounted for a total CO_2 capture cost of 39.84 €/ton CO_2, in comparison to 44.35 €/ton CO_2 for the WGS enhanced case. It is evident that while the WGS-enhanced approach significantly improves the overall capture capacity of CCS systems, the extra components required, as well as the increased energy consumption, accounted for increased operational costs.

Considering the economic sustainability of CCUS, the benefits of this approach lie in the mitigation of emissions related costs (i.e., carbon taxation) and the creation of revenues from the production of MeOH. In 2020, Centi et al. performed a techno-economic analysis for the production of MeOH by CO_2 hydrogenation [86]. The study compared the cost of MeOH production in relation to an average value of 400 €/ton MeOH in Europe. In addition, the study examined the MeOH production cost considering the case study of renewable energy production from remote areas, as proposed by Barbato et al. [87]. Both infrastructure and operation for CO_2 capture and hydrogenation were considered and a 20 €/ton CO_2 benefit of avoided emissions was estimated. The study concluded a net MeOH production cost of 294 €/ton MeOH. The costs of the different CCUS technologies are summarized in Table 10.

Table 10. Cost of different CCUS technologies.

CCUS Components	Solvent	Carbon Capture Efficiency (%)	Cost for CCS (€/ton CO_2)	Cost for CCU (€/ton MeOH)	Reference
CCS	MEA	90	63.2	-	[82]
CCS	MEA	90	43	-	[83]
CCS	MEA	90	76.7	-	[84]
CCS	Potassium taurate	90	64.7	-	
CCS	MDEA	46.5	39.84	-	[40]
CCS (WGS-enhanced)	MDEA	83.8	44.35	-	
CCU	-	-	-	294	[86]

6. Summary and Future Perspectives

CCUS technologies are considered an efficient and promising approach to increase the environmental sustainability of energy-intensive industries such as extractive industries. With an average CO_2 capture efficiency of approximately 90%, CCS technologies have the potential to significantly reduce CO_2 emissions. CCU technologies for the production of MeOH facilitate the circular economy of the industrial sector, creating revenues from green fuel production.

The cost to capture and store post combustion emissions is very high, even though the technology has been around for a long time. The capital and high operational costs of such systems pose a major challenge towards their commercialisation. The development of CCUS faces a number of barriers lying mostly on the economic feasibility. The high energy requirement of the conventional CCS units, mostly due to the required regeneration energy for the solvents, significantly affects economic sustainability. In addition, incorporating CCS systems in industrial plants results in decreased plant efficiencies and increased water use, adding to the operational costs [88]. Transportation of the captured CO_2 requires significant amounts of energy for the compression, as well as maintaining the high pressure in the pipelines. The pipelines themselves must be specially designed, as existing oil and gas pipelines are not suitable. Impurities in CO_2, such as water, can damage pipelines, leading to leaks or even explosions due to the rapid expansion of the compressed fluid.

According to the literature, CCS solutions and suppliers have increased steadily, which means that the general technical availability is no longer a hurdle in itself for industrial applications. Nevertheless, further investigation can be performed, focusing on reducing the costs of CO_2 capture for investors as well as boosting technological progress. In this regard, it is clear that the next developments of CCS will have a focus on increasing their capture rates and the overall efficiency.

Captured CO_2 can also be used as a feedstock for a variety of industrial purposes (e.g., methanol). CO_2 hydrogenation for MeOH production is one of the most common practices. Current operational conditions significantly rise the energy demand for the compression and heating of the reactants, while achieving CO_2 conversions of less than 16% in a single pass, except in cases of extremely high pressures, which are mostly limited in lab-scale studies due to safety reasons. To tackle these issues, incorporation of recycling streams is necessary to achieve conversion efficiencies high enough to justify the high energy consumption. In addition to recycling, the use of RWGS reactors prior to CO_2 hydrogenation can significantly increase the overall CO_2 conversion of CCU, by converting CO_2 to CO. A key factor towards achieving the optimal and sustainable use of these technologies is the development of efficient CO_2 reactors. CCU comprises technologies at different levels of maturity and process complexity. In this respect, the development of highly selective, low cost and long-term recyclable catalysts for the production of MeOH with CO_2 in absence of CO is also investigated.

The production of the necessary H_2 for the CO_2 hydrogenation is another aspect requiring considerable research. Most commonly used H_2 production methods rely on the reforming of hydrocarbons. While production by water electrolysis has seen a constant rise over the years, the high energy costs result in scaling-up difficulties. Energy optimisation and scaling up of water electrolysis technologies are key to increase the sustainability of CCUS and promote the development of CO_2 hydrogenation technologies. A promising solution finding ground is the production of H_2 through photochemical water splitting with CO_2 reduction. Complimenting CO_2 hydrogenation systems with energy-efficient H_2 production solutions is a major challenge towards increasing the technologies' sustainability and promoting its future development.

In general, limited awareness is reported regarding CCS and CCU technology advantages. It is necessary to communicate how CCUS affect the everyday life of EU citizens and consumers' choices. To this end, the coaction of policymakers at national and European level with companies and other societal actors, such as trade unions, will facilitate the further deployment of CCUS projects and raise awareness about their climate and economic benefits.

To sum up, existing barriers to the industrial scale deployment and development of CCS and CCU have to be addressed. The European climate targets provide better prospects to overcome economic barriers in order to enable the take-off of CCS and CCU projects and to introduce incentives for the technologies. At the same time, technical constraints are also essential for creating new CCU decarbonisation pathways (e.g., ensuring symbiotic infrastructure among CO_2 emitters and CO_2 converters in clusters of concentrated industrial activity).

7. Conclusions

The CCUS technologies provide a path towards a carbon neutral society through mitigating CO_2 emissions from fossil fuels by capturing and converting them into fuels and chemicals. Among the CCS, chemical absorption using amine-based solvent is one of the most commercially used, being suitable for the low CO_2 containing flue-gases deriving from the processes of extractive industries. Evidence suggests that tertiary amines in particular, such as MDEA, account for carbon capture efficiencies of up to 90% while having significantly lower regeneration energy demands than other amines. When mixed with promoters such as PZ, such systems are able to achieve absorption capacities as high as 0.99 mol CO_2/kg solvent.

To achieve even higher CO_2 capture, amine-based CCS processing flue-gases of high CO content, such as the flue-gases of steel mills, often utilise WGS reactors for the transformation of CO to CO_2, in the WGS-enhanced CCS. WGS-enhanced CCS has received great attention with several projects focusing on the development of such systems to be used in extractive industries. C4U project examined the CO_2 capture from BFG gases in the steel industry. Compared to conventional amine-based systems, the WGS-enhanced CCS accounted for an increase of approximately 80% in CO_2 production.

CCU for carbon transformation to MeOH is widely applied with MeOH finding commercial applications in various sectors. CO_2 hydrogenation for MeOH production takes place in catalytic reactors, typically Lurgi reactors. The most commonly used catalysts are $Cu/ZnO/Al_2O_3$ and $CuO/ZnO/Al_2O_3$ catalysts, being able to achieve CO_2 conversion and MeOH selectivity up to 65.8% and 77.3%, respectively, when operating at high H_2:CO_2 ratios and pressures. In addition to CO_2 hydrogenation, production of MeOH from CO is also applied. Such systems utilise RWGS reactors for the conversion of CO_2 to CO for the Fischer-Tropsch reaction. This approach allows for higher CO_2 conversion, but lower MeOH selectivity.

Overall, CCUS has already seen several applications in the extractive industry, with CCS especially already being considered a commercial technology. Utilisation of these technologies is key for the decarbonization, with new projects and plants constantly arising. For these reasons, further investigation for maximizing the CCUS capture and transformation efficiency is required. The optimization of these technologies is a major factor towards the mitigation of GHG emissions, for the extractive industries to play their part in the net zero emissions milestone of 2050.

Author Contributions: Conceptualization, A.P.; methodology, A.P., C.P., S.K. and P.M.A.; validation, C.P. and P.M.A.; investigation, S.K.; writing—original draft preparation, S.K.; writing—review and editing, C.P.; visualization, S.K.; supervision, A.P. All authors have read and agreed to the published version of the manuscript.

Funding: This research has received funding from the European Union's Horizon Europe research and innovation programme under grant agreement with number 101058696, project HEPHAESTUS.

Institutional Review Board Statement: Not applicable.

Informed Consent Statement: Not applicable.

Data Availability Statement: All data sources used are cited and all data produced are reported in the manuscript.

Conflicts of Interest: The authors declare no conflict of interest.

eng

Article

Geomechanical Feasibility Analysis of Salt Cavern Gas Storage Construction in Sanshui Basin, Guangdong Province

Haitao Li [1,*], Qiqi Wanyan [1], Guosheng Ding [1], Kang Li [1], Yanxia Kou [1], Song Bai [1], Lina Ran [1], Jianan Wu [1] and Jingen Deng [2]

1 PetroChina Research Institute of Petroleum Exploration & Development, Beijing 100083, China
2 College of Petroleum Engineering, China University of Petroleum, Beijing 102249, China
* Correspondence: lht19902010@163.com

Abstract: Salt cavern gas storage has become the key project of current and future underground gas storage (UGS) facilities construction due to their efficient peak-shaving and supply assurance capacities. The Sanshui Basin in Guangdong Province, China, is rich in salt resources and has high-purity salt rock, which is a potential area for the construction of salt cavern underground gas storage. To speed up the large-scale construction of underground gas storage in China and promote the sustainable development of the natural gas market, it is very necessary to analyze the geomechanics of the target salt layer and study the feasibility of gas storage construction. Based on comprehensive experiments of rock mechanics and thermodynamics, the strength, creep and temperature-sensitive mechanical properties of the target rock in Sanshui Basin were studied. Then, according to the geological conditions of Sanshui salt formation, a three-dimensional geological model was established to analyze the stability of salt cavern gas storage under the injection-production operation. The results show that the average tensile strength and uniaxial compressive strength of salt rock are 1.51 MPa and 26.04 MPa, respectively, showing lower strength. However, under triaxial compression, the compressive strength of salt rock increases significantly, and there is no obvious shear failure phenomenon observed. Moreover, after the peak strength, the salt rock still has a large bearing capacity. In addition, under the confining pressure of 30 MPa, the strength of salt rock decreases by 8.3% at a temperature of 60 °C compared with that at room temperature, indicating that the temperature has a low, modest effect on the mechanical properties of salt rock. The stability analysis shows that, under an injection-production operating pressure of 10–23 MPa, the displacement, plastic zone range and volume convergence rate of single cavity and cavity group are small, and the cavity shows good stability. Overall, the target salt formation in Sanshui Basin, Guangdong Province, presents a good geomechanical condition suitable for the construction of underground salt cavern gas storage. This study can provide a reference for the development and design of salt cavern UGS.

Keywords: Sanshui Basin; salt rock; mechanical property test; temperature; gas storage

Citation: Li, H.; Wanyan, Q.; Ding, G.; Li, K.; Kou, Y.; Bai, S.; Ran, L.; Wu, J.; Deng, J. Geomechanical Feasibility Analysis of Salt Cavern Gas Storage Construction in Sanshui Basin, Guangdong Province. *Eng* **2022**, *3*, 709–731. https://doi.org/10.3390/eng3040048

Academic Editors: Reza Rezaee and Yujie Yuan

Received: 23 November 2022
Accepted: 15 December 2022
Published: 16 December 2022

1. Introduction

Salt rock is recognized as a good medium for underground energy storage [1,2]. The unique properties of salt rock were discovered abroad early, and is utilized for the storage of oil, gas and various industrial wastes. In the early 1940s, Canada first used salt caverns to store liquefied petroleum gas. In the 1950s, North America and Europe began to use salt cavities to store light hydrocarbons. After the 1960s, the use of salt caverns for compressed air energy storage and hydrogen energy storage has been developed in the United States, Germany and other countries [3,4]. In 1978, Germany took the lead in building the first commercial salt cavern compressed air energy storage (CAES) power station (Huntorf) in the world, with a transmission power of 321 MW. The second commercial salt cavern CAES power station (McIntosh in Alabama) was constructed in the United States in 1991, forming a 110 MW transmission power scale [5]. With continuous exploration and practice, there are more and more types of salt cavern storage, which are also used to store hydrogen, helium, etc. An example is the Helenburg salt cavern helium storage in Russia [6]. Among many storage media, natural gas is the energy medium with the largest storage capacity in salt cavern storage [7]. Up to now, more than 108 underground salt cavern gas storages have been built and operated in the world, including more than sixty in Europe, forty-seven in North America and one in China, most of which exist in the United States and Germany [8]. During more than 60 years of development of salt cavern gas storage in salt dome formation, mature theories and technologies have been formed abroad in terms of water solution cavity construction, large wellbore cavity building, stability evaluation and injection-production operation [9,10].

China's rich salt mineral resources are distributed throughout 15 provinces. Making full use of salt mineral resources to build underground gas storage will greatly promote the sustainable development of China's natural gas market, which is also one of the key contents of China's "Fourteenth Five Year Plan" [11] on energy development planning. With the successful development and utilization of Jintan Salt Mine's salt cavern gas storage, in order to ensure the stable supply of natural gas in China's market and relieve the pressure of transportation and storage brought by the rapid growth of gas demand, China is stepping up efforts to develop existing salt mineral resources around the main cities, accelerating the construction of salt cavern gas storage, and forming a series of infrastructure supporting the supply and marketing of natural gas in large- and medium-sized cities. At present, Hubei Yunying Salt Mine, Jiangsu Huai'an Salt Mine, Henan Pingdingshan Salt Mine and other salt cavern gas storage areas have been built or are planned to be built, and the feasibility studies of gas storage construction, such as site selection and rock mechanical properties, have been carried out successively. In the geomechanical analysis of the construction of salt cavern gas storage, the mechanical parameters of the target salt layer are the key and foundation for the construction of gas storage [12–15]. Scholars have carried out a lot of research on the mechanical properties of salt rock. Ma [16], Liu [17] and Alkan [18] et al. studied the yield characteristics and failure criteria of salt rock based on a large number of core uniaxial-triaxial and tensile tests. Yang [19], Li [20] and Mansouri [21] et al. carried out a large number of indoor creep tests and loading and unloading tests to study the creep characteristics of salt rock and its deformation characteristics. Wanyan [22] et al. carried out triaxial compression creep tests on salt rocks from salt mines in Yunying, Hubei province, Huai'an, Jiangsu province and Pingdingshan, Henan province, and compared the creep characteristics of salt rocks in different regions. Liu [23], Jiang [24] and Wang [25] et al. studied the mechanical properties of layered salt rock and the influence of interlayer on the mechanical properties of salt rock. Gao [26] and Li [27] et al. studied the influence of temperature on the mechanical properties of salt rock by using core high temperature tests, and obtained the understanding that the mechanical properties of salt rock deteriorate due to temperature rise. At present, the research on mechanical properties of salt rock is mainly based on a combination of experiment and theoretical analysis, taking into account the influences of tensile and compressive strength, yield, creep and temperature on the characteristics of the rock. This provides the necessary basic data for the construction of gas storage cavity and the long-term stability of the cavity.

A salt mine in Sanshui Basin, Guangdong Province, is rich in salt rock resources with high purity and medium buring depth of 1200–1300 m. It is a potential mining area for building salt cavern UGS. At present, there is little research on the mechanical properties of rock in this mining area. In order to fully evaluate the feasibility of the construction of salt cavern gas storage in this area, it is necessary to carry out experimental research on the mechanical properties of salt rock in the main salt layer, recognize the deformation law and rheological properties of salt rock and evaluate the stability of salt cavern under operation pressure, so as to provide a theoretical basis for the construction of salt cavern gas storage in Sanshui Basin, Guangdong Province.

2. Experiment Analysis and Discussion

2.1. Experiment Method

The experimental salt rock core used in this study is taken from the main salt layer in Sanshui Basin, Guangdong Province, with a buried depth of about 1200 m. Some core columns collected by the coring barrel are shown in Figure 1a. A core from the core column was drilled and processed into standard rock sample for experiments (Ø50 ∗ H100 mm), as shown in Figure 1b. As the salt rock is easily dissolved in water, in order to avoid large damage to the core caused by the conventional coring method, the wire cutting process was adopted for this coring. During processing, the wire cutting fluid did not contain water or react with the core, which ensured the core is not damaged and polluted.

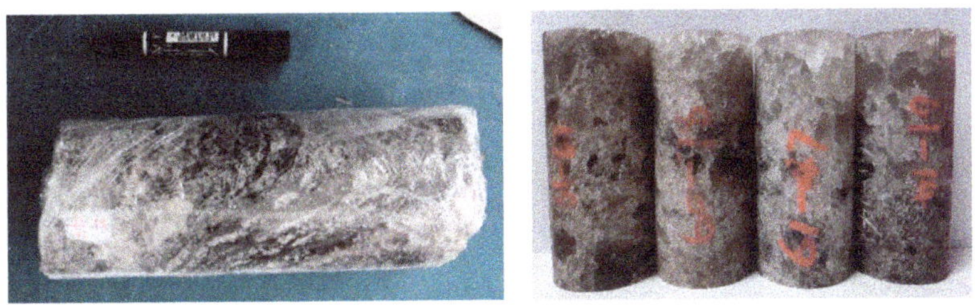

(**a**) (**b**)

Figure 1. Core pillar and standard rock sample for experiment (**a**) for core pillar, (**b**) for standard core.

The experimental salt rock core was grayish black due to impurities, and some salt rock cores contained an obvious mudstone interlayer. This research mainly studies the mechanical behavior and deformation law of salt layer in the target area through direct tension, Brazilian splitting, uniaxial, triaxial and creep tests, as well as strength and creep tests at high temperatures. The self-designed direct tensile apparatus, high-temperature and high-pressure triaxial apparatus and servo-controlled testing machine of the Rock Mechanics Laboratory of China University of Petroleum (Beijing) were used to conduct the salt rock mechanics test. See Table 1 for basic physical parameters and the experimental scheme of salt rock core. From the perspective of physical parameters, the salt rock of the target salt bed in Sanshui Basin contains impurities, with a density range of 2.15–2.43 g/cm^3 and a large distribution range, with an average density of 2.24 g/cm^3. Acoustic data show that the longitudinal wave velocity is about twice that of the shear wave.

Table 1. Physical parameters and experimental scheme of salt rock.

No.	Density /g/cm^3	Diameter /mm	P-Wave Slowness-Time /us/ft	S-Wave Slowness-Time /us/ft	Temperature/°C	Methods
1	2.24	50.09	/	/	25	Indirect Tension
2	2.20	50.05	/	/		
3	2.25	50.06	/	/		
4	2.23	50.12	/	/		
5	2.23	49.76	77.25	146.86		Direct Tension
6	2.21	50.04	78.16	152.35		
7	2.43	49.99	68.96	113.93		Uniaxial Compression
8	2.22	50.25	68.86	121.04		
9	2.41	50.17	78.23	142.16		
10	2.21	49.68	69.09	122.06		
11	2.23	50.15	70.04	127.80		

Table 1. *Cont.*

No.	Density /g/cm^3	Diameter /mm	P-Wave Slowness-Time /us/ft	S-Wave Slowness-Time /us/ft	Temperature/°C	Methods
12	2.20	49.96	65.13	131.06	25	Triaxial Compression
13	2.29	50.03	65.93	106.04		
14	2.34	50.01	69.56	123.93		
15	2.17	49.98	75.56	135.67		
16	2.21	49.89	71.000	125.24		
17	2.33	50.09	76.88	147.35		Uniaxial Creep
18	2.36	50.09	77.79	152.84		
19	2.15	50.30	70.44	129.59		
20	2.23	50.15	72.86	130.47		
21	2.17	50.15	68.46	132.35		
22	2.16	50.03	70.20	118.21		Triaxial Creep
23	2.15	50.05	69.15	125.08		
24	2.17	50.06	70.22	118.79		
25	2.17	50.04	75.19	142.20		
26	2.23	50.03	76.23	146.41		
27	2.15	49.98	68.25	129.46	60	High Temperature Triaxial Compression
28	1.97	49.86	71.92	138.38		
29	2.46	50.12	111.59	184.04		
30	2.20	49.80	67.6	124.82		
31	2.27	50.07	69.83	135.27		
32	2.19	20.16	69.47	128.66		High Temperature Triaxial Creep

Note: The size of indirect tensile core is small, and no acoustic test was conducted. For simplicity and clarity, the core numbering in the table is rearranged.

2.2. Analysis of Experimental Data

2.2.1. Tension Tests

The direct tensile test adopts the self-designed core direct tensile loading device, which can directly conduct the direct tensile test on the press, as shown in Figure 2a. All the cores were processed into an I-shape for the experiments. The Brazil splitting test method was used for indirect tension. The tensile properties and parameters of salt rock were determined by comparing the two methods. Figure 3 shows that, in the direct tension test, the salt rock fracture occurs at the neck (as shown in Figure 3a, the red dotted line), and the fracture surface is basically perpendicular to the axial direction. However, in the indirect tension test, the core is split into nearly symmetrical halves. The results of both experiments are shown in Table 2. The tensile test shows that there is little difference between the core strength measured by the two methods, indicating that both methods can be used as the core tensile strength measurement mode. The experimental results show that the tensile strength of salt rock in Sanshui Basin is not high, ranging from 1–2 MPa with an average of 1.51 MPa. For salt cavern gas storage, the tightness of the cavity is one of the key parameters for its safety assessment. In the process of construction and operation, it is necessary to control the pressure in the cavity to avoid tensile damage in key parts, which may lead to the risk of cavity instability or sealing failure.

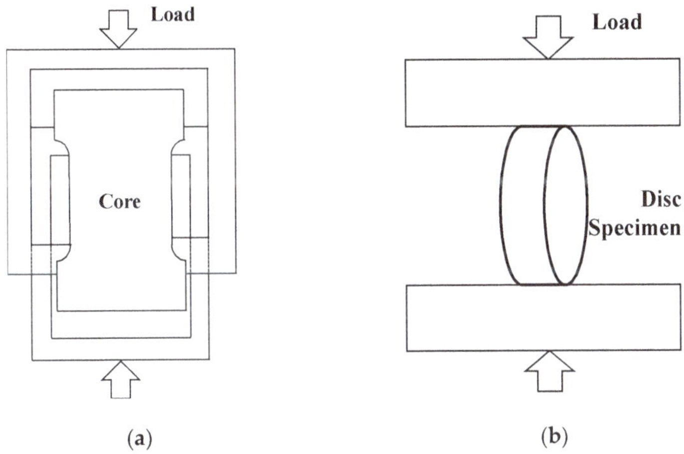

Figure 2. Schematic diagram of tensile test (**a**) for direct tensile, (**b**) for indirect tensile.

Figure 3. Tensile test photos of salt rock. (**a**) #5 direct tension (**b**) #1 indirect tension (left Before experiment, right after experiment).

Table 2. Tensile test data.

No	Density (g/cm^3)	Maximum Load (N)	Uniaxial Tensile Strength (MPa)	Average Strength (MPa)
1	2.24	2505	1.27	
2	2.20	2690	1.36	
3	2.25	4000	2.02	1.51
4	2.23	N/A	N/A	
5	2.23	2130	1.54	
6	2.21	1940	1.38	

Note: N/A indicates that the core is damaged at the beginning of test.

2.2.2. Uniaxial/Triaxial Test at Room Temperature

The core length used in the uniaxial and triaxial compression tests of salt rock in Sanshui Basin was about 100 mm. It can be seen from some core photos after the experiment that, under uniaxial compression, tensile cracks along the axial direction were produced in the salt rock (as shown in the red dotted line at the lower part of Figure 4a), while under triaxial compression, no obvious cracks or damage were produced in the salt rock samples, and the salt rock samples were in a radial expansion state. Core photos show that the crystal grains of salt in the salt rock fabric of Sanshui Basin are large (as shown in the red dotted line circle in Figure 4b above), visible to the naked eye, and the size is inlaid with an irregular distribution. There are also randomly distributed impurity grains of different sizes inside. The experimental results show that under the action of stress, there is dislocation, sliding and fragmentation between the salt rock crystal particles. Large salt rock crystal particles are broken into many small crystal particles, and the boundaries between the particles are denser and clear (as shown in the red dotted circle below Figure 4a,b). Uniaxial compression test results in Table 3 show that the average peak compressive strength of salt rock is 26.04 MPa, which is about 17.2 times of its uniaxial tensile strength, and the average elastic modulus and Poisson's ratio are 6.16 GPa and 0.296, respectively. Under uniaxial compression, the stress–strain curve of the salt rock in the salt mine has obvious elastic–plastic behavior. The elastic stage is short. It then goes through a long plastic flow stage. Before the peak failure, the axial strain and radial strain both increase significantly, and the volumetric strain also shows that the dilatancy of salt rock core under uniaxial compression test is obvious.

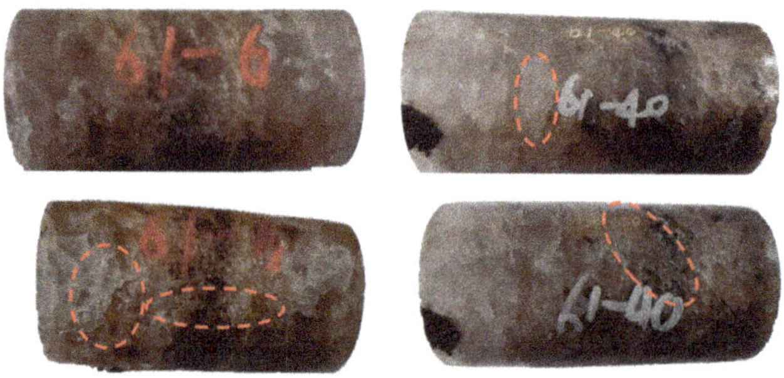

(**a**) #11 (**b**) #15

Figure 4. Photos of some rock cores before and after uniaxial and triaxial compression test. The upper part is before the test, and the corresponding lower part is after the test.

Under the triaxial compression test, the peak strength of the rock generally increases with the increasing of confining pressure. Compared with the uniaxial compressive strength, the strength of the core under triaxial compression has been greatly improved, and the elastic modulus also significantly increased, as shown in Table 4. The stress–strain curves (Figure 5) show that the strain of salt rock increases significantly when it reaches failure under triaxial compression test. The axial strain of the core is larger than the radial strain under the confining pressure, indicating that the confining pressure has a greater impact on the radial deformation of the core salt rock. Volume strain shows that at the initial stage of the experiment, the core is mainly compacted, which is obvious from the triaxial compression experiment. Then, the damage occurs in the rock when microcracks are generated and expand, and volume expansion happens at the beginning of the experiment. The core compaction is significant, and under the influence of confining pressure, the volume expansion is significantly weakened compared with the uniaxial compression condition. In Figure 6, axial stress and strain under different confining pressures show that, with increasing of confining pressure, the peak strength of salt rock does not increase by the same factor, but rather increases slightly. After reaching peak strength, the salt rock still has a high bearing capacity.

Table 3. Resultant for the uniaxial compression test.

No	Density (g/cm^3)	Peak Strength (MPa)	Elastic Modulus (GPa)	Poisson's Ratio
7	2.43	25.74	8.89	0.33
8	2.22	28.55	7.63	0.3
9	2.41	28.18	6.48	0.23
10	2.21	22.87	6.23	0.3
11	2.23	24.86	1.56	0.32
Average	/	26.04	6.16	0.296

Table 4. Resultant data for the triaxial compression test.

No	Density (g/cm^3)	Confining Pressure (MPa)	Peak Strength (MPa)	Elastic Modulus (GPa)
12	2.20	30	72.87	10.93
13	2.29	10	55.91	8.46
14	2.34	15	66.94	8.04
15	2.17	30	76.27	8.67
16	2.21	15	65.65	7.37

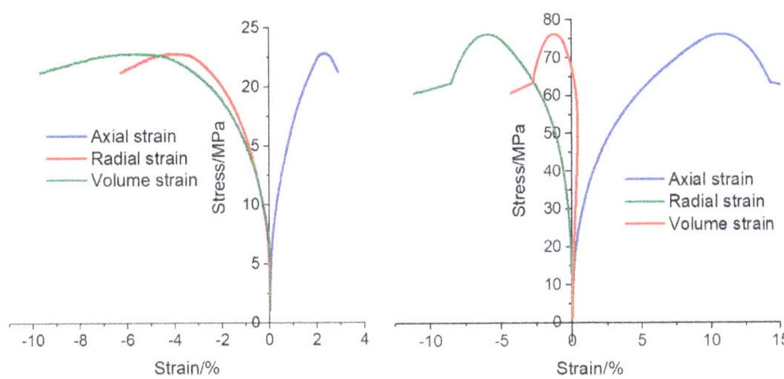

(**a**) #10 Uniaxial compression test (**b**) #15 Triaxial compression test

Figure 5. Uniaxial and triaxial stress-strain curves of salt rock in Sanshui Basin.

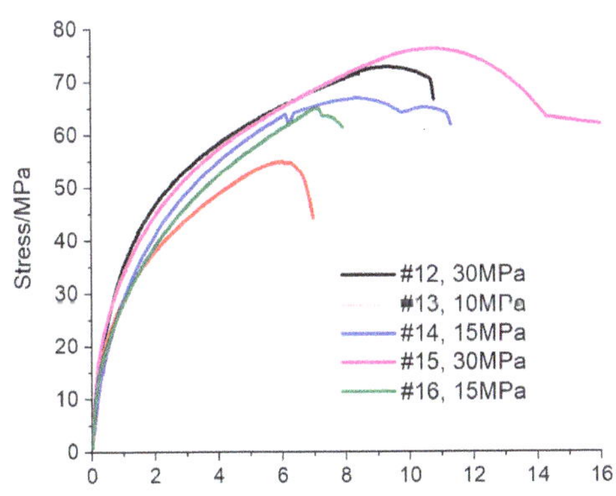

Figure 6. Relationship between stress and axial strain under different confining pressures in triaxial compression test.

2.2.3. High Temperature Triaxial Test

Liang et al. [28] discussed the influence of temperature on the mechanical properties of salt rock and concluded that the uniaxial compressive strength and axial strain of salt rock increase with increasing temperature, while the elastic modulus decreases. According to the temperature distribution characteristics of Sanshui Basin, this section studies the mechanical deformation behavior of salt rock at 60 °C. The comparison of the experimental data in Tables 4 and 5 shows that the peak strength decreases slightly with the increase of temperature under 30 MPa confining pressure, and the average decrease is about 8.3% compared with that under normal temperature, indicating that temperature has a certain influence on the mechanical parameters of salt rock. With the increase of temperature, the mechanical parameters of salt rock show a tendency to weaken. Gao et al. [26] believed the thermal stress caused by temperature makes it easy to produce stress concentrations at the boundary of mineral particles which exceed the strength limit of rock, thus causing the connection to fracture between mineral particles at the boundary surface. Thus, microcracks are generated and extend to form a network, finally leading to macroscopic deterioration of rock mechanical properties. Chen [29] believed that the thermal damage of salt rock was mainly caused by intergranular cracks caused by uneven grain expansion. Figure 7 shows the triaxial test curve of some cores under high temperature, and that the stress–strain changing characteristics of cores under different confining pressures are basically the same. Under the action of thermal stress, the higher the confining pressure, the greater the peak load required for the same deformation. Figure 8 shows that the axial deformation required by the salt rock core under the high-temperature environment is smaller when entering the volume expansion stage under the confining pressure of 30 MPa, which also shows that temperature has an impact on the deformation of the salt rock. From the microscopic point of view, with the increase of temperature, the thermal movement of molecules inside the salt rock is enhanced, the cohesion between molecules is weakened, and the indirect contact surfaces of grains are more prone to slippage. In addition, the different thermal expansion coefficients between grains lead to more damage cracks, deteriorated mechanical properties of the salt rock. It is easier to achieve damage under load compared with the normal temperature state. As a whole, the salt rock in Sanshui Basin, Guangdong Province is not significantly affected by temperature.

Table 5. Experimental data of salt rock under triaxial compression at high temperatures.

No	Density (g/cm^3)	Confining Pressure (MPa)	Peak Strength (Mpa)	Elastic Modulus (Gpa)
27	2.15	30	69.28	7.21
28	1.97	30	67.45	9.98
29	2.46	10	53.81	5.23
30	2.20	15	61.37	7.63
31	2.27	15	63.02	8.12

2.2.4. Creep Test

The rheological properties of salt rock are of great significance to the long-term operational safety of underground projects. Figure 9 shows the pre and post creep test photos of some salt rock cores in Sanshui Basin under uniaxial and triaxial conditions. The creep test did not cause large apparent damage to the salt rock core. After the test, the rock core appears to be complete, and there were no fractures or fragmentation. The only difference was in the boundary between particles, which appeared clearer and more obvious (as shown in the red dotted ellipse in Figure 9a below). Under the uniaxial creep test, the creep rate increases with the increase of axial pressure, as shown in Figure 10a. Table 6 shows that the accelerated creep stage occurs in uniaxial creep of individual cores, such as core #18 under an axial pressure of 24 MPa and core #20 (as shown in Figure 10b) under an axial pressure of 20 MPa, while individual cores (such as #17) still do not enter the accelerated creep stage under an axial pressure of 25 MPa, reflecting the heterogeneity of the stratum, indicating that there are differences in properties of salt rocks even in the same mining area.

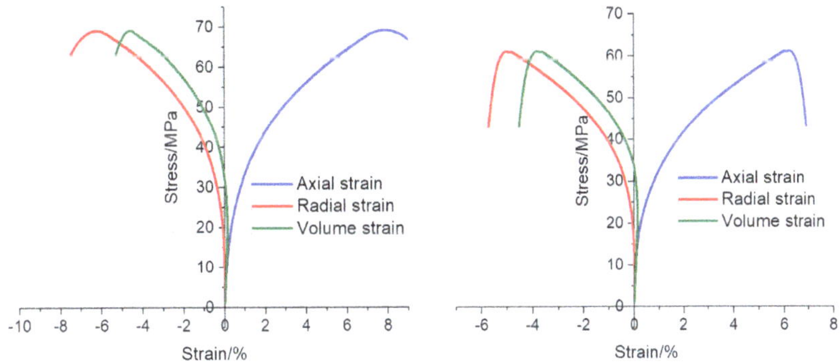

(**a**) #27, confining pressure of 30 MPa (**b**) #30, confining pressure of 15 MPa

Figure 7. Triaxial compression experimental data of salt rock at high temperature.

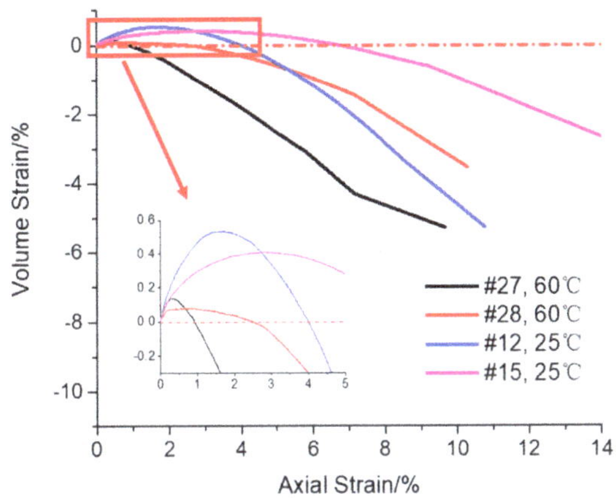

Figure 8. Comparison of axial strain and volumetric strain of salt rock at room temperature and high temperatures under confining pressure of 30 MPa.

The confining pressure of the triaxial creep test was set according to the stress state of the coring formation depth. Under the triaxial creep test, the deviatoric stress is in direct proportion to the creep rate. The greater the deviatoric stress, the greater the creep rate (as shown in Table 6). When the confining pressure is fixed, the salt rock does not enter the accelerated creep stage with the increase of deviatoric stress. Even if the deviatoric stress reaches 35 MPa, there is no accelerated creep phenomenon, indicating that the confining pressure has a significant limiting effect on the salt rock creep. According to Figure 11, the fitting results of creep rate show that the creep rate is exponential with the deviatoric stress.

(**a**) #19 (**b**) #24

Figure 9. Photos of some salt rocks before and after creep tests under uniaxial and triaxial conditions.

Table 6. Creep test results of salt rock in Sanshui Basin.

No	Density (g/cm^3)	Confining Pressure (MPa)	Stress Difference (MPa)	Creep Rate (h^{-1})
17	2.33	0	10	0.000186
			18	0.000418
			25	0.000696
18	2.36		12	0.000315
			18	0.000664
			24	/
19	2.15		8	0.00023
			13	0.000676
			18	0.000978
20	2.23		15	0.000954
			20	/
21	2.17		10	0.00036
			15	0.000962
			20	0.001523
22	2.16	30	15	0.000539
			25	0.001419
			35	0.00279
23	2.15		12	0.00026
			20	0.000921
			28	0.001782
24	2.17		10	0.000187
			20	0.000606
			30	0.001672
25	2.17		15	0.000518
			25	0.000857
			35	0.001533
26	2.23		10	0.000424
			20	0.000579
			30	0.000846

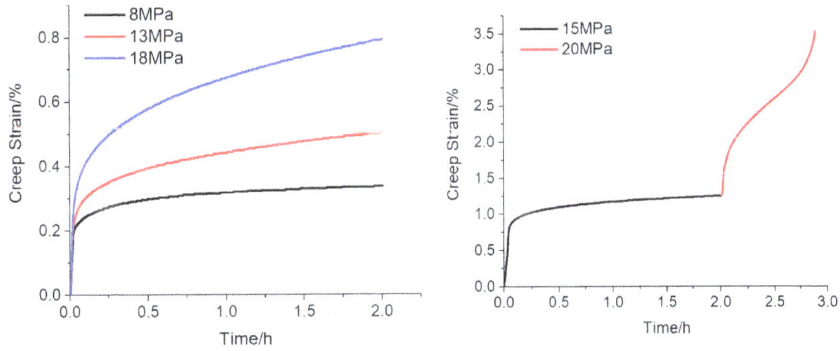

(**a**) #19 (**b**) #20

Figure 10. Results of uniaxial creep tests.

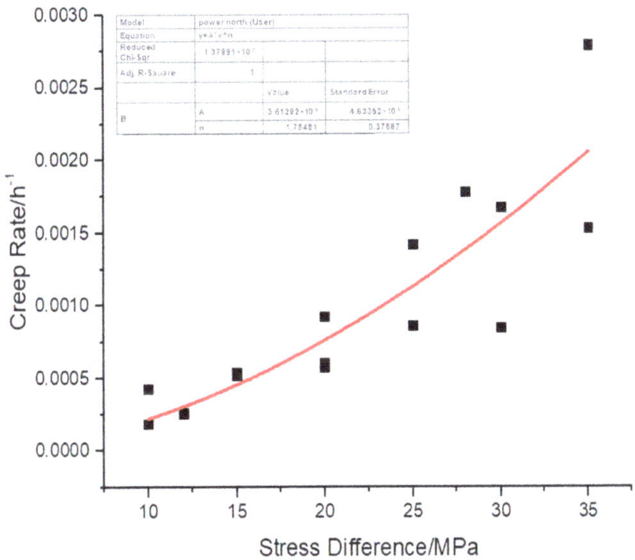

Figure 11. Fitting of creep rate dependence on stress difference under confining pressure of 30 MPa.

Seen from the comparison of triaxial test results at room temperature and high temperature, the target formation temperature of 60 °C in Sanshui Basin has a slight impact on the strength properties of the salt rock. In order to explore the influence of the temperature on the creep property of salt rock, a creep test (core No. #32) was conducted at 60 °C and compared with triaxial creep data under the same confining pressure and deviatoric stress at the room temperature, as shown in Figure 12. The compared results of creep rate show that, under the same deviatoric stress and confining pressure, the effect of temperature on creep rate is not significant. The creep rate of salt rock, under the confining pressure of 30 MPa, deviatoric stress of 25 MPa and environmental temperature of 60 °C in the target salt layer of Sanshui Basin, has little difference from the rate under the same conditions at room temperature. Therefore, the influence of temperature can be ignored when analyzing the creep problem.

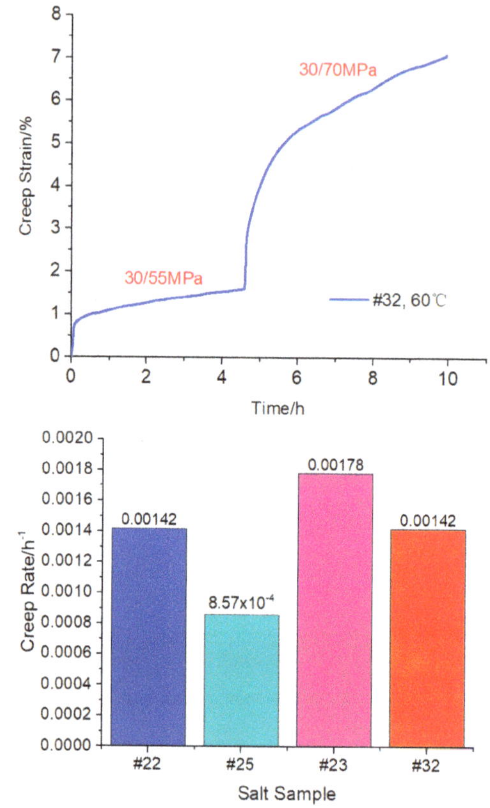

Figure 12. Experimental curve of triaxial creep of salt rock samples at high temperature, and the comparison of its creep rate with that at normal temperature.

2.3. *Result Analysis and Dicussion*

2.3.1. Strength Model Parameters of Salt Rock

The results of uniaxial and triaxial compression tests of salt rock in Sanshui Basin show that the failure mode of salt rock under uniaxial compression is tensile failure, while the failure mode is radial expansion failure without obvious shear fracture under triaxial compression. A lot of research [30,31] shows that the strength of salt rock can be expressed by the Mohr–Coulomb strength criterion:

$$\frac{1}{2}(\sigma_1 - \sigma_3) = C\cos\varphi - \frac{1}{2}(\sigma_1 + \sigma_3)\sin\varphi \tag{1}$$

where σ_1, σ_3 are maximum and minimum principal stresses, respectively; C is cohesive force; φ is internal friction angle.

Based on the uniaxial and triaxial compression test results in Sanshui Basin, the cohesion and internal friction angle of salt rock at room temperature are 8.32 MPa and 28.1°, respectively. By comparing Yingcheng Salt Mine and Jintan Salt Mine [32], as shown in Table 7, the cohesion of salt rock in Sanshui Basin is slightly greater than those of Yingcheng Salt Mine and Jintan Salt Mine, while the internal friction angle is slightly lower than those of the latter two. Cohesion reflects the strength of material. Here, the smaller the cohesion, the lower the rock strength. The influence of internal friction angle on rock strength is greater than that of cohesion [33]. From the comparison results of strength parameters, the strength of salt rock in Sanshui basin is between that of the Yingcheng and the Jintan salt mines in Hubei.

Table 7. Comparison of strength parameters of salt rock from Sanshui, Yingcheng and Jintan salt mines.

Core from	Cohesion (MPa)	Internal Friction Angle (°)
Sanshui	8.32	28.1
Yunying	4.36	39.9
Jintan	6.29	28.6

The high temperature compression test of salt rock shows that the temperature has a slight deterioration effect on the mechanical properties of salt rock. In the design and application of engineering parameters, the temperature factor can be ignored.

2.3.2. Creep Model Parameters of Salt Rock

Yang et al. [34], based on a large number of studies on the creep properties of salt rock, have come to the conclusion that the steady creep rate of salt rock is closely related to the deviatoric stress and confining pressure. Without considering the influence of temperature, steady state creep model of salt rock conforms to the Norton exponential function:

$$\dot{\varepsilon}^{cr} = A(\sigma_1 - \sigma_3)^n \tag{2}$$

where $\dot{\varepsilon}^{cr}$ is the creep strain rate; A, n are the material characteristic parameters of salt rock.

By fitting the steady-state creep rate of salt rock samples under 30 MPa confining pressure in Sanshui Basin with the Norton steady-state creep model (as shown in Figure 11), the creep model parameters A and n were calculated to be 3.61×10^{-6} and 1.78, respectively. The creep parameters of the salt rock of Jintan Salt Mine in Jiangsu Province and Yunying Salt Mine in Hubei Province [35] are compared (both fitted with the Norton creep model) and shown in Table 8. The results show that the higher the value of n, the higher the sensitivity of steady creep rate to the variation of deviatoric stress. The smaller the A value, the lower the sensitivity of steady creep rate to the change of confining pressure. The creep model parameter A of Sanshui basin salt rock is much higher than that of Yunying and Jintan salt rock, while n is lower. Figure 13 shows the comparison of creep strain rates under different deviatoric stresses. When the deviatoric stress is lower than 20 MPa, the creep rate of salt rock in Sanshui Basin, Guangdong Province is the highest under the same confining pressure. When the deviatoric stress exceeds 20 MPa, it is between the creep rates of Yunying in Hubei Province and Jintan in Jiangsu Province.

Table 8. Comparison of creep model parameters of salt rock from Sanshui, Yingcheng and Jintan salt mines.

Core from	A (MPa^{-n}/h)	n
Sanshui	3.61×10^{-6}	1.78
Yunying	3.56×10^{-10}	3.68
Jintan	1.98×10^{-8}	3.50

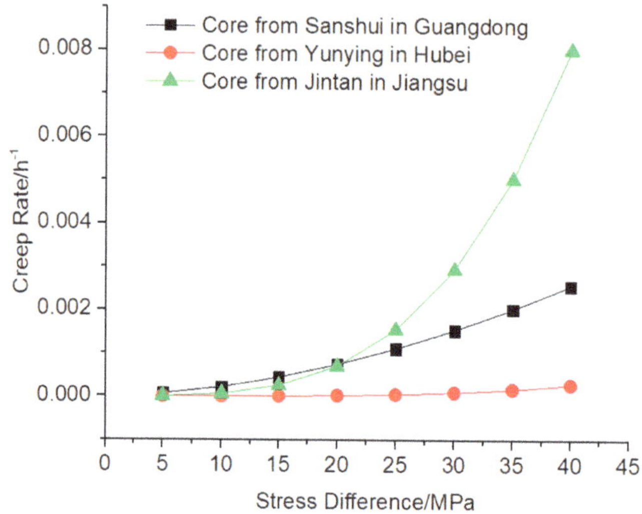

Figure 13. Comparison of creep strain rates under different stress difference.

3. Mechanical Stability Evaluation

The rock strata in the salt cavern area of the reservoir rock and its overburden and floor rock strata can be considered as an approximately horizontal distribution, so its tectonic stress is weak. The vertical component of the original geostress of the stratum in this area is calculated according to the average density of the overlying strata and the relationship between the average density and the burial depth of the rocks in each layer of the calculation area, and according to the calculation method of self-weight stress. It is thought that after a long period of tectonic evolution, the three components of the stresses are equal in the salt formation. According to the above rock mechanics analysis, the stability evaluation and analysis are carried out for the salt cavern gas storage to be built in the salt layer in this area. The finite difference software Flac^{3D} [36] (Flac^{3D} vs. 6.0, Itasca Consulting Group Inc., Minneapolis, MN, USA) was used to establish the geomechanical model and perform the simulation, which is widely used to handle rock and soil engineering problems for its outstanding ability and advantage to solve large displacement problems. The Cpower model was introduced into the simulation to describe the mechanical behavior of salt cavern gas storage during injection-production operation, which is composed of the Mohr–Coulomb criterion and the Norton Power creep model and widely used in creep engineering of underground excavation.

3.1. Geomodel of Salt Cavern Gas Storage

Based on the geological conditions of target strata and previous research results [37–39], a pear-shaped cavity was used to construct the salt cavern gas storage in Sanshui Basin. The morphological structure of the salt cavity had a height of 30 m at the top, a diameter of 60 m in the upper hemisphere and 70 m in the lower hemisphere. The whole height of the cavern was 95 m. There were 10 mudstone interlayers distributed in the target salt formation. The corresponding geological model is shown in Figure 14. The hexahedral mesh was used in the model and the sensitivity analysis of its mesh density was done. The model was meshed with a total number of 244,044 elements and 52,112 nodes.

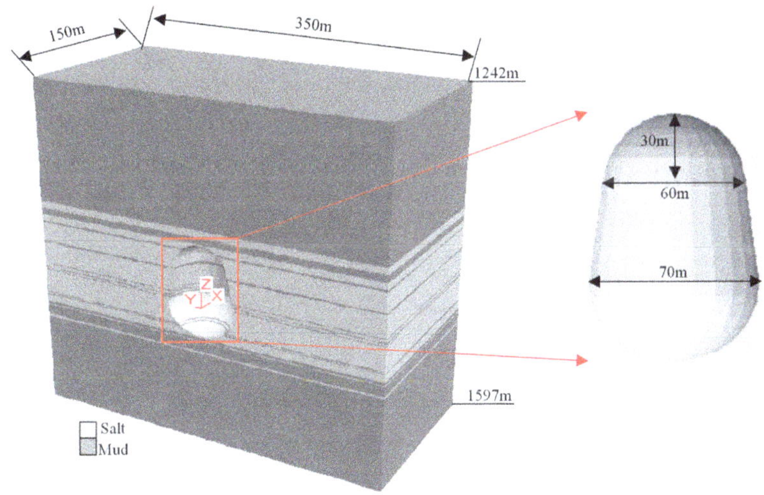

Figure 14. Geomechanical model of salt cavern gas storage (half model).

The upper surface of the model (1242 m) was the stress boundary condition and the average density of the overlying strata was 2.5×10^3 kg/m^3. Thus, the vertical load component of the upper boundary (z = 1242 m) of the model was 31.05 MPa. The bottom surface of the model (1597 m) was constrained in the Z-direction, and the four sides in the horizontal direction of the model were constrained by the simple support in the corresponding normal direction. That is, the front, back, left, right and bottom surfaces of the model were considered to have normal constraints, and movement in normal direction was not allowed, and the influence of the cavity mining process on them could be ignored.

3.2. Injection and Production Pressure Scheme

According to the depth of the last casing and domestic and foreign experience, the upper limit pressure of the designed and constructed salt cavern gas storage was 23 MPa, and the minimum internal pressure is 10 MPa. The injection-production operation simulation is carried out with a service life of 30 years. One cycle consists of four stages of gas injection, well shut in, gas production and well shut in, as shown in Figure 15.

3.3. Stability Analysis

Based on the above rock mechanical parameters and the constitutive model, combined with the injection-production operation scheme, a mechanical stability analysis of single cavity and cavity group was carried out to determine the feasibility of the construction of salt cavern gas storage. The parameters of rock mass layers are listed in Table 9.

Table 9. The parameters of rock mass layers.

Lithology	Elastic Modulus (GPa)	Poisson's Ratio	Cohesion (MPa)	Friction Angle (°)	Tensile Strength (MPa)	A (MPa^{-n}/h)	n
Rock salt	6.16	0.3	8.32	28.1	1.51	3.61×10^{-6}	1.78
Mudstone	16	0.27	5	26.4	3.23	2.5×10^{-7}	1.5

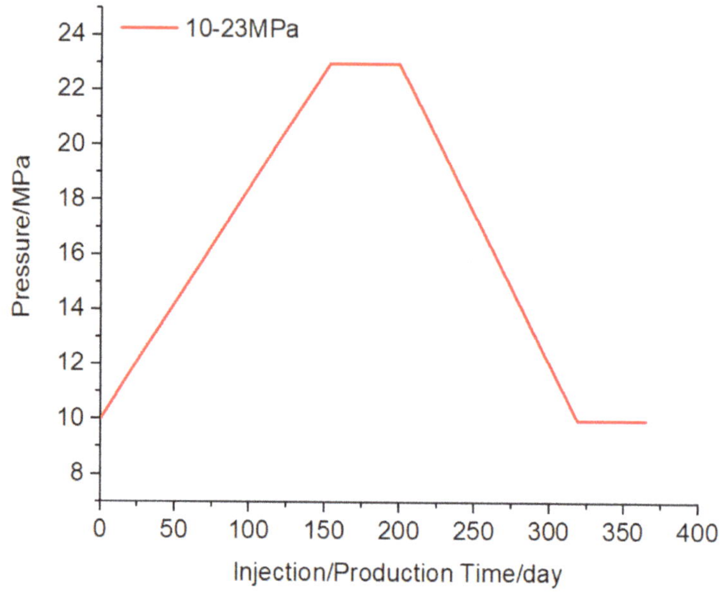

Figure 15. Pressure scheme of injection and production operation (one cycle).

3.3.1. Stability Evaluation Criterion

An index system composed of displacement, plastic zone, volume shrinkage and vertical stress was summarized to evaluate the safety of salt cavern UGS during the stability analysis. The displacement of cavity surrounding rock increases with the operation time caused by the rock salt creep. The allowed displacement was not over 5% of the maximum cavern diameter [40]. The Mohr–Coulomb criterion, one of the most extensively used criteria in geotechnical engineering, was used to judge whether the cavity surrounding rock enters the plastic failure state. The expression is shown in Equation (1). Volume shrinkage was used to characterize the reduction ratio of cavity volume loss to its original volume. The allowed ratio is 30% after operating 30 years according to much research on salt cavern gas storage and the geological data of Sanshui Salt Mine.

3.3.2. Single Cavity Stability Analysis

Figures 16 and 17 show the deformation and plastic zone of surrounding rock in the first year and the 30th year of salt cavity creep after gas production in that year. The roof displacement in the first year is relatively large, about 0.32 m. The deformation of the lower sphere of the cavity is small. The plastic zone is mainly concentrated on the positions of each interlayer and the cavity top and bottom, and the depth of the plastic zone of the interlayer and bottom area in the middle of the cavity is large. In the 30th year, the top and bottom parts both deform towards the interior of the cavity and the large displacement is concentrated near the top and bottom position, with maximum displacement of about 1.57 m. The plastic zone is essentially the same as the first year, and the volume of the plastic zone changes slightly.

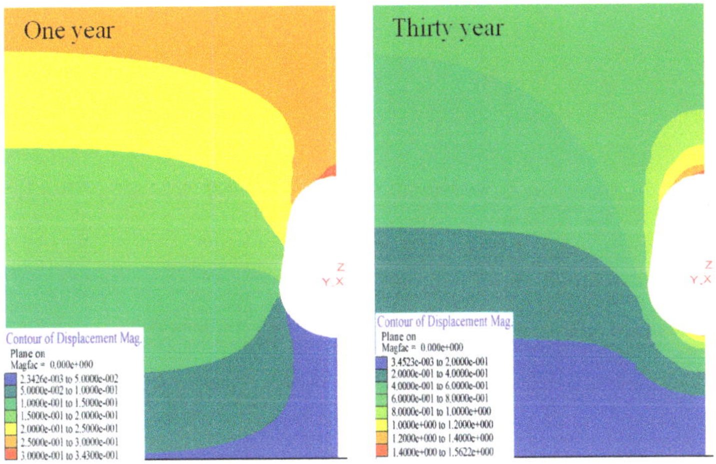

Figure 16. Displacement distribution of salt cavern under the different operation times.

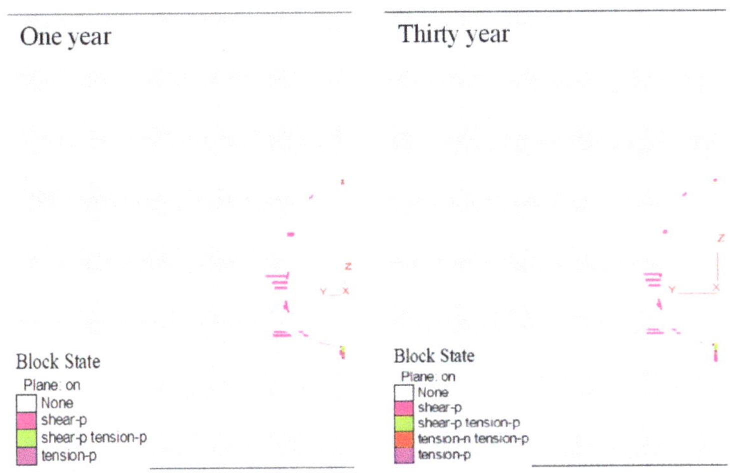

Figure 17. Plastic zone distribution of salt cavern under the different operation times.

According to Wang's research [40], the displacement of the cavity top should not exceed 5% of the maximum diameter of the salt cavity. After 30 years of operation, the cavity top settlement displacement is about 1.57 m (as shown in Figure 18), which is 2.2% of the largest diameter (70 m) of the cavity, indicating that the stability of the cavity is good under the operating pressure. Figure 19 shows the volume convergence of the chamber under different operation times. After 30 years of operation, the volume convergence rate of the cavity is 6.55%, which meets the stability requirement that the volume convergence rate should not exceed 5% during 5 years of operation. Similarly, it reflects the good stability characteristics of the cavity.

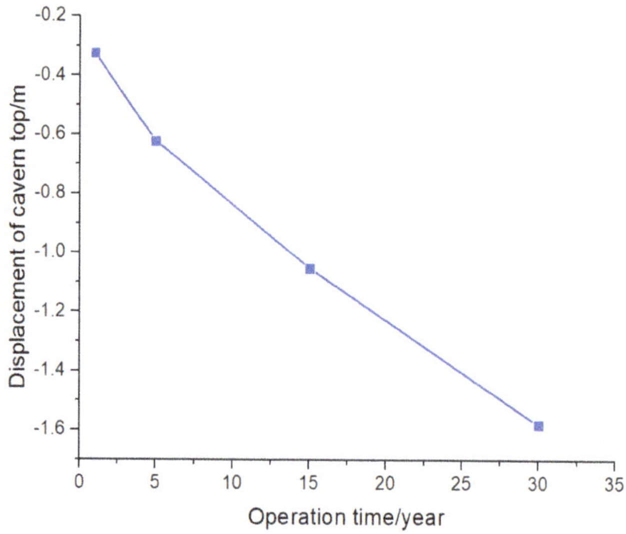

Figure 18. Displacement change of cavern top over the 30 year operational time.

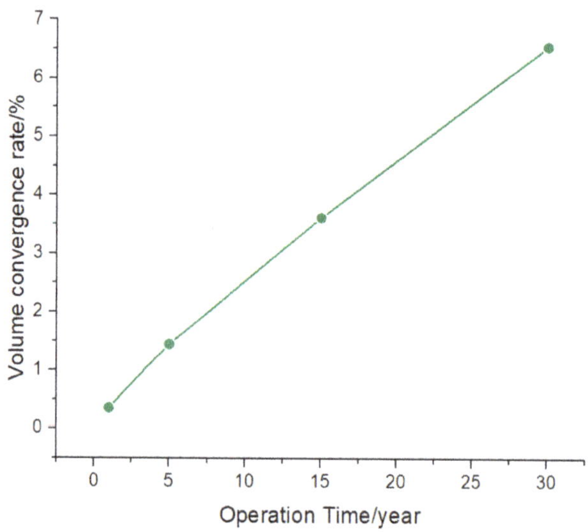

Figure 19. Volume convergence rate of salt cavern over the 30 year operational time.

3.3.3. Stability Analysis of Two Caverns

For the construction of salt cavern storage cavern group, the minimum safety distance of adjacent caverns should be considered. The layout of the multi cavern is recommended to be in a positive triangle layout, and the pillar design of each adjacent cavern should be equal [41]. Therefore, the problem can be simplified to only analyze two adjacent caverns. Figure 20 shows the design width of adjacent storage pillars, where D is the maximum diameter of any cavity. The designed well spacing of Sanshui Basin is 250 m, and the pillar width is 2.14D, so the pillar width is considered as a 2D working condition. The maximum operating pressure for the numerical simulation of injection-production operation of tunnel group is 23 MPa, the minimum is 10 MPa, and the operating time is 30 years.

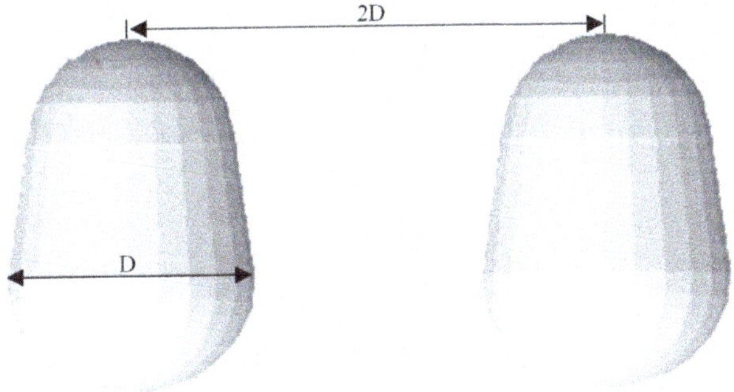

Figure 20. Pillar width of two caverns.

Figures 21 and 22 show the vertical stress and plastic zone of surrounding rock in the first year and the thirtieth year after gas production. The stress distribution is basically symmetrical along the vertical centerline of the pillar. In the first year, the stress at the top and bottom of the cavity is relatively small, while the stress at the minimum spacing is relatively large. The plastic zone is distributed asymmetrically. The plastic zone of the surrounding rock of the roof of the right cavity is significantly larger than that of the surrounding rock of the left cavity. The plastic zone is mainly concentrated in each interlayer, the cavity top and bottom. The failure mode is mainly shear failure, and the plastic zone of the pillar is not connected. The extreme value of vertical stress decreases in the thirtieth year. The distribution of plastic zone is similar to that of the first year, and the volume of plastic zone changes slightly. During the overall operation, the convergence rate of two cavities meets the requirement that the convergence rate is less than 10% in ten years [42], shown in Figure 23, and the pillar design is reasonable.

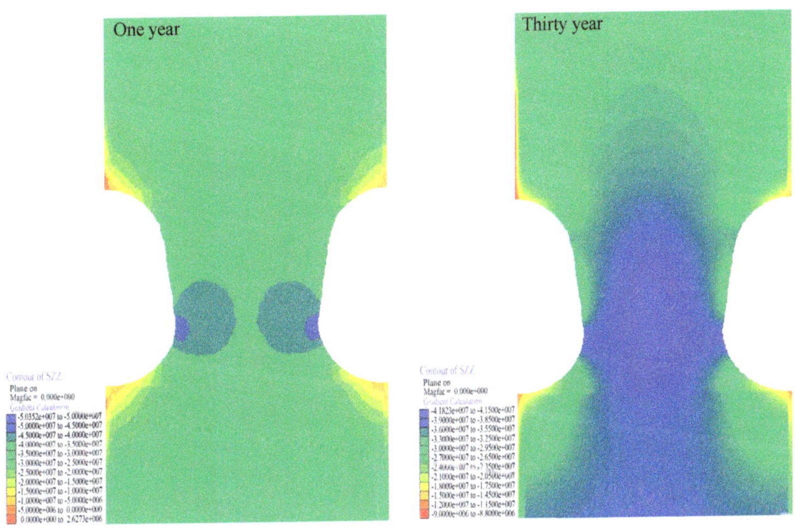

Figure 21. Vertical stress distribution under different operation time.

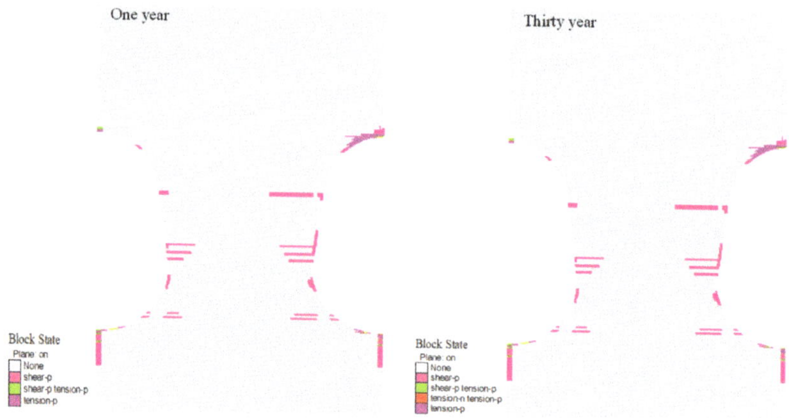

Figure 22. Plastic zone distribution under different operation time.

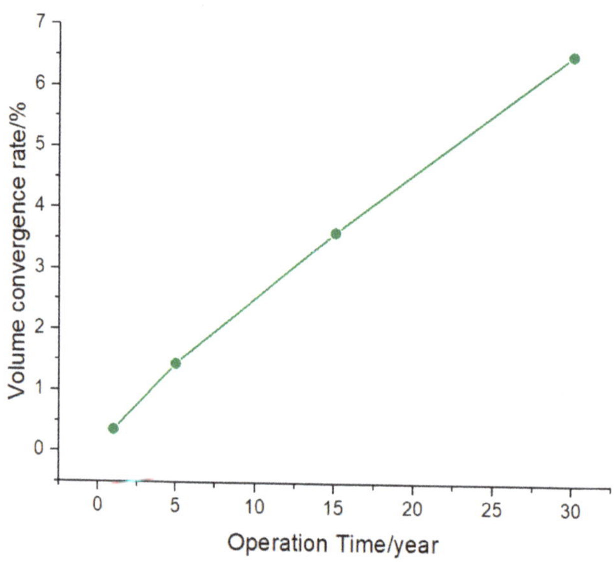

Figure 23. Volume convergence rate of cavern group over the operational time of 30 years.

4. Conclusions

In this paper, the salt rock of a salt mine in Sanshui Basin of Guangdong Province was taken as the subject of research. Direct and indirect tensile, uniaxial compression, triaxial compression and creep tests were carried out. At the same time, the influence of temperature was considered to study the mechanical properties of the salt rock of the salt mine. A stability evaluation of salt cavern gas storage under injection-production operation conditions was conducted based on actual formation conditions. The main conclusions are as follows:

1. The salt rocks in Sanshui Basin generally contain impurities. Both tensile test methods show that the tensile strength of the salt rock is low, less than 2 MPa.
2. Under uniaxial compression, the circumferential tensile stress of the core is greater than its own tensile strength. The failure mode of salt rock presents as radial tensile failure and axial splitting failures. However, under triaxial compression, the salt

rock shows a radial expansion state without obvious shear splitting. The confining pressure significantly improves the compressive strength of salt rock by limiting its radial deformation.

3. The mechanical properties of salt rock are slightly influenced by temperature. With the increase of temperature, the peak strength and elastic modulus of salt rock decrease, and axial deformation at peak failure also reduce. Under the confining pressure of 30 MPa, the strength of salt rock decreases by 4.5% at the temperature of 60 °C compared with that at room temperature.
4. Whether under uniaxial or triaxial compression condition, the salt rock after creep test has no obvious collapse failure. With the increase of axial load, the accelerated creep of salt rock only occurs under uniaxial compression operation, indicating that the confining pressure has a certain inhibition effect on the creep of salt rock. The greater the deviatoric stress, the greater the creep rate of salt rock, which conforms to the power exponent relationship. An operating temperature of 60 °C has no obvious influence on the creep rate of the salt rock. On the whole, the salt rock of salt bearing strata in Sanshui Basin, Guangdong Province, shows good mechanical properties.
5. Under the operating pressure of 10–23 MPa, the roof settlement displacement and the volume convergence rate of the cavity are small. Moreover, the plastic area between two caverns is unconnected, and the overall stability of the salt cavity is good. Combined with the results of mechanical experiments and a stability analysis, the salt formation of Sanshui Basin in Guangdong Province has a good geomechanical condition, and it is suitable for the construction of underground salt cavern gas storage.

Author Contributions: Conceptualization, J.D.; Methodology, J.D.; Validation, K.L.; Formal analysis, Y.K., S.B. and J.W.; Investigation, H.L., K.L. and L.R.; Data curation, S.B.; Writing—original draft, H.L.; Writing—review & editing, Q.W. and Y.K.; Visualization, L.R. and J.W.; Supervision, G.D. All authors have read and agreed to the published version of the manuscript.

Funding: This research was funded by Major Scientific and Technological Projects of China, grant number No2022DJ8303.

Institutional Review Board Statement: Not applicable.

Informed Consent Statement: Not applicable.

Data Availability Statement: The study did not report any data.

Conflicts of Interest: The authors declare no conflict of interest.

References

1. Ding, G.; Zhang, B.; Yang, C.; Xie, P.; Tian, G. The Creep Deformation Rule of Gas Storage Salt Cavern. *Nat. Gas Ind.* **2007**, *27*, 94–96.
2. Crotogino, F.; Schneider, G.-S.; Evans, D.J. Renewable energy storage in geological formations. *Proc. Inst. Mech. Eng. Part A J. Power Energy* **2017**, *232*, 100–114. [CrossRef]
3. Ran, L.N.; Zheng, D.W.; Han, B.J.; Wang, Y.; Kou, Y.X. Overview and Development of Energy Storage Using Salt Caverns in the World. In Proceedings of the 2013 National Natural Gas Academic Annual Conference, Kunming, Yunnan, China, 14–15 August 2013; pp. 440–445.
4. Zheng, Y.L.; Zhao, Y.J. General Situation of Salt Cavern Gas Storage Worldwide. *Oil Gas Storage Transp.* **2010**, *29*, 652–655.
5. Li, N.N.; Zhao, Y.Q.; Wang, T.T.; Yang, C.H. Trends Observation: Strategy and Development of International Salt Cavern Energy Storage Research. *Bull. Chin. Acad. Sci.* **2021**, *36*, 1248–1252.
6. Khan, S.A.; Igoshin, A.I.; Sokhranskiy, V.B. Some Technological Aspect of Compressed Helium Storage in Salt Caverns. In Proceedings of the SMRI Spring 2011 Technical Conference, Galveston, TX, USA, 18–19 April 2011.
7. IEA U.S. Underground Natural Gas Storage Capacity 2020 [EB/OL]. Available online: https://www.eia.gov/dnav/ng/ng_stor_cap_dcu_nus_a.htm (accessed on 25 December 2020).
8. Zhang, B.; Lv, B.L.; Wu, Y.H.; Cui, L.H.; Zhou, S.Q. Development and Trend of Salt-Cavern Gas Storage in Domestic and Abroad. *China Well Rock Salt* **2021**, *52*, 21–24.
9. Sedaee, B.; Mohammadi, M.; Esfahanizadeh, L.; Fathi, Y. Comprehensive modeling and developing a software for salt cavern underground gas storage. *J. Energy Storage* **2019**, *25*, 100876. [CrossRef]

10. Moghadam, S.N.; Nazokkar, K.; Chalaturnyk, R.J.; Mirzabozorg, H. Parametric assessment of salt cavern performance using a creep model describing dilatancy and failure. *Int. J. Rock Mech. Min. Sci.* **2015**, *79*, 250–267. [CrossRef]
11. Stern, N.; Xie, C. China's new growth story: Linking the 14th Five-Year Plan with the 2060 carbon neutrality pledge. *J. Chin. Econ. Bus. Stud.* **2022**, *2022*, 1–21. [CrossRef]
12. Bruno, M.S.; Durseault, M.B. Geomechanical Analysis of Pressure Limits for Thin Bedded Salt Caverns. In Proceedings of the SMRI Spring Technical Meeting, Banff, AB, Canada, 29–30 April 2002.
13. Bruno, M.S. *Geomechanical Analysis and Design Considerations for Thin-Bedded Salt Caverns: Final Report*; Terralog Technologies: Arcadia, CA, USA, 2005.
14. Bruno, M.; Dorfmann, L.; Han, G.; Lao, K.; Young, J. 3D geomechanical analysis of multiple caverns in bedded salt. In Proceedings of the SMRI Fall Technical Meeting, Nancy, France, 1–5 October 2005.
15. B'Erest, P.; Brouard, B. Safety of salt caverns used for underground gas storage blow out; mechanical instability; seepage; cavern abandonment. *Oil Gas Sci. Technol.* **2003**, *58*, 361–384. [CrossRef]
16. Ma, H.; Yang, C.; Li, Y.; Guo, E.; Liu, J. Experimental and theoretical research on yield and failure characteristics of salt rock. *Chin. J. Rock Mech. Eng.* **2012**, *31*, 3747–3756.
17. Liu, X.R.; Guo, J.Q.; Wang, J.B.; Li, P.; Zhang, Q.Q. Investigation on mechanical properties and failure criterion of salt rock based on energy principles. *Rock Soil Mech.* **2013**, *34*, 305–310, 315.
18. Alkan, H.; Cinar, Y.; Pusch, G. Rock salt dilatancy boundary from combined acoustic emission and triaxial compression tests. *J. Rock Mech. Min. Sci.* **2007**, *44*, 108–119. [CrossRef]
19. Yang, C.; Zeng, Y.; Wu, W.; Chen, F. Constitutive relationship of deep salt rock and its application to petroleum drilling engineering. *Chin. J. Rock Mech. Eng.* **2003**, *22*, 1678–1682.
20. Li, Y.; Jiang, W.; Liu, J.; Chen, J.; Yang, C. Direct shear tests for layered salt rocks of Yunying salt mine in Hubei province. *Chin. J. Rock Mech. Eng.* **2007**, *26*, 1767–1772.
21. Mansouri, H.; Ajalloeian, R. Mechanical behavior of salt rock under uniaxial compression and creep tests. *Int. J. Rock Mech. Min. Sci.* **2018**, *110*, 19–27. [CrossRef]
22. Wanyan, Q.Q.; Wu, J.P.; Wang, Z.Y.; Zheng, Y.L. Experimental study on creep mechanical behaviors of salt rock in gas storage. In Proceedings of the Advances in Rheology—The 11th National Academic Conference on Rheology, Langfang, China, 17–18 October 2012; pp. 281–285.
23. Liu, J. *Experimental Investigation and Theoretic Analysis on the Mechanical Properties of Layered Salt Rock*; Institute of rock and soil mechanics, Graduate University of Chinese Academy of Sciences: Wuhan, China, 2006.
24. Jiang, D.; Ren, T.; Chen, J.; Ren, S.; Yang, C. Experimental study of mechanical characteristics of molded salt rock with weak interlayer. *Chin. J. Rock Mech. Eng.* **2012**, *31*, 1797–1803.
25. Wang, A.M.; Li, X.G.; Yang, C.H.; Huang, Z.Q. Study of interaction between creep deformation of bedded salt rock. *Rock Soil Mech.* **2010**, *31*, 3964–3970.
26. Gao, X.P.; Yang, C.H.; Wu, W.; Liu, J. Experimental studies on temperature effect of mechanical properties of salt rock. *Rock Soil Mech.* **2005**, *26*, 84–87.
27. Li, Z.; Ma, H.L.; Yao, Y.F. A preliminary study on basic mechanical properties of salt rock at high temperature and high pressure. *Chin. J. Undergr. Space Eng.* **2013**, *9*, 981–985.
28. Liang, W.G.; Zhao, Y.S.; Xu, S.G. Testing study on physical and mechanical properties of heated salt rock within 240 °C. *Chin. J. Rock Mech. Eng.* **2004**, *23*, 2365–2369.
29. Chen, J.W. *Studies on Temperature Effect of Mechanical Properties and Micro Mechanism of Salt Rock*; Institute of rock and soil mechanics, Graduate University of Chinese Academy of Sciences: Wuhan, China, 2008.
30. Wang, T.; Yang, C.; Ma, H.; Daemen, J.; Wu, H. Safety evaluation of gas storage caverns located close to a tectonic fault. *J. Nat. Gas Sci. Eng.* **2015**, *23*, 281–293. [CrossRef]
31. Yang, C.; Wang, T.; Li, Y.; Yang, H.; Li, J.; Qu, D.; Xu, B.; Yang, Y.; Daemen, J.J.K. Feasibility analysis of using abandoned salt caverns for large-scale underground energy storage in China. *Appl. Energy* **2015**, *137*, 467–481. [CrossRef]
32. Liu, J.; Yang, C.; Wen, W.U.; Yinping, L.I. Experiment study on short-term strength and deformation properties of salt rocks. *Chin. J. Rock Mech. Eng.* **2006**, *25*, 3104–3109.
33. Liu, D.K.; Li, K.G. The discussion on effect of cohesion and internal frictional angle on the peak strength of rock. *Conserv. Util. Miner. Recourses* **2015**, *3*, 16–19.
34. Yang, C.H.; Li, Y.P.; Qu, D.A.; Cheng, F.; Yin, X.Y. Advances in researches of the mechanical behaviors of bedded salt rocks. *Adv. Mech.* **2008**, *38*, 484–494.
35. Wang, A.M. *The Deformation Mechanism and Nonlinear Creep Constitutive Model of Bedded Salt Rock*; Institute of rock and soil mechanics, Graduate University of Chinese Academy of Sciences: Wuhan, China, 2008.
36. Itasca Consulting Group Inc. *FLAC3D Version 6.0 Users' Manual*; Itasca Consulting Group Inc.: Minneapolis, MN, USA, 2019.
37. Cyran, K.; Kowalski, M. Shape Modelling and Volume Optimization of Salt Caverns for Energy Storage. *Appl. Sci.* **2021**, *11*, 423. [CrossRef]
38. Cyran, K. Insight into a shape of salt storage caverns. *Arch. Min. Sci.* **2020**, *65*, 363–398.
39. Onal, E. Stability Analyses of Differently Shaped Salt Caverns for Underground Natural Gas Storage. Master's Thesis, The Penn-Sylvania State University, State College, PA, USA, 2013.

40. Wang, T.; Yang, C.; Chen, J.; Daemen JJ, K. Geomechanical investigation of roof failure of China's first gas storage salt cavern. *Eng. Geol.* **2018**, *243*, 59–69. [CrossRef]
41. Wang, T.T.; Yan, X.Z.; Yang, H.L.; Yang, X.J. Stability analysis of pillars between bedded salt cavern gas storages. *J. China Coal Soc.* **2011**, *36*, 790–795.
42. Li, H.; Deng, J.; Wanyan, Q.; Feng, Y.; Lenwoue, A.; Luo, C.; Hui, C. Numerical Investigation on Shape Optimization of Small-Spacing Twin-Well for Salt Cavern Gas Storage in Ultra-Deep Formation. *Energies* **2021**, *14*, 2859. [CrossRef]

eng

Review

DFN: An Emerging Tool for Stochastic Modelling and Geomechanical Design

Peter Kolapo [1,*], Nafiu Olanrewaju Ogunsola [2], Prosper Munemo [3], Damilola Alewi [4], Kayode Komolafe [5] and Ahmid Giwa-Bioku [6]

1 Department of Mining Engineering, College of Engineering, University of Kentucky, Lexington, KY 40506, USA
2 Department of Mineral Resources and Energy Engineering, College of Engineering, Jeonbuk National University, Jeonju-si 54896, Republic of Korea
3 Department of Mining and Mineral Processing Engineering, Faculty of Engineering, Manicaland State University of Applied Sciences, Mutare P.O. Box 7001, Zimbabwe
4 Department of Materials Science and Engineering, College of Engineering, University of Kentucky, Lexington, KY 40506, USA
5 Department of Mining Engineering, College of Engineering, Colorado School of Mines, Golden, CO 80401, USA
6 Production Engineering Department, Chorus Energy Limited, Lagos 105102, Nigeria
* Correspondence: kolapopeter@gmail.com or peter.kolapo@uky.edu

Abstract: The discrete fracture networks (DFN) have become indispensable tools for geomechanical modelling of jointed rock masses. The technology creates a three-dimensional (3D) representation of fracture geometry used in the construction of surface and subsurface engineering projects in mining, civil engineering, and fracturing of the reservoir in the oil and gas industry. The approach depends on the accuracy of the data obtained during site investigation to create models that represent the fracture geometry of the structure. The better the acquired information available, the better the stochastic analysis that determines the engineering applications and designs that can be carried out. Therefore, it is important to use instruments that can capture fracture distribution characteristics such as fracture intensity, fracture orientation, spatial distribution, fracture length, fracture aperture, and size. This study provides a detailed review of the recent advances in the application of a DFN for modelling jointed rock masses in different engineering applications. The paper shows the principles of modelling in a DFN, including various data-capturing methodologies, and the general application of DFN in various fields. Several case studies where the DFN method was applied are presented in the paper. These include evaluation of slope in an open pit mine, modelling of discontinuity in tunneling, stability evaluation of coal seam longwall, the design of high-level radioactive waste, prediction of groundwater flow, fracturing of petroleum reservoirs, and geothermal cracking of shale gas in the coal bed. However, despite the versatility of the DFN technique, there are still some limitations and challenges to the integration of complexities encountered in rock masses within DFN models.

Keywords: discrete fracture network; 3D laser scanner; fracture geometry; numerical modelling; hydraulic fracturing

check for updates

Citation: Kolapo, P.; Ogunsola, N.O.; Munemo, P.; Alewi, D.; Komolafe, K.; Giwa-Bioku, A. DFN: An Emerging Tool for Stochastic Modelling and Geomechanical Design. *Eng* **2023**, *4*, 174–205. https://doi.org/10.3390/eng4010011

Academic Editors: Reza Rezaee and Yujie Yuan

Received: 25 November 2022
Revised: 29 December 2022
Accepted: 30 December 2022
Published: 5 January 2023

1. Introduction

Numerical modelling of rocks begins with setting up a model that is a representation of a rock mass and can be investigated to predict its behaviour upon being subjected to several external conditions. However, rock masses are rarely intact but contain discontinuities such as joints, faults, and bedding planes which affect their mechanical behaviour. When modelling rock masses, it is critical to incorporate the inherent complex discontinuities. Discrete fracture network (DFN) models have proved to be versatile in numerical simulation of such discontinuities, in that they explicitly capture the discontinuity properties and fracture geometries. The concept of the DFN involves the mathematical modelling of fracture sets which represent the discontinuities' network structure in a real rock mass.

Crustal rocks are the fundamental building blocks of the earth's crust and are important to the successful implementation and construction of many surface and subsurface engineering structures, including nuclear waste repositories, enhanced oil recovery, enhanced geothermal energy systems, and other underground structures. However, these surface and subsurface rock masses are characterized by a network of complex geological structures (fractures), including faults, bedding planes, unconformities, joints, etc. Accurate characterization of the variability and distribution of these fracture systems is essential to the optimal functioning of these constructed engineering structures. A discrete fracture network (DFN) is one of the numerical modelling approaches used to describe joint properties and variability in a rock mass based on available data from mapping techniques such as the laser scanning technique, geophysical mapping, borehole mapping, and photogrammetry technology. The DFN method presents a realistic three-dimensional (3D) geometry, joint properties, and the variation in the joint fracture system within a rock mass based on the acquired data. DFN modelling specifically defines the geometrical properties of an individual fracture and the topological relationship between the rock fracture such as joints, faults, veins, and bedding planes [1]. The development of DFN models requires sourcing and integrating into a unit of a vast amount of fracture network data from both surface and underground bedrocks [2]. The DFN software generates the 3D representation of fractures by making use of the fractures' characteristics, such as size and orientation, which are treated as random variables with given probability distributions. The development of the DFN model is categorized into four main components, namely the consistent structural and stratigraphic model, integrating fracture characterization, static DFN model construction, and DFN dynamic validation and upscaling.

However, collecting fracture data seems a herculean task, particularly at several kilometres below the Earth's crust. Even with continuous advancements in technology, the bulk of high-resolution observations of fracture patterns is still restricted to surface mapping and borehole measurements, which raises difficulties with under-sampling and stereotyping [3]. A DFN model is basically a statistical ensemble, and because it is challenging to collect data underground, most DFNs rely on statistics of fracture networks, and it was anticipated that the target site will be a manifestation of this statistical ensemble [4]. The result of the DFN approach serves as advanced tools in mining operations and other engineering projects such as dam construction, underground excavations, caverns, tunneling, design of underground support systems, fragmentation analysis in blasting [5], and fluid flow in a rock mass. Similarly, DFNs are extensively used in the oil and gas industry to predict the hydraulic behaviour and fracture connectivity in a rock mass [6]. The hydraulic behaviour and fracture parameters are used to establish the zones with high pore pressure during the construction work such as underground hydrocarbon reservoirs and nuclear waste. The DFN is not a novel technology, and the use of the DFN numerical approach can be traced to the early 1970s when it was used to characterize and model fluid percolation in natural fractures [7]. It was adopted and gained wider practice in the mineral extraction industry, most especially in rock engineering and rock mechanics professions. It has also been widely used in geological engineering and hydrogeological and petroleum applications. Currently, the DFN is an integral tool used in various studies and engineering designs for several civil and mining projects [8]. For instance, the DFN can be a useful tool for an in-depth analysis of rock mass strength in a jointed rock mass. It is predominantly used in a rock mass with the presence of discontinuities or highly fractured rock mass.

This study thus aims at reviewing the recent advances in the application of the DFN for modelling jointed rock masses in different engineering applications. The purpose is to catalogue all the principles of modelling in the DFN, including various data-capturing methodologies, different computer programming codes for DFN generation, and the general application of the DFN in various fields, and some specific case studies of DFN models in different engineering applications are elicited. Following that, the observed challenges with the already developed DFN models were pinpointed with the aim of offering purposeful solutions. The review concluded by outlining future research directions in this field. It is, however, pertinent to note that this review is not totally comprehensive on the subject matter, and for the sake of concision, just a small number of credible sources are referenced. For a comprehensive review on specific studies of coupled hydromechanical (HM) characteristics of fractured rocks using the DFN, readers are directed to consult the reviews by [1,9].

2. Background Study

2.1. Overview of Discrete Fracture Network in Mining

There has been an increase in the use of the DFN in providing solutions to mining and geotechnical engineering problems. DFN models have proven to be an effective technology for rock masses characterization using statistical distribution to represent the fracture network [10]. The availability of high computing technologies (hardware and software) has increased the performance of the DFN serving as one of the most versatile tools for analyzing complicated rock mass states and interactions of the fracture network [11].

The method produces a 3D representation of the fracture network by utilizing the observed field data to define the orientation, size, intensity, and spatial model [12]. The success of the DFN technology depends on the quality and quantity of the fracture network acquired from the field. Hence, the method relies on the accuracy of the fracture data collection techniques. After a successful collection of quality data from the site visit, the DFNs then make use of the stochastic analyses to give a range of possible models which will indicate the real mapping area of the slope or tunnel with their stable performance. Alghalandis et al. [13] defined the DFN as a stochastic approach that represents only one possibility derived from statistical distributions of dip directions, dip angles, intensities, and persistence of each fracture set.

Miyoshi et al. [10] stated that the DFN makes use of the statistical distribution principle to describe the fracture parameters such as the fracture location, orientation, intensity, and size. The acquired data from mapping and site investigation are used to create a 3D stochastic model presenting the statistics that describe properties of the specific discontinuities such as faults and persistence of fracture in the rock mass. The basic fracture properties that are required to generate a DFN model are classified into primary and secondary properties. According to Elmo et al., [14], the primary properties entail the geometry of the fracture network while the secondary properties of the specific applications of the DFN model were discussed in Elmo [15]. Rogers and Booth [16] modified the primary and secondary properties of the DFN by adding the sources of the data tabulated in Table 1.

However, to obtain a great amount of persistence during mapping, a Bayesian bootstrap method can be applied to generate rational results [17]. The bootstrap technique is a statistical approach that focuses on the random sampling from an original sample used to create a pseudo-replicate sample of fracture orientation [10], that is, the method provides a sum of the DFN model to simulate highly dispersed orientation data. Similarly, Dijkstra's shortest-path algorithm can also be used to evaluate the linear discontinuity persistence based on the 3D DFN [17], that is, the extent to which the size of the discontinuities persists in rock masses. Although the shortest-path algorithm approach is relatively difficult to use due to the complexity of 3D spatial analysis, it is considerably easier to estimate the persistence of joints and trace the length of discontinuities in 2D. Generally, the stochastic DFN model approach also has its limitations. Mathis [18] stated the limitations of using the stochastic approach include the uncertainties in statistical parameters and oversimplifications of fracture geometries and topologies. Likewise, the DFN model often led to complex geometric configurations, that is, the modelling of DFN involves thousands of fractures [19]. Recent studies by [20–22] showed that the establishment of a hybrid DFN approach can be accounted for by the advantages of both stochastic and geological approaches to create a better 3D fracture network that can present a realistic geological system. Moreover, Hyman [23] stated that the main challenge of using the DFN approach to simulate fluid flow is the creation of an efficient and scalable workflow.

Table 1. List of required primary and secondary properties for DFN modelling and the sources of these properties (modified after Rogers and Booth, 2014 [16]).

S/N	Primary Properties	Sources of Data
1	Fracture intensity distribution	Core sample orientation, borehole image interpretation, and mapping
2	Orientation distribution	Core sample orientation, borehole image log, and mapping
3	Fracture size distribution	Mapping, ideally at multiple scales
4	Spatial variation	Interpretation of borehole or mapping data
	Secondary Properties	
5	Fracture stiffness properties	Shear testing techniques from literatures
6	Fracture shear properties	Core logging, mapping, and shear testing
7	Aperture distribution	Core logging, mapping, and hydraulic testing
8	Storativity distribution	Packer testing, well testing
9	Termination percentage	Mapping
10	Fracture transmissivity distribution	Packer testing

The framework for DFN was discussed by Davy [4]. In their study, a five (5) step framework, whose core components are a set of deterministic data and statistical properties, is shown in Figure 1.

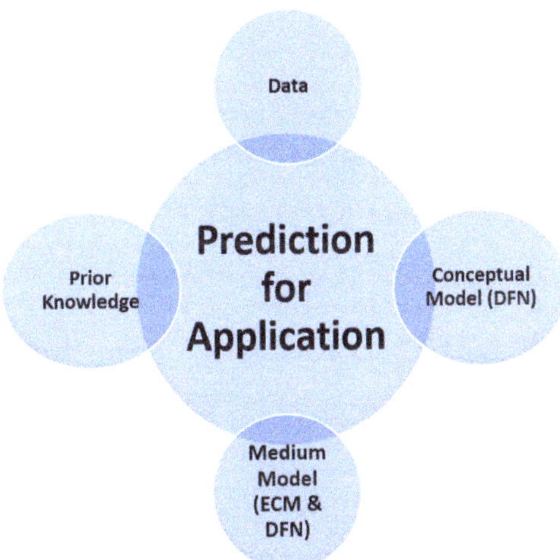

Figure 1. The five (5) basic steps to DFN framework (adapted from Davy [4]).

These properties include the black circle arrow which is developed from the blue arrow and the prior knowledge which is the yellow arrow. The outcome of the model can either be empirical or theoretical relationships, that is, the DFNs are used to establish a stochastic model which is denoted by the green arrow, and the prediction which is the brown arrow continuum model can either be a discrete DFN model or an equivalent continuum model (ECM). However, Weir and Fowler [12] argued that it is imperative to develop a network model statistical and spatial measurement that describes the actual rock mass. This developed DFN model must represent the actual ground condition. Notably, the DFN model must contain a structural domain that accurately describes the fracture characteristic such as the joint orientation, degree of fracturing, and joint spacing. A workflow on how to create a DFN modelling process and the path to establish an accurate engineering model is presented in Figure 2.

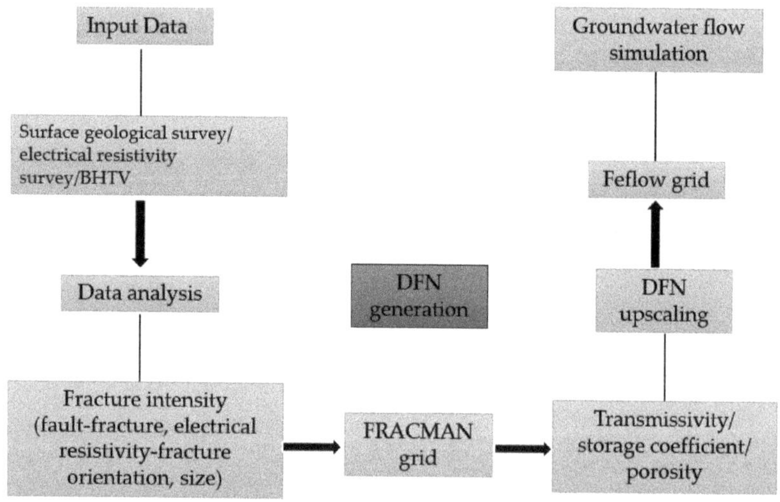

Figure 2. Description of how to build hydraulic properties of fracture rocks using DFN generation and upscaling (adapted from Cheong [24]).

2.2. Principles of Modelling in DFN

Modelling in DFN entails statistically created networks that represent the major structure of the rock mass. The statistically developed DFN model can be created based on the acquired data during the site investigation. These data can be borehole and mapping of the structural domain of the rock mass. According to Dershowitz [25], the modelling of DFN is performed in different forms and shapes, and these include the flat polygon shapes that are the most flexible, realistic, and trending shape representation. Although, modelling of DFNs in a polygonal shape can be complex and requires some specialized stochastic tools. For instance, Alghalandis [26] made use of an enhanced brute-force search and RANSAC approach to generate a DFN from micro seismic data to an enhanced geothermal system. Similarly, Alghalandis's [13] study used Alghalandis discrete fracture network engineering (ADFNE) to proffer solutions to the complexity in geometry and spatial characteristics of rock masses.

Moreover, the choice of selecting input parameters for the DFN modelling also depends on the complexity of the geological structure of the site, the scale, and the purpose of modelling [12,27]. In most cases, modelling of a DFN requires the structural mapping data recorded during site investigation. The acquired data are grouped into four main input parameters, namely fracture orientation, intensity, spatial location, and fracture size [12,28]. The current technological advancement brought about the introduction of digital photogrammetry for mining operations and geotechnical mapping of outcrops. This technology can provide joint properties of the rock surface and other data sets required for the DFN modelling, including large-scale waviness, defect orientation, persistence, joint spacing, block size, and geological structure. Unfortunately, the technology cannot provide the key characteristic of the geological discontinuities, including the joint roughness properties, joint wall strength, and properties of joint infill material [16]. These parameters serve as guidelines to generate a 3D stochastic model representation of the site.

2.2.1. Fracture Orientation

Fracture orientations in rock mass are mostly determined by the stress field conditions of the rock and other factors such as the geochemical properties and local and regional scales [26]. Unfortunately, it is impossible to have a 3D representation of the fracture or map the fracture present in the subsurface. The common way of generating information related to fracture distribution (orientations, intensities, apertures, and lengths) in the subsurface can be achieved by taking measurements from the outcrop or carrying out well-loggings. From this, a stochastic fracture network describes a single fracture property in both 3D and 2D, such as fracture shape (in 3D), length (in 2D), orientation, aperture, and position of the fracture centre [29].

Tuckwell [30] emphasized that fractures with different orientations possibly belong to different fracture sets because of different stress conditions. Mostly, in each fracture set, the orientation is highly concentrated with a high value of the degree of concentration of the distribution around the mean direction [31]. Hence, the fracture orientation is the only important factor to differentiate various fracture sets and the neighbouring and overprinting criteria between fractures on an outcrop map [32]. The fracture orientation describes the attitude of a discontinuity in space, which measures the dip and the dip of the joint. The orientation of the fracture network in the DFN model can be created by the statistical approach of aggregate or disaggregate [28]. The statistical approach can be represented as Fisher, Bingham bivariate Fisher, and bivariate Bingham. Alghalandis [26] argued that the probability distributions approach is the most convenient and widely used technique for summarizing fracture orientations.

Similar to the statistical method, the common probability distribution approach for summarizing fracture orientation includes the von-Mises (the circular normal) for 2D and the Fisher distribution and uniform distribution for 3D. The von-Mises–Fisher distribution is usually used to describe the orientation of the fractures [31,33,34]. According to Staub [35], of all the above-mentioned approaches, the Fisher distribution is the most commonly used. This is because it presents an analogue method of normal distribution for fracture data, and the parameters from the mapping can be easily derived. Likewise, Miyoshi [10] added that in cases where the orientation of the data is dispersed, it would be difficult to define the fracture set. Although, the bootstrap method can be applied to generate an aggregate DFN model. The bootstrap will create a pseudo-replicate sample of fracture orientation. According to Alghalandis [26], the orientation of fractures can be defined as oriented, semi-oriented, and fractures with no orientation. The description of the orientation depends on the generated value of k from the von-Mises distribution. For instance, the von-Mises distributions set k to be equal to 1000 for oriented fractures, k is 10 for semi-oriented, and k is 0 for randomly oriented fractures, as shown in Figure 3.

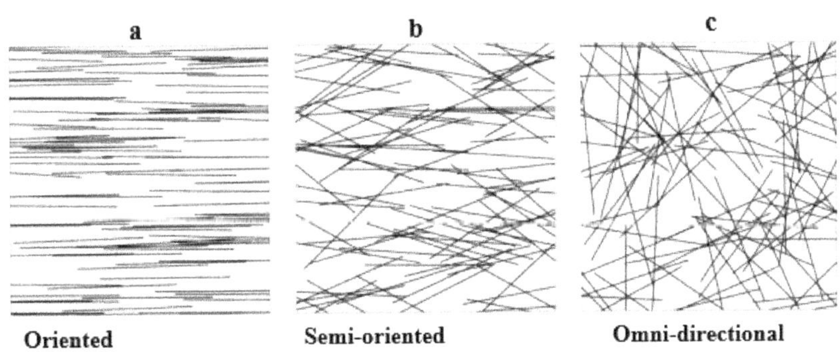

Figure 3. Different fracture orientations: (**a**) fractures oriented in E–W direction (von-Mises distribution = 0 and k = 1000); (**b**) fracture partially oriented towards E–W (von-Mises distribution = 0 and k = 10); (**c**) fractures that are randomly oriented (von-Mises distribution = 0 and k = 0).

2.2.2. Fracture Intensity

Fracture intensity is one of the factors that describes the geometrical properties of the rock mass. Zhang and Einstein [36] defined fracture intensity as the surface area of discontinuities per unit volume. Lu [37] stated that the fracture intensity is the ratio of the total area of discontinuities and the volume of the rock mass considered. The intensity of jointed rock mass plays an important role in determining the mechanical properties and hydraulic conductivity of discontinuous rock mass. Therefore, it is important to accurately determine the fracture intensity in a rock mass to be able to realistically model the 3D representation of the rock mass [12]. Although, a rock mass demonstrates spatial variability in fracture intensity due to the variation in geometric constituent in a rock mass [37]. In estimating the fracture intensity, the fracture distribution size and the number of discontinuities must be known. However, the fracture tensor can be used to quantitatively describe the intensity and the orientation of the fracture. Kachanov [38] quantified the geometry of microcracks in rocks by introducing a tensor α_{ij}, that is, the geometry of cracks in the rock mass can be defined by Equation (1):

$$\alpha_{ij} = \frac{1}{v}\sum_{k=1}^{m(v)}\left[S^{(k)}\right]^{3/2} n_i^{(k)} n_j^{(k)} \tag{1}$$

where V denotes the volume of the rock mass, $S^{(k)}$ represents the kth discontinuity, and $m^{(V)}$ is the number of discontinuities in volume V, while $n_i(^k)$ and $n_j(^k)$ $(i,j = x, y, z)$ are the components of the unit normal vector of the kth discontinuity with respect to orthogonal reference axes i and j $(i, j = x, y, z)$. Dershowitz [39] used the FracMan discrete fracture code to estimate the intensity of the discontinuities. This is carried out by using the discrete fracture code to generate discontinuity networks with known discontinuity size distribution and intensity to simulate the field sampling process. According to Weismüller [32], the fracture intensity is denoted by p_{ij}. In the p_{ij} system, the subscript i represents the dimension of the sample, while the second index subscript j refers to the dimension of the measurement. Lu [37] explained the use of the two-dimensional fracture intensity (P_{21}) method to predict the fracture intensity in a rock mass. This method makes use of Poisson distribution to model the uncertainty of a P_{21} measurement. The outcome of their study showed that the uncertainty of P_{21} measurement can be exhibited by the size of the sampling window, fracture diameter, and fracture intensity. Weir and Fowler [12] argued that the first step in characterizing fracture intensity is the identification of zones of rock mass where the degree of fracturing remains constant over intervals significant to the scale of the model (typically 10 s to 100 s of meters).

2.2.3. Spatial Distribution

The spatial distribution of fractures is determined from the spacing measurements along the sampling lines such as a borehole or the 2D rock face mapping. Zhu [29] stated that it is a complex procedure to characterize the position of the fracture centre. In the studies of Berkowitz and Adler [40], Bour and Davy [41], and Berkowitz [42], they emphasized the importance of fracture centre. That is, when fractures are uniformly distributed from the centre, the fracture orientation and number of fractures per unit volume can be determined. Although, the spatial density distribution brings the clustering effects and might be closer to reality [43]. Over the years, one of the significant changes that has taken place in DFN technology is the evolution from a simple geometric model where the spatial distribution of fracturing is constant within a volume.

However, the commonly used approach to record the spatial distribution of fracture networks is the one-dimensional sampling technique. This approach measures the spacing between fractures in a given fracture set through the map of fracture traces exposed on the rock mass. In Table 1, Rogers and Booth [16] stated that the source of data for spatial variation for DFN modelling can be taken from the analysis of borehole or mapping data. The data were obtained from drilling through the subsurface, which provides quality data detailing the geotechnical logging, core photographs, and rigorous quality-control program. The core samples from the drilling will highlight parameters of the fracture orientation, intensity, and spatial model for the drilling surface. An example of the spatial model from drilling is presented in Table 2 by [12]. The table summarizes the spatial input data and resultant value for the DFN modelling.

Table 2. Sample of spatial data for DFN in shale unit [12].

Domain	Data Sources	Spatial Model	Mean P_{10}	Sizes (m)		Total Number of Models
				Min	Max	
1	11 boreholes	Enhanced Baecher	0.007	25	400	35
2	7 boreholes		0.018	75	400	24
4	Recent underground decline and 4 boreholes		0.014	25	400	26

From the drilling, the stereographic orientation was recorded where the fracture orientation and Fisher dispersion were obtained for each of the data sets. Thereafter, an enhanced Baecher spatial distribution from the Poisson process model was applied for each model. The enhanced Baecher model works on the principle of using a fracture centre located uniformly in space and applying the Poisson process to create a disk with a given radius and orientation. According to Yue [44], the enhanced Baecher model makes use of fracture shape to generate a polygon with three to sixteen sides. The established polygon can be equilateral or elongated with the ratio of major to minor axis size and orientation being defined.

2.2.4. Fracture Size and Length

The fracture size in the DFN model is regarded as the most challenging parameter in the DFN to quantify, particularly from borehole parameters [12]. Most studies depend on literature to determine the size and the length of fractures of the same type of rock at different locations. Alternatively, the size and the length are estimated from a back-analysis approach, that is, the size and the length are estimated using an empirical relationship between the fracture length and aperture of the rock sample [45]. However, before the empirical method can be used, the rock mass must have the same or similar properties to the rock at hand.

Several researchers have discussed several methods of determining the fracture size using different models. Li [46] described the fracture size in six forms; these include orthogonal model, Baecher disk model, Veneziano model, Dershowitz model, Mosaic block tessellation and modified tessellation models, and lastly, the finite element method (FEM). In the orthogonal model, the model assumes three mutually orthogonal sets of parallel fractures. The fracture size can be determined as the space in between the fracture plane of a rectangular shape fracture. The Baecher disk model assumes that the fracture shape has a circular structure, and the fracture size can be determined by the diameter of the fracture. In the Veneziano model, the fracture is polygonal, and the fracture size is defined by the intensity of the Poisson line processed and the proportion of polygons marked as fractures. In Dershowitz models, the fracture shape is a polygon, the fracture size is also determined by the intensity of the Poisson plane processes, and the proportion of the polygon is marked as a fracture. Both mosaic block tessellation models and modified tessellation models depend on Poisson–Voronoi tessellation models where the fracture size is determined by the density of Poisson points from the Voronoi polygon. Lastly, the finite element method's (FEM) model approach discretized the fracture into triangular meshes for hydromechanics analysis [47–50].

During DFN modelling, the user must be able to differentiate between the size and the length of fractures from the rock mapping data. It was also mentioned by Elmo [28] that it is critical to differentiate between the fracture length and size as input parameters for the DFN modelling. Miyoshi [10] defined the fracture size as an equivalent radius if the fracture under investigation is assumed to be polygonal, that is, the fracture size measured from the field is traced with the polygon fracture with a free surface equivalent to the measuring plane in the field. The fracture size or length is usually obtained from bench or outcrop trace mapping. It is necessary to convert the fracture length to an equivalent radius distribution to be incorporated into the DFN model [12]. The length of the fracture in the DFN model for characterizing fluid flow in fractured rock masses plays a key role in generating fracture model networks. An increase in fracture length causes an increase in the permeability of the fracture network. In a study by Zhang [51], the fracture length was increased from 3 m to 13 m, causing an exponential increment in the value of permeability by approximately two orders of magnitude. Likewise, an increment in fracture length also caused an increment in the intersection length and number of intersections. The result of the study by Zhang [51] showed that the fracture length influenced the permeability of the fracture more than that of the geometric parameters, including the intersection length and the number of intersections. Similarly, Liu [52] acknowledged that longer fractures usually have higher permeabilities and larger apertures than shorter fractures. Schnabel's [53] study made use of the RANSAC shape detection tool to obtain the fracture size from the point cloud data using the cloud compare software. The dip and dip angle are automatically generated using the plane-patch filter RiSCAN Pro software. Li [46] emphasized the importance of the shape and sizes of polygonal fractures. The study stated that a better way of estimating the size is to focus on the vertices of the polygon.

3. Data Capturing for DFN Modelling

Data capturing in a DFN is the most critical process that influences the outcome of the DFN modelling. The quality of the input data determines the outcome of the fracture network. At the end of the modelling, the 3D representation of the discontinuous surface would depend on the quality of field data acquired during site investigations. According to Weir and Fowler [12], it is difficult to conceptually describe the geological interpretation of a rock mass, that is, the relationship between various fracture sets and their properties. It is important to interpret the fracture distribution and properties in terms of their characteristics such as the joint roughness, joint infill materials, and termination of joints. Therefore, to develop a robust DFN model that represents the ideal geometry of fractures, there is a need to generate large statistically and spatially located data that can be georeferenced. For instance, the construction of a tunnel in a highly fractured rock mass requires details of a subsurface fracture network to evaluate the mechanical response. However, reliable mapping data can be used to characterize and establish a sound 3D representation of the fracture geometry.

Over the years, there have been various methods of recording characteristics of fracture networks from a surface. The approaches to map rock surface have been evolving using traditional and modern technologies to capture fracture properties of the rock mass. Modern technologies present digitalized fracture geometry with the advantage of capturing a huge amount of data within a short period of time. Modelling with a DFN heavily depends on the quality of the input parameters from the structural properties of the rock mass [8]. The traditional approach of measuring the spatial data such as the dip and strike is generally performed by hand with a compass or inclinometer, which is dangerous when collecting data in an unstable area. Moreover, the approach is time-consuming. Zhang and Einstein [54] stated that the traditional approach for acquiring input parameters for the DFNs is derived from the manual method of mapping rock faces such as the use of the scanline sampling method, circle sampling, area or window sampling, and use of the geological compass. Although, the manual approach was reported to be time-consuming and thus not often used in mining operations [8]. Similarly, Weir and Fowler [12] stated that the challenge of using the traditional method of kinematic analyses is that assumption is made about the infinite joint length intersecting the excavation plane. Ovaskainen [55] argued that the source of uncertainties in using a manual approach can be attributed to the availability and size of the outcrops and variable visibility of fractures on the outcrop surface caused by censoring, such as the limitation in sample sizes and sample area, which mask the true length of fractures. The data obtained from the mapping will provide information on the strength of the rock mass and its deformability properties, the geological structure, rock mass structure, and the properties of major planes of weakness [56]. Consequently, the absence of quality and comprehensive data from the field can impact the stability analysis of excavation in rock, that is, the better the information available from the mapping field, the better the outcome of the engineering analysis.

The advancement of technology has brought about accuracy and precision in field data capturing. Apart from the accuracy and precision, the introduction of digital mapping instruments such as the terrestrial laser scanner, borehole televiewers, and photogrammetry can generate high-density 3D point cloud data from rock surface mapping [57]. According to Cawood [58], the laser scanning system provides a complete and detailed mapping of a rock mass in mining. Several researchers [59–63] have discussed the strengths and flaws of using the laser scanner instrument and photogrammetry approach in rock surface mapping.

3.1. DFN Data Acquisition with 3D Laser Scanning Technologies

The 3D laser scanning technology is an instrument used for mapping, surveying, and monitoring rock mass movement in mining. The instrument is a non-contact technology that captures high-density data by utilizing a transmitted and reflected laser beam to determine the distance of the reflected objects. This principle of measurement and recording data is known as light detection and ranging (LiDAR). The outcome of the scanning generates millions of point cloud data that shows the scan surface and discontinuity data such as joint orientation, fracture density, and length. Kolapo [64] mentioned that laser scanner technologies can digitize a rock face in 3D with a fast scanning speed and high resolution of up to millimetre accuracy. The 3D laser scanning technologies have greatly improved the methods of collecting in situ data due to the convenient mode of operation. The instrument provides a uniform data format and mapping algorithm that makes it possible to extract accurate and precise data regarding the discontinuities [65]. The accuracy and speed of data acquisition over wide areas enable the creation of 3D models of geological bodies.

Moreover, after scanning tasks, the result is 3D point cloud data. The point cloud data provide a 3D representation of the scan surface made up of single points that are densely populated. The raw point cloud needs to be processed to eliminate human mapping errors or blunders that may be present. The efficiency and accuracy of the acquired point depend on the skill and experience of the operator. The knowledge of the source of errors in using laser scanning technologies is essential in generating reliable point cloud data. The acquired point cloud data can be transferred into computer software packages for processing and further analysis; although, the time taken for processing the data is more than the time required to manually collect the data [66].

The processing software will improve the display performance such as the feature extraction and fusion with imagery data [64]. Tracing of fractures in the mapping of a rock surface using point cloud data was carried out. For instance, Maerz [67] performed the extraction of the geometric parameters of a discontinuous rock mass from laser scans. Several researchers have introduced various computer software packages to extract the geological features from laser scanner point cloud data. Riquelme [68] applied the discontinuity set extractor (DSE) method to estimate the discontinuity spacing in the rock mass. Deweza [69] used the FACETs function on cloud compare software to perform planar extraction and to calculate the azimuth (dip and dip direction) of fractures. Recently, there have been developments in various intelligence algorithms that can identify discontinuity in laser scans. These include the random sample consensus (RANSAC) discussed in Ferrero [70] and the principal component analysis (PCA) approach by Gomes [71]. Likewise, Guo [72] used 1D truncated Fourier series and a curvature-weighted Laplacian-like smoothing method to trace and identify discontinuity directly from 3D point cloud data; Gigli and Casagli [73] applied a MATLAB tool called discontinuity analysis (DiAna) to analyze the rock mass discontinuities from laser scanning point cloud data; Ge [66] proposed a modified region growing (MRG) algorithm to identify discontinuities from the point cloud data. Similarly, Zhang [65] discussed the five (5) steps presented in Figure 4 used to trace the intersection line of two adjacent discontinuity planes.

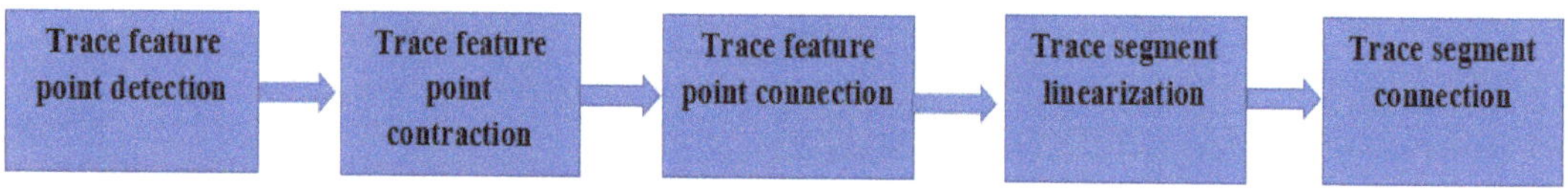

Figure 4. Steps taken to trace discontinuity from point cloud data (Adapted from [65]).

In Vazaios's [74] study, polygonal models were used to assess the structural feature of a rock mass such as joint orientation and fracture length. In this approach, the orientation of fractures present is determined by fitting planes to the selected areas of the polygon models after visual inspection. Thereafter, the direction cosines of the normal of these best-fit planes are used to evaluate the dip and the direction of fractures as shown in Figure 5a,b.

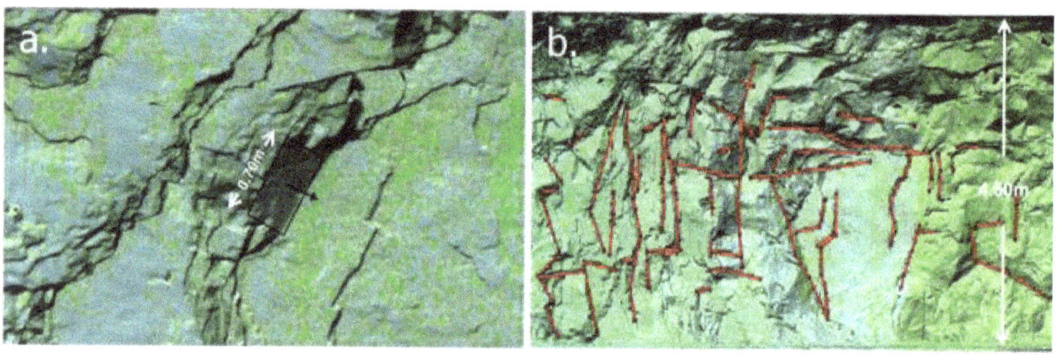

Figure 5. (**a**) Virtual mapping of discontinuities; (**b**) tracing fracture on 3D surface of a rock mass (adapted from [74]).

3.2. Photogrammetry Approach of Collecting DFN Data

Like 3D laser scanning instruments, the photogrammetry approach also produces digital elevation models (DEM) at high densities. The method makes use of digital photography of a surface captured from two or different locations to determine the depth and the dimension of space [16]. The mode of operation is by overlapping multiple photographs to generate DEM that can create 3D models of the surface. This approach makes use of motion video of surfaces or objects with cameras to generate 3D space to estimate X, Y, and Z coordinates for each pixel of an image capture. The photogrammetric approach provides a combination of 3D and spectral remote sensing data in a flexibly and cost-effective manner [75]. The method has been described as the main tool to compute the volume of ore in an open pit mine [76].

In the early 20th century, photogrammetry principles were introduced with the use of stereoscopic instruments for the construction of topographic maps. The advancement in technologies has brought about the integration of spectral remote-sensing data to generate DEM. The emergence of modern and high-resolution digital cameras proffers solutions to the restriction of the number of photographs that can be taken during data collection and provides quality pictures that can easily be assessed during data collection. Bemis [77] discussed that the noticeable barrier in using this approach is achieving the optimal positioning for the camera relative to the object of interest due to topography, vegetation, etc. Although, the barrier is being reduced over the years with the advent of high flight height digital photography from UAVs that provides improved synoptic views from overhead. Rogers and Booth [16] stated that a noticeable limitation in the use of digital photogrammetry for geotechnical mapping is that it cannot record measurements of certain key characteristics such as the thickness and type of infill materials, joint wall strength properties, and small-scale roughness. This is the reason why the remotely taken measurement by photogrammetry mapping is commonly supplemented with conventional outcrop mapping data.

The photogrammetry technique has been widely used in mapping, monitoring, and evaluating the stability of rock mass in mining operations. Traditionally, it is difficult to survey and collect geological data from high walls in open pit mines. The geotechnical engineers and geologists must walk up to the high walls to physically capture data needed for high walls stability analysis and to plan for the optimal mineral extraction. The modern photogrammetry approach can provide engineers with a user-friendly tool for rapid data acquisition. It provides a remote geotechnical mapping of rock slopes without physical access to the rock mass. The photogrammetry method of data acquisition works perfectly with DFN modelling as it can provide large and spatially located geotechnical data sets required to produce the 3D representation of the surface. The technique permits the use a probabilistic method for evaluating kinematic stability and rock block size distributions [16]. Application of photogrammetry approach in open pit mine was presented by Patikova [78] where high-resolution aerial images with low flight altitude are taken for open pit mining purposes. High-speed computers equipped with photogrammetric software are used to generate the digital terrain modelling (DTM) of the mine for volume calculation, as shown in Figure 6a,b.

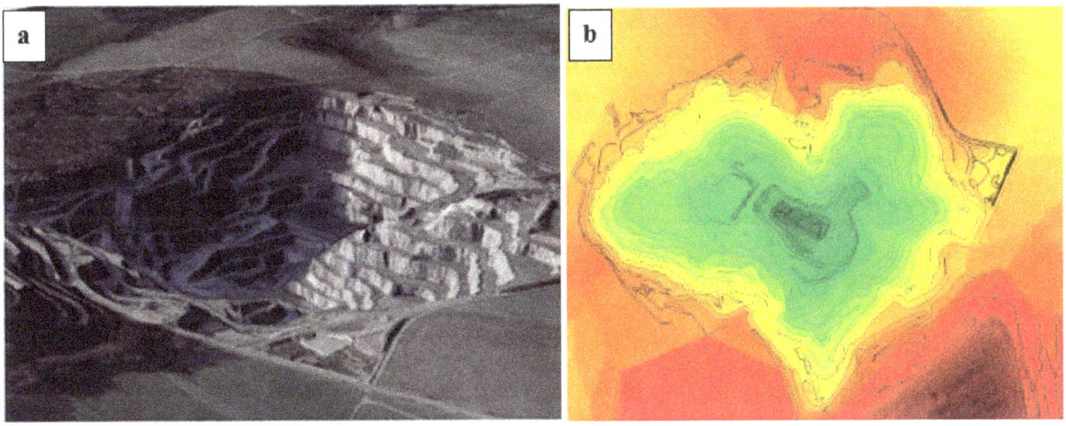

Figure 6. (**a**) Three-dimensional view of the open pit; (**b**) DTM representation of the open pit (adapted from [78]).

Moreover, Rogers [79] explained the application of the photogrammetric method to acquire fracture orientation for DFN modelling. The study stated the importance of photogrammetry in generating large, spatial geotechnical data sets required to produce a robust 3D modelling of the surface.

This approach produces large volumes of data, which makes it possible to develop DFN models for the actual fracture observation from the field for various geotechnical and geomechanical evaluations. This is one of the advantages of the photogrammetric approach in the DFN modelling process. The generation of the high density of data makes it possible for defining the effects of fracture properties on both the global and local DFN properties. Availability of technologies integrated with multiple sensors allows acquisition of fractures from mapping and photogrammetry approaches to produce exact fracture distribution of the mapped surface. The workflow for DFN model development from conditioned to actual mapping data is presented in Figure 7.

Likewise, the method has the potential to develop DFN models directly from photogrammetric surveys for evaluation of kinematic analysis from a structural description that accurately reflects the scanned location, providing a significant step forward in modelling capability. According to Rogers [79], the result of the analysis has the ability to make a sound decision on underground excavation techniques and ground support designs as a result of the heterogeneity of the rock mass. However, several studies have also used the photogrammetric approach for DFN modelling in the underground mining environment. For instance, Grenon [8] used photogrammetry tools to provide input parameters for generating a series of DFN models at the Éléonore underground mine in Canada. Three images were taken from three rock faces (Amn-0354, Gmn-0354-1s, and Gmn-0357-1n) at the development stage. These images were used to construct the 3D numerical modelling of the rock mass with orientations of 88°/252°, 87°/185°, and 86°/161° for the Amn-0354, Gmn-0354-1s, and Gmn-0357-1n drifts, respectively. The availability of the data facilitates the calibration of the DFN models at the mine. Similarly, Rogers and Booth [16] used the photogrammetric method to provide input parameters required for the DFN of rock mass in a tunnel. The study explained that the instrument can generate 3D fracture models driven by the statistical structural data and conditioned to the features observed by the photogrammetry to yield a synthetic rock mass on which improved design methods can be applied. The application of the photogrammetric approach for DFN modelling in underground mines was recorded by Benton [80]. Photogrammetric surveys were carried out in the 54 ramps at the depth of 2.1 kilometres at the Lucky Friday mine to produce measurements to aid in the interpretation of measurements from conventional instruments.

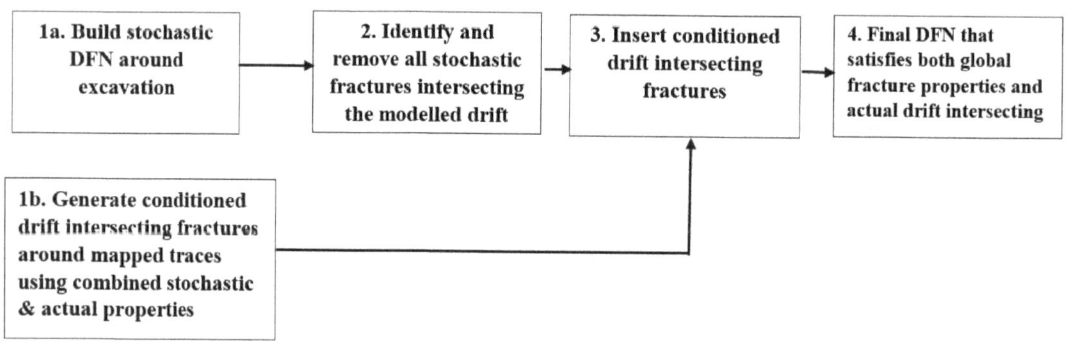

Figure 7. Interface for the development of DFN model from conditioned to actual mapping data (adapted from [79]).

4. Application of DFN Approach in Predicting Stability Analysis of Jointed Rock

DFN has been widely applied to study the stability of important engineering structures and utilities in jointed rocks at the surface, underground, small-scale, and large-scale. The areas of deployment of this tool include the stability of the fractured rock slope, coal seam longwall face stability, highway rock slope stability, high-level radioactive waste disposal wells, spent fuel reprocessing, hydrocarbon reservoirs, hydropower projects, and tunneling [1,81–87]. Table 3 provides a summary of DFN models developed for different engineering applications and structures. For each summarized area of application, the characteristics of the rock mass, model properties, figures, and drawbacks of the model were briefly discussed. Further explanation of the applications of the DFN model in these areas is provided in Section 5.

Table 3. Summary of developed DFN models in various engineering applications.

Reference	Summary of Developed DFN Models
Adapted from Li [81]	Geological/geotechnical characteristics: bedrock sliding body fracture and rock matrix fracture. 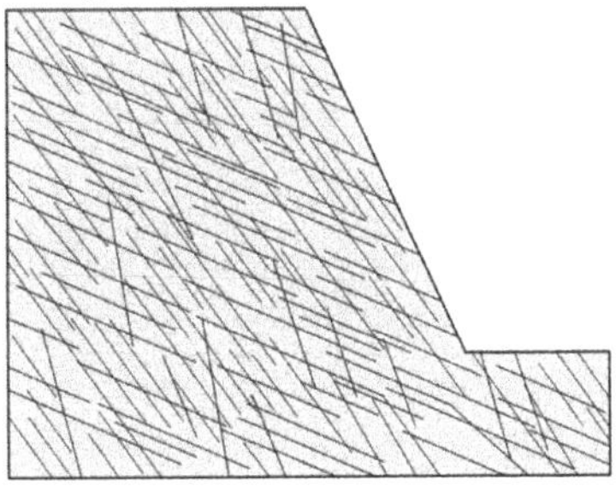 Figure: Numerical fractured rock slope model with regularly distributed fractures. Model properties: geometry of length, height, and slope angle as 96 m and 125 m, 50 m, and 95 m and 30° and 40°, for single and multi-fractures containing slopes, respectively. Sliding volume as 447.84 m^3 per unit width and the interface or fracture length as 69.282 m for a slope containing one single fracture. Field of deployment: stability of the fractured rock slope. Remarks: The developed DFN model is constrained by the employment of simplified and assumed material parameters and constitutive relations which may affect the accuracy of the model. The unavailability of groundwater flow within the slope may also limit the accuracy of the model.
Adapted from Le and Oh. [82]	Geological/geotechnical characteristics: Average sedimentary thickness of 540 m which mainly consists of sandstone, siltstone, gritstone, and coal seams. Two main mine-scale fault sets whose strikes are in parallel and meridian directions. The parallel-oriented strike faults plunge to the north or north–northwest, and their dip angle ranges from 55 to 75 degrees. The mine contains many small-scale faults whose strike directions are like those of the main faults. The panel has an average cover depth of 200 m, extraction thickness of 7.98 m, and seam dip angle of less than 10 degrees. The immediate roof of the coal seam is mainly siltstone with an average thickness of 8.48 m. The main roof is mainly sandstone with a thickness of 16.72 m. 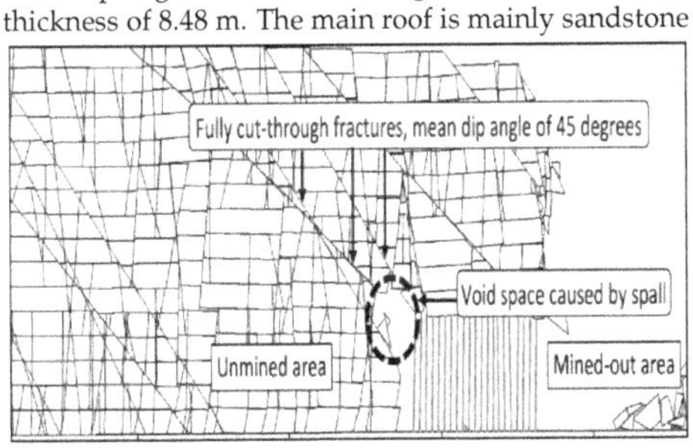 Figure: The impact of a 45-degree fracture dip angle on the stability of the longwall face. Model properties: The model used plane-strain condition because the face advance in panel length was much greater than that in panel width. The model dimensions in vertical direction were 113 m, 250 m in strike direction, and 1 m in the out-of-plane direction. DFN domain of length 50 m and thickness of 8 m. The siltstone floor coal seam, siltstone immediate roof, sandstone main roof, and siltstone overburden measured 40, 8, 8.5, 16.5, and 40 metres in thickness, respectively, from bottom to top. Bedding planes spacing of 0.5 m, mean dip angle of 115, and 90 degrees for the first and second fracture sets, respectively. Field of deployment: coal seam longwall face stability. Remarks: Due to limited field mapping and laboratory testing, a few assumptions were made to determine the inputs for the model; nevertheless, their impact on the model's outputs is minimized by a comprehensive calibration and validation procedure versus field measurement.

Table 3. *Cont.*

Reference	Summary of Developed DFN Models
Adapted from Singh [83]	Geological/geotechnical characteristics: Upper Tal Formation dominantly consisting of orthoquartzite and arkosic sandstone and partial partings of mudstone. Cretaceous anikot shelly limestone. The rock group belongs to a regional-scale syncline known as Garhwal Syncline. 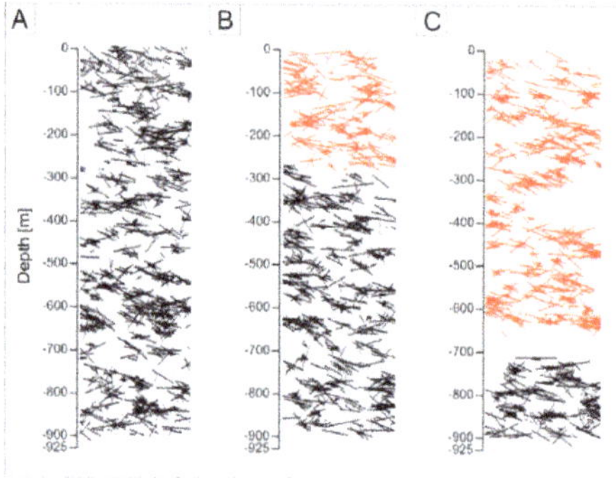 Figure: Developed DFN for site 2 having three joint sets coloured yellow, pink, and dark green (not to scale). Model properties: Three persistent joint sets were considered per each of the studied sites in the DFN model. The length of the joints range between 0.9 m ± 1.23 and 4.75 m ± 0.87 for the three sites. Both log-normal and power-law were the length distributions adopted. The average Fisher dip/dip direction ranged between 35/30 and 80/80, and the average fracture intensities (P_{32}) were between 0.451 and 4.5. Field of deployment: rock slope stability along national highway. Remarks: Due to the broad variability of rock mass properties, engineering professionals are encouraged to study each rock mass independently in the proximity of structurally damaged zones.
Tóth [84]	Geological/geotechnical characteristics: The host rock is the Boda Claystone Formation (BCF). The formation consists of well-compacted reddish-brown claystone, siltstone, and albitolite (authigenic albite >50%) with dolomite and sandstone intercalations. The formation maximum thickness is 1000 m and with a distribution area of 150 square kilometres. The average dip direction of the bedding in the well is SE–SSE, and the dip is 40° based on acoustic borehole televiewer (BHTV) observations. Four classes of veins including branched veins, straight veins, *en echelon* vein arrays, and breccia-like veins characterized the deepest well of the formation (BAF-2). In the BAF–2 well, the average dip values of the vein types are 42° (branched veins), 70° (straight veins), and 22° (*en echelon* vein arrays). 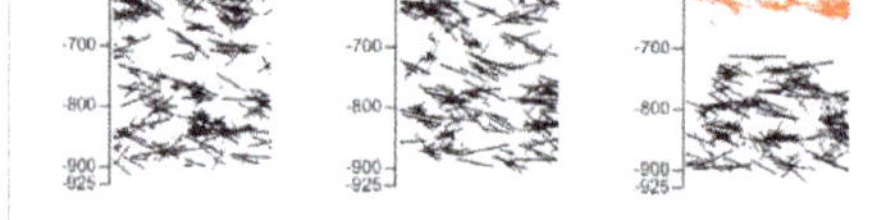 Figure: Vertical sections of simulated fracture network geometry patterns (**A–C**) of the BAF–2 well based on 20 independent runs. Model properties: the DFN model domain was 150 × 150 × 925 m. Field of deployment: High-level radioactive waste repository deep well. The average fracture density (P10) was 9–10 m^{-1}. The average fractal dimension (D) was 1.37 ± 0.05 while the single length exponent (E) value was −0.90 and the F parameter value was 10.00. The borehole planes dip in SSE direction with an average orientation of 162° N—60° E. Remarks: the study is basic and could be improved upon to provide detailed hydrodynamic modelling of the well and its environment.

Table 3. *Cont.*

Reference	Summary of Developed DFN Models
Adapted from Gao [85]	Geological/geotechnical characteristics: The lithology of the formation is made up of biotite granodiorite in the basement and gravel sand and breccia in the upper layers. Gravel sand and breccia are mixed, and their thickness typically ranges between 0 and 16.8 m from northeast to southwest. The formation is characterized by several structural and weathering fractures that are relatively straight, and the apertures are generally ~0.1–2 mm. 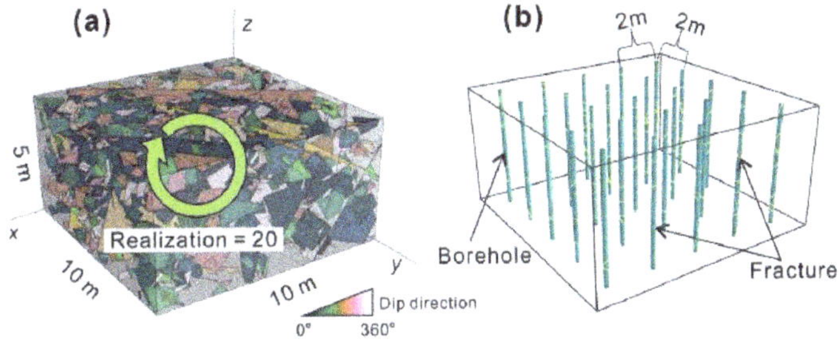 Figure: (**a**) A finite volume DFN model; (**b**) P10 obtained by virtual drilling at an interval of 2 m. Model properties: The rock mass is characterized by at least three fracture sets striking orthogonally towards northeast and northwest directions. The DFN model domain was a cuboid model (length = width = 10 m, height = 5 m). The P32 values range between 0.1 (m^2/m^3) and 4 (m^2/m^3). The fracture size was d_{min} = 2 m, d_{max} = 25 m, b = 1.9343. Field of deployment: spent fuel reprocessing (SFR) site fracture characterization. Remarks: By effectively capturing the variability in fracture spatial intensity, the framework employed in this study addresses the primary drawback of the conventional DFN models. The DFN model can, however, be improved upon and optimized inside this framework in the future by including more important information.
Adapted from Zhu [86]	Geological/geotechnical characteristics: Shale reservoirs with natural fractures and hydro-fractures. 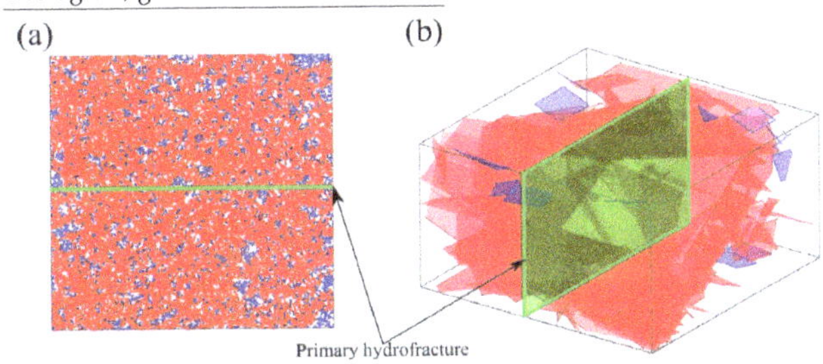 Figure: DFN model: (**a**) 2D fracture network; (**b**) 3D fracture network. Model properties: Stochastic DFN domain with 2D and 3D fractures represented by a line segment (100 m × 100 m) and square plate (50 m × 50 m × 50 m), respectively. The minimum and maximum fracture length used in the power-law distribution is 1 m and 100,000 m. The fractal dimension (*FD*) varies between 1.1 and 2.0 and between 2.1 and 3.0 for 2D and 3D fracture networks, respectively. Fracture intensity (FI) for 2D and 3D fracture networks range between 0.8 and 2.6. Injected fluid pressure of hydraulic fracturing (P_f) = 1; maximum horizontal stress (Sh_{max}) = 1.3 P_f; minimum horizontal stress (Sh_{min}) = 0.8 P_f; vertical stress (Sv) = 1.1 P_f; reservoirs pressure (P_p) = 0.5 P_f; segment length (L_{se}) from between 1 m and 0.2 m. Fracture roughness (JRC) and strength (JCS) and fracture orientation (k) vary between 0 and 20, 0.5 P_f and 18.5 P_f, and 0 and 20, respectively. Field of deployment: sensitivity analysis of the impacts of the fracture network on the formation and development of stimulated reservoirs' volume. Remarks: It is crucial to precisely measure the fracture sealing degree, fracture intensity, and fracture orientations of the subsurface fracture networks to accurately estimate SRV or have a good production prediction.

Table 3. *Cont.*

<table>
<tr><th>Reference</th><th>Summary of Developed DFN Models</th></tr>
<tr><td>Adapted fromWu [87]</td><td>Geological/geotechnical characteristics: The rock slope rock mass environment is complex and characterized by three major fracture groups with orientations of 25.98/235.15, 28.16/351.34, and 85.80/94.73, respectively. The trace length and spacing of the three fracture groups are 3.42/1.35, 3.12/1.70, and 2.96/0.69, respectively.
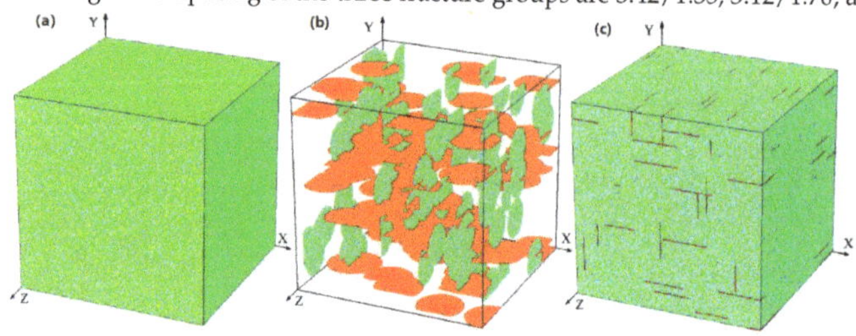

Figure: The three-dimensional equivalent DFN model's construction process (a) rock model, (b) DFN model, and (c) equivalent DFN model.
Model properties: The DFN domain was 15 m × 15 m ×15 m and the model of fracture rock masses was 8 m × 8 m × 8 m. Model edges of between 2 m and 15 m were established to investigate scale effects and anisotropy of the fractured rock masses. Displacement load of $2e-5$ times the model edge per step was applied to the model. The heterogeneity, UCS, elastic modulus, friction angle, and Poisson's ratio of the rock and fracture for the numerical model were, 5 108.9, 37.6, 56, and 0.24; and 2, 5.45, 1.88, 30, and 0.39, respectively. A water head (initial head H1 = 0) with an increase (ΔH) of 0.3 mm/step was applied to the surface in the positive direction of the Z-axis, and the water head (H2 = 0) in the corresponding negative direction was zero in the model.
Field of deployment: hydropower station project.
Remarks: Nevertheless, the suggested paradigm has some drawbacks. The dynamics, impacts of stress changes, fracture interactions, significant displacement, and rotation issues of natural fracture systems cannot be fully considered.</td></tr>
<tr><td>Adapted from Lei [1]</td><td>Geological/geotechnical characteristics: The rock mass is a crystalline rock with incompressible grains and networks of fractures of varying lengths and intensities. The intact rock is characterized by density (P), young modulus (E), Poisson's ratio (v), internal friction angle (φi_{nt}), tensile strength (f_t), cohesion (c), mode I energy release rate (G_I), mode II energy release rate (G_{II}), and matrix permeability (K_m) values of 2680 kg.m^{-3}, 60.0 GPa, 0.25, 35.0°, 15.0 MPa, 55.0 MPa, 74.3 J.m^{-2}, 88.4 J.m^{-2}, and 1 × 10−15 m^2, respectively. The fracture is characterized by residual friction angle (φr), laboratory sample length (L_0), joint roughness coefficient (JRC), and joint compressive strength (JCS) values of 31.0°, 0.2 m, 150.0 MPa, and10.0, respectively.
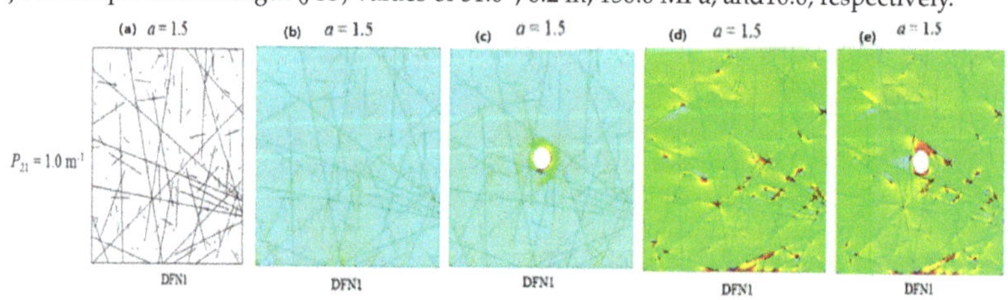

Figure: Generated DFN (a) length exponent (a) = 1.5 and fracture intensity (P_{21}) = 1.0 m^{-1}; (b) isotropic in situ stress condition before excavation; (c) isotropic in situ stress condition after excavation; (d) anisotropic in situ stress condition before excavation; (e) isotropic in situ stress condition after excavation.
Model properties: The DFN domain was 20 m × 20 m. The fracture length bounds were defined as $lmin$ = L/100 = 0.2 m and $lmax$ = 100L = 2000 m. Length exponent (a) of 1.5, 2.5, and 3.5 and fracture intensity (P_{21}) values of 1.0, 2.0, and 3.0 m^{-1} were adopted. Overburden stress ($S'v$) = 8.0 MPa, $S'_H/S'v$ =1.0 and 2.0, i.e., S'_H = 8.0 and 16.0 MPa, respectively. Average element size (h) = 5.0 cm, diameter of tunnel (d) = 2 m, penalty term (p) = 600 GPa, and damping coefficient (η) = 6.34 × 105 kg/m·s.
Field of deployment: tunnel excavation in fractured rock masses.
Remarks: The presence and distribution of natural cracks in the subsurface were found to have a substantial impact on the parameters of the excavation damaged zone (EDZ).</td></tr>
</table>

5. General Applications of DFNs

The DFNs have been extensively used in geoengineering projects such as mining operations, civil engineering projects, and the petroleum industry. The application of DFN modelling has become the most versatile, powerful, flexible, and useful tool for advanced engineering analysis that requires the evaluation of fracture geometry in jointed rock masses. It becomes valuable in projects where the fracture network controls the behaviour of the system, such as in stability analysis of a jointed rock mass in a mining rock slope, excavation stability in fractured rock, petroleum cracking, and hydrocarbon and groundwater flow. Alghalandis [13] argued that it is important to establish and analyze the rock mass response, interaction, and complexity of the fracture networks. The outcome of the evaluations will assist in making a sound judgement on the design that will guarantee safety and enhance productivity. Over the last decades, a great number of studies have extensively used DFN modelling as a solution in various engineering projects. Some of the general applications of DFN modelling are described in the next subsections.

5.1. Stability Analysis in Mining Operations

DFN models are widely used in both open pit and underground mining environments. The DFN serves as a tool for modelling the fracture geometry of a rock mass that can influence the stability of high walls in underground and bench configuration in open pit mines. Modelling rock mass structures has been a challenge due to the variability of material properties and uncertainty [88]. The observed variability in material properties is significant because it influences performance in engineering applications [89]. Despite the variation in material properties of the ground, the DFN produces a 3D representation of the rock mass that is captured during data collection. DFN technology poses advantages over other conventional tools in the area of key block stability. The tools can evaluate the fracture system, rock wedge formation, and monitor the deviation of excavation when compared with the actual design. The technology can also be used as a risk assessment tool for geotechnical engineers and geologists in underground stability assessment [79].

In an open pit mining operation, the DFN computer tools were used to create the 3D representation of rock mass structure from statistical distribution. The software integrates fracture data input such as the fracture length, height, spacing, orientation, and aperture to generate a discrete fracture network, compartment and fluid migration pathway analysis, and visualization. The software includes FracMan, SiroModel, FracSim, Alghalandis Discrete Fracture Network (ADFNE), PFLOTRAN, DFNFLOW, DFNSIM, FRANEP, SDFN, MOFRAC, and HATCHFRAC. Kuppusamy [88] used FracMan to generate the bench scale DFN model using a geometric network construction approach from an open pit phosphate-bearing ore mine located in Limpopo province in South Africa. Both scanline and window mapping surveys were conducted across various lithological units. The discontinuity type, persistence, material infilling, joint roughness, termination index aperture, rock type, and groundwater condition were recorded from the scanline while the photogrammetry technique generated the digital map structure where the areas were inaccessible or too dangerous to complete physical mapping surveys. The result of the digitally mapped area of the outcrop is presented in Figure 8 from terrestrial laser scanner point cloud data. Style [90] used the DFMN model to provide stress paths that described the instability of a large slope. The outcome of the study showed that the variation in the mass strength between different fracture orientations has a greater influence on the DFN mass strength in low-strain environments. Foster [91] applied the DFN approach to investigate the hydrological significance and its impact around the open pit. Both DFN technology and the conventional approach of limit equilibrium analysis (LEA) were used to design a bench face slope in an open pit mine by [92,93]. In the study, the strength of the approach lies in the ability of the DFN to adequately capture the characteristics of the rock mass. Similarly, Bonilla-Sierra [94] used the combination of DFN technology and the discrete element method (DEM) to examine and predict unstable wedges and high walls in open pit mines. In the DFN models, the areas that are prone to failures are highlighted. Rogers [95] applied the DFN model to evaluate the volume of unstable material that occurred as a result of three different closure options that led to the formation of key blocks, which was the source of instability at the Chuquicamata open pit mine in Chile. Similarly, Kong [96] used the combination of a 3D terrestrial laser scanner and UAV photogrammetry to identify and measure the fracture properties in a rock mass based on the multi-source fusion of point clouds shown in Figure 8.

Figure 8. Digitalized outcrop created from TLS point cloud data (adapted from Kong et al., 2022) [96].

Most underground mine applications of DFN models are used in the area of evaluating the stability of underground excavation. DFNs are used to examine the direction and extent of fragmentation in mining operations [97]. More accurate geometric models of in situ fracturing can be developed, and the result of the modelling can serve as an integral tool for advanced geomechanical simulations. The in situ stress conditions of the ground have a greater influence on the stability of an underground excavation. The DFN can be used to describe the in situ fragmentation of the ground conditions. Similarly, Junkin [98] used the DFN approach to derive fracture geometry for a rock engineering system (RES) for an interaction matrix. The study incorporated the fracture geometry from the DFN into the RES framework for blast fragmentation modelling. In massive ore bodies, a hybrid DFN and block caving fragmentation (BCF) was applied to assess fragmentation in block caving projects and mines [99]. Xiao [100] constructed a drift of 40.32 m with a 3D high point cloud data of Jianshan Underground Mine in Panzhihua, Sichuan, China, shown in Figure 9. The result of the point cloud data can be used to make a sound decision on the stability analysis of the mine by importing it on DFN codes.

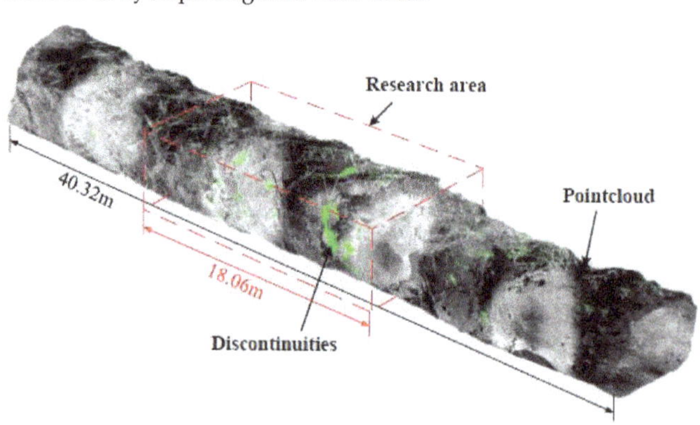

Figure 9. Three-dimensional point cloud data of an underground mine (adapted from [100]).

Similarly, Graaf [101] applied the DFN approach to accurately generate the 3D fracture network geometry of the Callie underground gold mine, which is used to develop a guideline to assist geotechnical engineers to know the variability and how the rock mass will behave under different loading conditions.

Moreover, the DFN can be used as a tool to determine the block size and volume of the unstable key block during underground excavation. The key blocks are formed as a result of the intersection of joints. In a study by Starzec and Tsang [102], they applied probabilistic calculations of unstable blocks around a circular tunnel to develop a relationship between the volume of blocks and fracture spacing and fracture intersection density. The study applied the volumetric feature intersection density as a means of quantifying the effect of feature parameter estimates on block prediction. In another application in the underground mining environment, Rogers [79] used the DFN approach with photogrammetric data to examine rock mass movement in an underground excavation. Elmo [103] used the DFN approach to simulate rock mass behaviour and combine the effects of intact rock fracturing and failure occurring along the natural fractures. The approach entails the use of synthetic rock mass (SRM) modelling where the DFN serves as one of the three components of SRM. Esmaieli and Hadjigeorgiou [104] conducted synthetic rock masses of the four-rock mass domain which was developed in the DFN to investigate the influence-induced damage from the transiting material in a Canadian underground mine. Harthong [105] used a coupled DEM-DFN approach to characterize the influence of the clustering and size distribution of pre-existing fractures on the strength of fractured rock masses. The DFN was used to control water inflow into underground excavations during and after construction [106]. Water ingress influences the stability of underground excavation and can cause adverse environmental impacts, which can in turn influence the project schedule. The DFN was used to examine the possibility of water inflow into the underground excavation in heterogeneous fractured rock. Likewise, Li [107] used the DFN to investigate the impact of water seepage on the surrounding strata of the shaft, that is, the effects of pore pressure and water inflow on the shaft pillar stability. DFNs have been widely used in mining operations. Most especially in the area of stability analysis, the effect of water ingression and pore pressure on strata control, underground water flow in deep level mines, rock mass characterization, and slope stability in open pit mines.

5.2. Application of DFN in Petroleum Industry

DFNs have been extensively used in the petroleum industry to generate the stochastic realization of the fracture network. In oil and gas production, the description of the reservoir fracture network is one of the parameters to be considered in reservoir geomechanics and production performance evaluation. The data acquisition mode in the petroleum industry for the DFN is different from that of solid minerals. During oil and gas production, data are collected from wireline logs, convention cores, small drilling mud loss analysis, and field-scale observation from outcrop analogue inspection [108]. Fracture data such as types, orientation, distribution, and properties are used in generating the stochastic model of the fracture reservoir.

However, the DFN model is used in prediction flow characteristics, which is a robust parameter in defining the dynamic performance of a naturally fractured reservoir (NFR). In Karatalov's [109] study, the work presents applications of fracture characterization parameters such as fracture length, aperture, and intensity to carry out sensitivity analysis of NFR modelling. The outcome of the study showed that the top three parameters of the fracture network are prioritized in the following order: fracture aperture, upscaling method, and fracture intensity. Likewise, Mahmoodpour [110] applied the DFN approach to mimic the reservoir permeability. Based on the acquired field data, a stochastic DFN was developed to evaluate the permeability behaviour of the reservoir. The simulation showed that a large number of fractures are involved in the stochastic model, and equivalent permeability fields are calculated to create a model which is computationally feasible. Similarly, Gentier [111] developed models based on a realistic fracture network from hydraulic systems to simulate the fluid flow that takes place within the fracture. A DFN model was built from the data collected from well imagery and estimation of fracture density using the Fisher distribution approach with the four major fractures set that constitute the reservoir. To develop an effective and efficient shale natural gas well, the designer or engineer must have a clear understanding of the role of natural fractures in shale gas production. In most cases, DFNs are applied in the area of hydraulic fracturing of the rock formation to stimulate the flow of oil and gas into the wellbore. This hydraulic fracturing approach has been in use for over 60 years and has promoted the shale gas boom in the past decade [112]. The primary purpose of the hydraulic fracturing approach is to increase the oil and gas yield in the well. This is carried out by injecting large quantities of fluid (chemical additives, water, or proppant) at high pressure into the targeted rock formation. Several studies have proven that hydraulic fracturing causes an increment in oil and gas production. Viegas [113] applied the DFN to provide a detailed analysis to highlight fluid flow paths and proppant placement in a shale reservoir. The method was used to calculate both static and dynamic parameters such as the fracture plane orientation, moment tensor, seismic efficiency, radiated energy, slip direction, and dynamic and static stress drop. Likewise, Zhang [112] used the DFN model to investigate four different shear deformation mechanisms. Similarly, Wang [114] used the DFN to model a unified shale gas reservoir to examine how each mechanism influences shale gas production and the corresponding rate transient behaviour. In the event of a tight reservoir, the flow in a fracture obeys a different pattern. Wang [115] applied an open-source numerical code (ShOpen) which comprises stochastic fracture modelling, multi-continuum modelling, and discrete fracture models for simulations of unconventional gas reservoirs. The numerical code can measure the fluid flow, absorption, transport, and indirect hydromechanical coupling in unconventional fracture reservoirs, as shown in Figure 10.

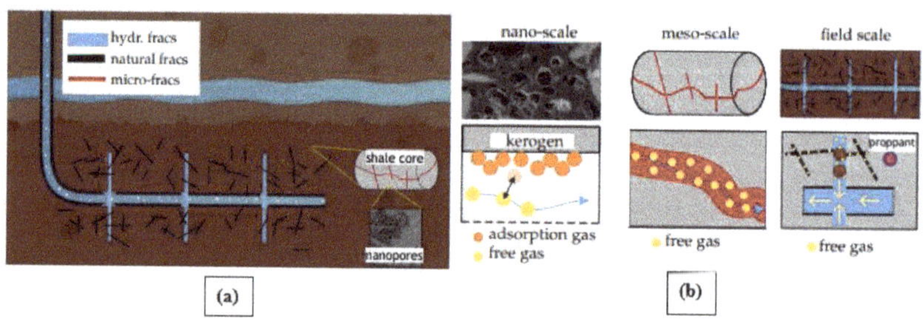

Figure 10. Description of shale gas production (**a**) gas transportation and (**b**) multiscale features (adapted from [115]).

Furthermore, the DFN is also applicable in geothermal projects. The understanding of fracture characteristics and properties is essential for effective geothermal reservoir management. A geothermal reservoir is a natural underground source of geothermal energy, typically consisting of hot water or steam often located near tectonic plate boundaries, where there is a high level of geothermal activity or in areas with a high concentration of volcanic activity. These reservoirs are formed when water seeps into the ground and is heated by the Earth's internal heat, whose temperature and pressure are dependent on the depth and geology of the area. Geothermal reservoirs are typically found at depths of a few hundred metres to several kilometres below the Earth's surface. Mahmoodpour [116] employed a DFN for modelling a reservoir and wellbore to simulate hydraulic and thermal processes involved in the geothermal energy extraction operation at Soultz Sous Forêts, France, with the aim to possibly extract higher amounts of geothermal energy from the existing industrial infrastructure. The validated hydro-thermal operational data were used for a numerical model to simulate the geothermal energy extraction operation for 3 years as a basis for the two operational scenarios for 100 years using four different injection wellhead temperatures of 70, 60, 50, and 40 °C. Mahmoodpour [116] asserted the feasibility of 100 years of geothermal energy extraction operation at Soultz-sous-Forêts with sufficient high-production temperature throughout the operation's duration.

Reservoir flow modelling technology was used to assign a constant heat flux at the bottom of the boundary of the domain while all boundaries were defined as no flow for both fluid and heat transmission. Discrete fractures within the domain were regarded as internal boundaries, implicitly considering the mass and energy exchange between the porous media and fractures or fault zones while the diameter of the injection well is represented by a line for simplicity. One of the earlier applications of the DFN in geothermal projects was recorded by Tester [117] to replicate the reservoir permeability behaviour. The outcome of their study stated the importance of creating a connected network of pathways between the injection well and production in a petrothermal system that lacks sufficient natural fluid flow and permeability for flow rates required for economic geothermal systems. These reservoirs' enhancement was achieved by creating new fractures or simulating naturally existing ones to achieve the permeabilities needed for economical fluid flow.

In a study by Aydin and Akin [118], the DFN model was used to create the fracture geometry, conductivity, and connectivity and to construct fracture networks that were instrumental to the characterization of the Alaşehir geothermal reservoir which is comprised of highly fractured marble and schist. Other fracture parameters that were considered in the fracture modelling include fracture permeability, aperture intensity, and fracture radius. Jambayev [119] constructed a DFN model for a free well section of naturally fractured "Field X" to improve reservoir characterization and to accurately predict the flow rate for the carbonate reservoir. Several researchers have performed fluid flow calculations by upscaling the DFN to an effective continuum model [120].

5.3. Application of DFNs in Tunnels and Underground Construction Works

DFNs are widely used in civil engineering projects such as tunnels, shaft sinking projects, and the construction of roads. A basic understanding of how fractures are distributed on a rock mass is important during the construction of underground projects, that is, the formation of key blocks and shape and size characterization are the fundamental properties and parameters to be considered in the engineering analysis of fractured ground. Although, for robust design, the strength of the rock mass, deformability properties, and permeability behaviour of such ground are also put into consideration. Moreover, the stress field conditions can also influence the stability of the tunnels and underground constructions. These parameters and point cloud data from laser scanners during field observation are used to develop a DFN model which highlights the area that needs a support system. It also provides tools to evaluate the stability of grounds around the excavation area, as the DFN would generate a model that represents the fracture geometry.

However, the application of DFN models in a tunnel is carried out by acquiring data from joint surfaces to create digital sections of the mapped area. The common method used to extract the plane orientation data and to determine the dominant joint sets from identified discontinuity planes is the use of polyworks [121]. The approach states the statistical sample of the joint dip and directions value can serve as input parameters for numerical modelling computer packages such as Dips from Rocscience. The exported data from the polyworks contain the original coordinates of the tunnel, which may not be linear but contain irregular geometry. The DFN technique has been applied by various researchers to solve geomechanical problems in civil engineering and stability analysis of underground construction projects. In Cacciari and Futai's [121] study, a computational approach for generating single and continuous DFNs was developed and applied to 324m of Monte Seco tunnel with foliated gneiss formation from Nova Venecia complex in Brazil. The result confirmed that it is possible to evaluate the variability of P_{32} (volumetric intensity). Notably, the variability of rock face orientations in the tunnel was considered by comparing the P_{32} results and taking the highest value as a reference for each tunnel sect. In Wang's [122] study, a larger convergence area was obtained in the circular opening model due to the effect of the fracture sets. Similar fracture patterns observed around openings with structural planes are shown in Figure 11. The DFN models created were able to provide a comprehensive overview of the rock mass behaviour which is used in predicting the stability and mechanical response of the ground condition to discontinuities present. The authors used a novel rough discrete fracture network (RDFN) characterization method based on the Fourier transform method. The outcome of the numerical simulation of the anisotropic mechanical properties was performed for the RDFN model with a complex joint network. Based on the results, the geometry of the joint network has a significant influence on the strength and failure patterns of jointed rock masses.

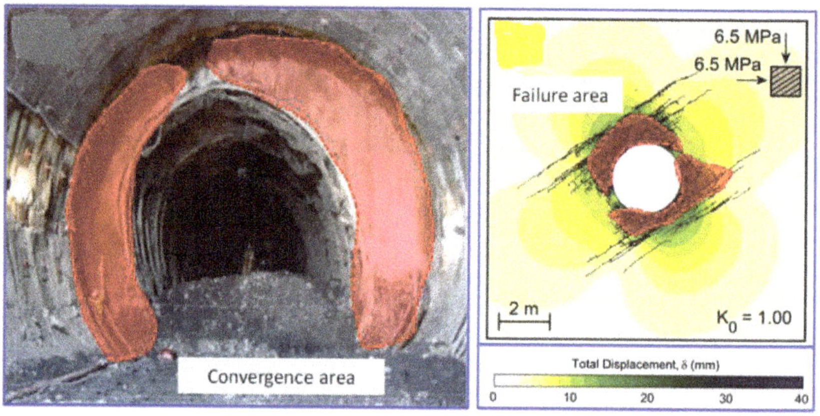

Figure 11. Construction of tunnel face from fracture pattern around circular openings (adapted from [122]).

Similarly, Ioannis [123] used a hybrid approach of both deterministic and statistical techniques to generate a DFN based on available data to simulate the rock mass behaviour and stability analysis of the Brockville tunnel in Ontario, Canada. Data from laser scanning data were used as input parameters in Morfa DFN software to construct the specific tunnel site.

5.4. Application of DFNs for Underground Flow Prediction

The presence of fractures plays an important role in the flow of fluid through the rock masses. The interconnected fracture network influences the channel flow which was developed through the conductivity of existing fractures or by new growth that connects existing fractures. The DFNs are used in modelling flow and fluid movement in the subsurface. The model is developed to simulate the movement of flow underground structures through interconnected fractures. This approach has been widely used in geoscience and engineering to control the movement of fluids such as in groundwater management, geothermal energy extraction (Saliu and Komolafe [124]), the aquifer recovery process, hydrocarbon extraction, carbon sequestration, and radioactive and toxic industrial waste [125]. Primary factors affecting groundwater flow include fracture density, fracture orientation, aperture width, and the rock matrix. According to Bordas [126], the fracture density and the fracture orientation are the most important parameters that define the interconnectivity of a fracture set, which are the contributing features of the hydraulic conductivity of a fractured rock mass. Moreover, this serves as vital input parameters for simulating the 3D representation of the subsurface fracture geometry. The fracture model created by the DFN describes the flow of the substance using data collected to characterize the fractures that influence the transportation and free flow of the substance. The main challenge of using the DFN for underground flow simulation has been the choice to include a detailed geometrical representation of the fracture and their respective fracture connectivity that is used in the predictive simulations of fluid flow in fractured rock. This makes it difficult to design an efficient and scalable workflow. If a well-defined fracture geometry of the rock mass can be acquired, the 3D representation of the subsurface fracture network can be determined. Moreover, the in situ stress and change in stresses of the rock mass have a significant influence on the flow, as this will cause changes in pore pressures within the rock mass.

Moreover, another common application of the DFN in underground flow is the evaluation of contaminated water-bearing rock. Contamination of groundwater flows occurs majorly at abandoned sites (Saliu and Komolafe [124]), where the fluid flows and where most contaminated mass flow occupies the much lower permeability matrix block between fractures [127]. The contamination of the fracture takes place when contaminated substances diffuse from the fracture where active groundwater flows occur into the permeability rock matrix. In traditional methods of investigating contamination, the rock matrix is disregarded as most of the fracture's active flows are ignored. The lack of data from borehole core sample contamination data makes it difficult for the conventional approaches to characterize the plume in a fractured rock mass. On the other hand, the DFN approach makes use of data from core rock samples from the subsurface to delineate the diffusion of contaminants on the fracture network. According to Parker [128], the application of the DFN method in this event is designed specifically for sedimentary formation, although it can work for all other rock types but must be adjusted to suit the nature of the matrix porosity, absorption rate, contaminant type, and time for diffusion.

5.5. DFN Application in the Classification of Rock Mass

The classification of rock mass is important for any engineering design The understanding of rock classification improves the quality of geoengineering designs as the geotechnical engineer will be provided with quality design input parameters [129–131], that is, the fundamental understanding of geological, structural, and mechanical classifications of the rock mass is required at the planning stage of excavation in mining and civil engineering projects. The classification properties of the rock material and the joint properties should be considered in the design. The result of the rock mass classification enables better engineering judgement most especially in underground construction when the subsurface fractures are not known. The absence of a true 3D representation of subsurface fracture geometry might be a challenge. The better the understanding of the fracture geometry of the rock mass, in quantifiable terms, the less the variability that needs to be considered in the input parameters for the classification. The DFN has the potential to describe the 3D representation of fracture geometry located in a rock mass in the subsurface, that is, the DFN model defines the geometrical properties of an individual fracture and its topographical relationship with the geological structure. For instance, a tunnel with a certain rock mass with a widely spaced geological structure may be unaffected by the structure, but the behaviour of an open stope in the same rock mass may be influenced by the jointed rock mass condition and major structural features such as faults.

The DFN arises to be an interesting tool in classifying rock mass structures. The data can be collected from the borehole core samples to produce the in situ fracture network. Since the DFN model can generate the in situ conditions of the subsurface rock mass, it is therefore important to continually update the DFN geometry as the excavation advances. The DFN approach has been widely used to simulate the classification of the rock mass. The method applies statistical and spatial measurement to examine the rock properties that determine the behaviour of the rock mass. The aim of the fracture geometry classification by DFN is to provide some geotechnical information that could be used for satisfactory planning and mining of safe excavations. It is critical to understand the variation in the properties of the rock mass, most especially in a deep underground mine where the properties of the rock mass change with depth. For instance, the DFN model can highlight the strength variation in rock mass from a small and strong intact specimen to a relatively intense fracture large slope [90].

6. Challenges of Using DFNs

DFNs have in recent times proved to be one of the veritable tools employed for geomechanical analysis for solutions in mining, geotechnical engineering, borehole mapping, and various applications in petroleum industries. Despite its versatility, there are still some limitations and challenges to the integration of complexities encountered in rock masses within DFN models. The inherent occurrence of geological discontinuities, such as joints, faults, shear zones, schistosity planes, and bedding planes [131], complicates the failure process observed during laboratory-scale experiments in comparison to the failure within the rock mass. As a result of these intrinsic uncertainties, the application of numerical modelling to the analysis of rock engineering problems poses a challenging task due to these discontinuities [132].

Most often, homogenization models are not a true representation of the real rock masses. For instance, in an underground cavern's rock mass deformation and failure analysis performed with a numerical tool for a blocky jointed rock mass, some pseudo-DFN models generated to simulate the joints in the rock mass explicitly having multiple joint sets, and many loose blocks are formed at the excavation boundary, making the cavern unstable; hence, the model does not converge. To stabilize the jointed model, supports are installed at the same stage as the excavation. This will underestimate the support capacity of the bolts and liners; hence, low deformation can be seen on the excavation boundary as expected in reality. As the k ratio increases, the induced stress on the crown pillar increases, creating a large deformation along the roof and floor. Martin [132] attested that 2D elastic analysis does not account for the effects of stress redistribution as failure progresses, which may be a result of discontinuities coupled with the kinematic constraints on the deformation and failure modes of structures in rocks [133] and causes stress and displacement redistributions to sensibly deviate from linear elastic, homogenous conditions [134]. Moreover, in all the model simulations, the maximum total displacement of the jointed model is always higher; hence, more robust support systems are required to stabilize the cavern along with the real-time stress and load monitoring in the field. It is therefore imperative to find a proper set of working plans and sections where a "creative thinking" process was applied, and geological and geotechnical concepts are tested prior to computerization [135].

Karimi and Elmo [136] stressed the need to improve engineering designs by integrating DFN numerical models within the geomechanical codes. There are challenges in coherently optimizing fractures in 3D and traces in 2D of the initially generated DFN models to minimize meshing issues while preserving the appropriate model's properties. The reliance on mesh sizes in continuum and discontinuum modelling methods dictates the mode and the types of failure mechanisms that are visible for capturing, and this must also be considered in the selection of the required input material parameters, which would depend on the degree of simplification adopted when building the model and the choice of a continuum or discontinuum model [137]. This limitation may occur due to inappropriate consideration of the subsequent mesh generation routines that are necessary for geomechanical simulation [138]. Considering the geometrical complexities of DFN conceptualization arising from many discontinuities intersecting with each other, there must be a compromise between mesh quality and the minimum element size, and this largely depends on the experience of the user for correct interpretations of how to correct or modify some specific discontinuities.

Of paramount interest to mining engineers in underground mine design is pillar strength, and this largely depend on the determination of rock mass strength with precision [55]. A more realistic depiction of the discrete fracture systems is required to accurately simulate the mechanical behaviour of mine pillars by numerical modelling [139]. There are challenges which may arise as a result of inadequate estimation and classification of the rock mass damage as a function of natural fracture network and loading conditions [135]. The accuracy and the inclusion of an impact of block to structural damage with respect to the initial block may not be properly accounted for as dilation in some models. Traditionally, there is an assumption that blocks are orthogonal, having mean volume, but the dissimilarity in joint conditions such as spacing, infillings, and orientation in the natural rock mass makes it rare to have an orthogonal block shape [140]. Effectively obtaining the size, shape, and other relevant block information of the exact blocks contributing to the structural damage is needed to define the relationship between the structure-damaging blocks and their parent blocks at the initial timestep for better investigation [135].

7. Summary and Conclusions

Fracture networks play a pivotal role in the quest to harness the critical minerals within the Earth system, hence the development of the DFN using various data-capturing methodologies and different computer programming codes for numerical modelling to enhance safe, smooth, economical, and sustainable exploitation of such minerals. The authors logically reviewed the recent advances in conceptualization of the DFN and its various engineering applications in the areas of mining, geotechnical engineering, borehole mapping, and petroleum industries.

The concept of the DFN model hinges on creating networks that replicate the major structure of the mapped rock mass using the data acquired during the site investigation while considering the geological structure of the site, the scale, and the purpose of modelling. The authors classified these data into fracture orientation, intensity, spatial location, and fracture size. The statistical method and probability distribution approach based on the principle of the von-Mises for 2D and the Fisher distribution and uniform distribution for 3D are used to represent fracture orientation (attitude of a discontinuity in space which measures the dip and the dip direction of the joints) using the data emanating from the outcrop measurement or well logging. Fracture intensity, which is the surface area of discontinuities per unit volume, plays an important role in determining the mechanical properties and hydraulic conductivity of a discontinuous rock mass, and this can be quantitatively determined by fracture tensor. The size and length of fractures in the DFN model, which are used to characterize fluid flow in fractured rock masses, were quantified by empirical and back analysis approaches before the advent of some modern advances in DFN modellings were developed to ease the process of delineating fracture size and length.

It has been established that the quality of the input data determines the outcome of the fracture network. The acquisition of a huge amount of data within a short period of time by modern data-capturing technologies has tremendously improved the solutions provided by DFN models for various applications. Besides the accuracy of the huge data captured, the advent of these cutting-edge technologies has brought about the introduction of digital mapping instruments such as the terrestrial laser scanner, borehole televiewers, and photogrammetry, which can generate high-density 3D point cloud data from rock surface mapping.

The institution of a DFN as solutions to many challenges, which in the past were unsurmountable in the mineral industries, has become one of the major technological innovations in recent times. Its versatility and flexibility made it a useful tool for advanced engineering analysis that requires the evaluation of fracture geometry in jointed rock masses. Its application in mining includes but is not limited to the stability of the fractured rock slope, coal seam longwall face, underground excavation, and investigating the possibility of water inflow into the underground excavation in heterogeneous fractured rock. In the petroleum industries, the DFN is applied in the area of hydraulic fracturing of the rock formation to stimulate the flow of oil and gas into the wellbore and to accurately predict the flow rate for the carbonate reservoir. The DFN also finds its wide expression in the civil engineering projects such as tunnels, shaft sinking projects, rock slope stability along highways, and the construction of roads. Ultimately, these applications are hinged on the accuracy of the rock mass characterization using the DFN models. The result of the rock mass classification enables better engineering judgement, most especially in underground constructions when the subsurface discontinuities are not known.

Despite the versatility of the DFN, there are still some limitations and challenges to the integration of complexities encountered in rock masses within DFN models. It cannot be convincingly said, however, that a model is a perfect imitation of reality, particularly at the initial stage when data availability is limited. For instance, forward modelling results cannot be said to be calibrated and validated at the prefeasibility or feasibility stage if the final shape of the stopes or caverns remains unknown. One of the greatest modelling challenges in caving geomechanics is the ability to develop a 3D caving model capable of incorporating fracture mechanics principles and other failure mechanisms to explicitly simulate fragmentation processes at the level required to depict secondary fragmentation while also maintaining a detailed representation of the rock mass structural physiognomies and replications of preconditioning scenarios [137].

The quest for valuable mineral resources to satisfy human wants, especially those identified in critical minerals lists for sustainable and economic success, has given rise to technological development in the mining industry. It is therefore necessary to develop technologies that can extract complex ore bodies with lower grade, at deeper depth, and sometimes in the challenging geological formation which necessitates the extension of life of the mine. These feats are made possible by the improvement of DFN models and their proper applications.

Author Contributions: Conceptualization and Manuscript Writing, P.K.; Manuscript Writing and Editing, N.O.O.; Review and Manuscript Editing, P.M.; Manuscript Administration, D.A.; Conceptualization and Manuscript Writing, K.K.; Manuscript Editing, A.G.-B. All authors have read and agreed to the published version of the manuscript.

Funding: This research received no external funding.

Conflicts of Interest: The authors declare no conflict of interest.

References

1. Lei, Q.; Latham, J.-P.; Tsang, C.-F. The use of discrete fracture networks for modelling coupled geomechanical and hydrological behaviour of fractured rocks. *Comput. Geotech.* **2017**, *85*, 151–176. [CrossRef]
2. Stephens, M.B.; Follin, S.; Petersson, J.; Isaksson, H.; Juhlin, C.; Simeonov, A. Review of the deterministic modelling of deformation zones and fracture domains at the site proposed for a spent nuclear fuel repository, Sweden, and consequences of structural anisotropy. *Tectonophysics* **2015**, *653*, 68–94. [CrossRef]
3. Le Goc, R.; Darcel, C.; Davy, P. Advanced DFN Models from Multi-Support Data for Underground Facilities. *Procedia Eng.* **2017**, *191*, 1015–1022. [CrossRef]
4. Davy, P.; Darcel, C.; Le Goc, R.; Munier, R.; Selroos, J.; Mas Ivars, D. DFN, why, how and what for, concepts, theories and issues. In Proceedings of the Seattle, Washington, USA, 2nd International Discrete Fracture Network Engineering Conference, Seattle, WA, USA, 20–22 June 2018.
5. Adepitan, R.O.; Owolabi, A.; Komolafe, K. Prediction of structural response to blast–induced vibration in Kopek Construction Quarry, Ikere–Ekiti, Ekiti State, Nigeria. *Int. J. Environ. Stud.* **2018**, *75*, 990–999. [CrossRef]
6. Esmailzadeh, A.; Kamali, A.; Shahriar, K.; Mikaeil, R. Connectivity and Flowrate Estimation of Discrete Fracture Network Using Artificial Neural Network. *J. Soft Comput. Civ. Eng.* **2018**, *2*, 13–26. [CrossRef]
7. Xu, C.; Dowd, P. A new computer code for discrete fracture network modelling. *Comput. Geosci.* **2010**, *36*, 292–301. [CrossRef]
8. Grenon, M.; Landry, A.; Hadjigeorgiou, J.; Lajoie, P. Discrete fracture network based drift stability at the Éléonore mine. *Min. Technol.* **2016**, *126*, 22–33. [CrossRef]
9. Huang, N.; Liu, R.; Jiang, Y.; Cheng, Y. Development and application of three-dimensional discrete fracture network modeling approach for fluid flow in fractured rock masses. *J. Nat. Gas Sci. Eng.* **2021**, *91*, 103957. [CrossRef]
10. Miyoshi, T.; Elmo, D.; Rogers, S. Influence of data analysis when exploiting DFN model representation in the application of rock mass classification systems. *J. Rock Mech. Geotech. Eng.* **2018**, *10*, 1046–1062. [CrossRef]
11. Jing, L.; Stephannson, O. *Fundamentals of Discrete Element Methods for Rock Engineering: Theory and Applications*, 1st ed.; Elsevier: Amsterdam, The Netherlands, 2007.
12. Weir, F.; Fowler, M. An Introduction to Discrete Fracture Network Modelling and its Geotechnical Applications. In Proceedings of the AUSROCK 2014: Third Australasian Ground Control in Mining Conference, Sydney, NSW, Australia, 5–6 November 2014.
13. Alghalandis, Y.; Elmo, D.; Eberhardt, E. Similarity analysis of discrete fracture networks. *arXiv* **2017**, arXiv:1711.05257. Available online: https://arxiv.org/pdf/1711.05257. (accessed on 24 August 2022).
14. Elmo, D.; Liu, Y.; Rogers, S. Principles of discrete fracture network modelling for geotechnical applications. In Proceedings of the 1st Conference on International Discrete Fracture Network Engineering, Vancouver, BC, Canada, 20 June 2018.

15. Elmo, D.; Eberhardt, E.; Stead, D.; Rogers, S. Discrete Fracture Network Modelling—Importance of Accounting for Data Uncertainty and Variability. In Proceedings of the Seventh International Conference & Exhibition on Mass Mining (MassMin 2016), Sydney, Australia, 17–19 September 2016.
16. Rogers, S.; Booth, P.W.; Brzovic, A. Integrated Photogrammetry and Discrete Fracture Network Modelling for Improved Rock Mass Characterization and Engineering Design. In Proceedings of the 15th Australasian Tunnelling Conference, Sydney, NSW, Australia, 17–19 September 2014.
17. Zhang, W.; Lan, Z.; Ma, Z.; Tan, C.; Que, J.; Wang, F.; Cao, C. Determination of statistical discontinuity persistence for a rock mass characterized by non-persistent fractures. *Int. J. Rock Mech. Min. Sci.* **2019**, *126*, 104177. [CrossRef]
18. Mathis, J.I. To DFN or not DFN—A question of circumstances. In Proceedings of the DFNE 2014, Vancouver, BC, Canada, 20–22 October 2014.
19. Shariff, L.; Elmo, D.; Stead, D. Dfncleaner: A Novel Automated Approach To Improve DFN Integration for Geomechanical Analysis. In Proceedings of the 2nd International Discrete Fracture Network Engineering Conference, Seattle, WA, USA, 19–21 June 2018.
20. Shevko, N.; Shaislamov, V.; Savelev, O. Application of Hybrid DPDP-DFN Modeling of Fractured Carbonate Reservoirs. In Proceedings of the SPE Russian Petroleum Technology Conference, Moscow, Russia, 22 October 2019.
21. Brenner, K.; Hennicker, J.; Masson, R.; Samier, P. Hybrid Dimensional Modelling and Discretization of Two Phase Darcy Flow through DFN in Porous Media. In Proceedings of the 15th European Conference on the Mathematics of Oil Recovery, Amsterdam, The Netherlands, 15–17 August 2016. [CrossRef]
22. Altundas, B.; Chugunov, N.; Corbett, L.; Brown, A.; Smith, V.; Thachaparambil, M.; Eliseeva, E.; Elghanduri, K.; Hartman, T.; Will, B.; et al. A Hybrid DFN with Elastic Properties to Construct a Seismic Forward Model. In Proceedings of the 15th European Conference on the Mathematics of Oil Recovery, Amsterdam, The Netherlands, 15–17 August 2016. [CrossRef]
23. Hyman, J.D.; Karra, S.; Makedonska, N.; Gable, C.W.; Painter, S.L.; Viswanathan, H.S. dfnWorks: A discrete fracture network framework for modeling subsurface flow and transport. *Comput. Geosci.* **2015**, *84*, 10–19. [CrossRef]
24. Cheong, J.-Y.; Hamm, S.-Y.; Lim, D.-H.; Kim, S.-G. Hydraulic Parameter Generation Technique Using a Discrete Fracture Network with Bedrock Heterogeneity in Korea. *Water* **2017**, *9*, 937. [CrossRef]
25. Dershowitz, W.; La-Pointe, P.; Doe, T. Advances in Discrete Fracture Network Modeling. In Proceedings of the US EPA/NGWA Fractured Rock Conference, Cambridge, MA, USA, 8–9 November 2000.
26. Alghalandis, Y. Stochastic Modelling of Fractures in Rock Masses. Ph.D. Thesis, The University of Adelaide, Adelaide, Australia, 2014.
27. Komolafe, K.; Achuenu, I.; Joro, J.D. Effect of Tantalum on Coltan in Cassiterite Using Mathematical Modeling Approach with Respect to Gibbs Energy. *Int. J. Sci. Technol.* **2021**, *9*, 1–17.
28. Elmo, D.; Stead, D.; Rogers, S. Guidelines for the quantitative description of discontinuities for use in discrete fracture network modelling. In Proceedings of the International Society for Rock Mechanics and Rock Engineering, Montreal, QC, Canada, 13 May 2015.
29. Zhu, W.; Khirevich, S.; Patzek, T.W. *HatchFrac: A Fast Open-Source DFN Modeling Software*; King Abdullah University of Science and Technology: Thuwal, Saudi Arabia, 2021.
30. Tuckwell, G.; Lonergan, L.; Jolly, R. The control of stress history and flaw distribution on the evolution of polygonal fracture networks. *J. Struct. Geol.* **2003**, *25*, 1241–1250. [CrossRef]
31. Kemeny, J.; Post, R. Estimating three-dimensional rock discontinuity orientation from digital images of fracture traces. *Comput. Geosci.* **2003**, *29*, 65–77. [CrossRef]
32. Weismüller, C.; Prabhakaran, R.; Passchier, M.; Urai, J.L.; Bertotti, G.; Reicherter, K. Mapping the fracture network in the Lilstock pavement, Bristol Channel, UK: Manual versus automatic. *Solid Earth* **2020**, *11*, 1773–1802. [CrossRef]
33. Song, J.-J.; Lee, C.-I.; Seto, M. Stability analysis of rock blocks around a tunnel using a statistical joint modeling technique. *Tunn. Undergr. Space Technol.* **2001**, *16*, 341–351. [CrossRef]
34. Whitaker, A.E.; Engelder, T. Characterizing stress fields in the upper crust using joint orientation distributions. *J. Struct. Geol.* **2005**, *27*, 1778–1787. [CrossRef]
35. Staub, I.; Fredriksson, A.; Outters, N. *Strategy for a Rock Mechanics*; Svensk Kärnbränslehantering AB, Swedish Nuclear Fuel and Waste Management Co: Stockholm, Sweden, 2021.
36. Zhang, L.; Einsten, H. Estimating the intensity of rock discontinuities. *Int. J. Rock Mech. Min. Sci.* **2000**, *37*, 819–837. [CrossRef]
37. Lu, C.; Tien, Y., Juang, C. Uncertainty of Fracture Intensity Measurement in Discontinuous Rock Masses—A Sampling Window Approach. In Proceedings of the 53rd U.S. Rock Mechanics/Geomechanics Symposium, New York, NY, USA, 20–22 June 2019.
38. Kasyanov, M. Continuum Model of Medium with Cracks. *J. Eng. Mech. Div.* **1980**, *106*, 1039–1051. [CrossRef]
39. Dershowitz, W.; Lee, G.; Geier, J.; Foxford, T.; La Pointe, P.; Thomas, A. *User Documentation. FracMan, Interactive Discrete Data Analysis, Geometric Modeling, and Exploration Simulation*, Version 2; Golder Associates Inc: Toronto, BC, Canada, 1998.
40. Berkowitz, B.; Adler, P.M. Stereological analysis of fracture network structure in geological formations. *J. Geophys. Res. Earth Surf.* **1998**, *103*, 15339–15360. [CrossRef]
41. Bour, O.; Davy, P. Connectivity of random fault networks following a power law fault length distribution. *Water Resour. Res.* **1997**, *33*, 1567–1583. [CrossRef]

42. Berkowitz, B. Analysis of fracture network connectivity using percolation theory. *J. Int. Assoc. Math. Geol.* **1995**, *27*, 467–483. [CrossRef]
43. Zhu, W.; Khirevich, S.; Patzek, T. Percolation Properties of Stochastic Fracture Networks in 2D and Outcrop Fracture Maps. In Proceedings of the 80th EAGE Conference and Exhibition, Copenhagen, Denmark, 10 June–15 July 2018.
44. Yue, G.; Wang, G.; Ma, F.; Zhu, X.; Zhang, H.; Zhou, J.; Na, J. Fracture Characteristics and Reservoir Inhomogeneity Prediction of the Gaoyuzhuang Formation in the Xiong'an New Area: Insights From a 3D Discrete Fracture Network Model. *Front. Earth Sci.* **2022**, *10*, 849361. [CrossRef]
45. Starzec, P.; Andersson, J. Probabilistic predictions regarding key blocks using stochastic discrete fracture networks—Example from a rock cavern in south-east Sweden. *Bull. Eng. Geol. Environ.* **2002**, *61*, 363–378. [CrossRef]
46. Li, M.; Han, S.; Zhou, S.; Zhang, Y. An Improved Computing Method for 3D Mechanical Connectivity Rates Based on a Polyhedral Simulation Model of Discrete Fracture Network in Rock Masses. *Rock Mech. Rock Eng.* **2018**, *51*, 1789–1800. [CrossRef]
47. Benedetto, M.F.; Berrone, S.; Scialò, S. A globally conforming method for solving flow in discrete fracture networks using the Virtual Element Method. *Finite Elements Anal. Des.* **2016**, *109*, 23–36. [CrossRef]
48. Ivanova, V.M.; Sousa, R.; Murrihy, B.; Einstein, H.H. Mathematical algorithm development and parametric studies with the GEOFRAC three-dimensional stochastic model of natural rock fracture systems. *Comput. Geosci.* **2014**, *67*, 100–109. [CrossRef]
49. Weiss, L. *The Minor Structures of Deformed Rocks: A Photographic Atlas*; Springer: Berlin, Germany, 1972.
50. Zhang, J.; Liu, R.; Yu, L.; Li, S.; Wang, X.; Liu, D. An Equivalent Pipe Network Modeling Approach for Characterizing Fluid Flow through Three-Dimensional Fracture Networks: Verification and Applications. *Water* **2022**, *14*, 1582. [CrossRef]
51. Liu, R.; Jiang, Y.; Li, B.; Wang, X. A fractal model for characterizing fluid flow in fractured rock masses based on randomly distributed rock fracture networks. *Comput. Geotech.* **2015**, *65*, 45–55. [CrossRef]
52. Schnabel, R.; Wahl, R.; Klein, R. Efficient RANSAC for point-cloud shape detection. In Proceedings of the 2007 Computer Graphics Forum, Warsaw, Poland, 17–20 August 2007; pp. 214–226. [CrossRef]
53. Zhang, L.; Einstein, H.H. Estimating the Mean Trace Length of Rock Discontinuities. *Rock Mech. Rock Eng.* **1998**, *31*, 217–235. [CrossRef]
54. Ovaskainen, N.; Nordbäck, N.; Skyttä, P.; Engström, J. A new subsampling methodology to optimize the characterization of two-dimensional bedrock fracture networks. *J. Struct. Geol.* **2022**, *155*, 104528. [CrossRef]
55. Kolapo, P. Investigating the Effects of Mechanical Properties of Rocks on Specific Energy and Penetration Rate of Borehole Drilling. *Geotech. Geol. Eng.* **2021**, *39*, 1715–1726. [CrossRef]
56. Kolapo, P.; Cawood, F. Factors to be considered in establishing a scanning laboratory for testing the accuracy of terrestrial laser scanning technologies. *Int. J. Min. Miner. Eng.* **2020**, *11*, 180–190. [CrossRef]
57. Cawood, F.; Yu, M.; Kolapo, P.; Qin, C. Development of a Laboratory for Testing the Accuracy of Terrestrial 3D Laser Scanning Technologies. *Int. J. Georesources Environ.* **2018**, *4*, 105–114. [CrossRef]
58. Tannant, D. Review of Photogrammetry-Based Techniques for Characterization and Hazard Assessment of Rock Faces. *Int. J. Geohazards Environ.* **2015**, *9*, 76–87. [CrossRef]
59. Birch, J.S. Using 3DM analyst mine mapping suite for rock face. In Proceedings of the Laser and Photo-Grammetric Methods for Rock Face Characterization Workshop, Minneapolis, MN, USA, 15–20 December 2006.
60. Haneberg, W.C. Using close range terrestrial digital photogrammetry for 3-D rock slope modeling and discontinuity mapping in the United States. *Bull. Eng. Geol. Environ.* **2008**, *67*, 457–469. [CrossRef]
61. Kemeny, J.; Turner, K.; Norton, B. *LIDAR for Rock Mass Characterization: Hardware, Software, Accuracy and Best-Practices*; Tonon, F., Kottenstette, J., Eds.; Laser and Photogrammetric Methods for Rock Face Characterization: Golden, CO, USA, 2006.
62. Sturzenegger, M.; Stead, D. Close-range terrestrial digital photogrammetry and terrestrial laser scanning for discon-tinuity characterization on rock cuts. *Eng. Geol.* **2009**, *106*, 163–182. [CrossRef]
63. Kolapo, P. Towards a Short-Range Laboratory for Testing the Accuracy of Terrestrial 3D Laser Scanning (TLS) Technologies. Master Thesis, University of Witwatersrand, Johannesburg, South Africa, 2019.
64. Zhang, K.; Wu, W.; Zhu, H.; Zhang, L.; Li, X.; Zhang, H. A modified method of discontinuity trace mapping using three-dimensional point clouds of rock mass surfaces. *J. Rock Mech. Geotech. Eng.* **2020**, *12*, 571–586. [CrossRef]
65. Ge, Y.; Tang, H.; Xia, D.; Wang, L.; Zhao, B.; Teaway, J.W.; Chen, H.; Zhou, T. Automated measurements of discontinuity geometric properties from a 3D-point cloud based on a modified region growing algorithm. *Eng. Geol.* **2018**, *242*, 44–54. [CrossRef]
66. Maerz, N.H.; Youssef, A.; Otoo, J.N.; Kassebaum, T.J.; Duan, Y. A Simple Method for Measuring Discontinuity Orientations from Terrestrial LIDAR Data. *Environ. Eng. Geosci.* **2013**, *19*, 185–194. [CrossRef]
67. Riquelme, A.J.; Abellan, A.; Tomás, R. Discontinuity spacing analysis in rock masses using 3D point clouds. *Eng. Geol.* **2015**, *195*, 185–195. [CrossRef]
68. Deweza, T.; Montaut, D.; Allanic, C.; Rohmer, J. FACETS: A CloudCompare plugin to extract geological planes from un-structured 3D point clouds. In Proceedings of the International Archives of the Photogrammetry, Remote Sensing and Spatial Information Sciences, Prague, Czech Republic, 12–19 July 2016.
69. Ferrero, A.M.; Forlani, G.; Roncella, R.; Voyat, H.I. Advanced Geostructural Survey Methods Applied to Rock Mass Characterization. *Rock Mech. Rock Eng.* **2008**, *42*, 631–665. [CrossRef]
70. Gomes, R.K.; Oliveira, P.L.; Gonzaga, L.; Tognoli, M.W.; Veronez, M.R.; Souzab, K.D. An algorithm for automatic detection and orientation estimation of planar structures in LiDAR-scanned outcrops. *Comput. Geosci.* **2016**, *90*, 170–178. [CrossRef]

71. Guo, J.; Wu, L.; Zhang, M.; Liu, S.; Sun, X. Towards automatic discontinuity trace extraction from rock mass point cloud without triangulation. *Int. J. Rock Mech. Min. Sci.* **2018**, *112*, 226–237. [CrossRef]
72. Gigli, G.; Casagli, N. Semi-automatic extraction of rock mass structural data from high resolution LIDAR point clouds. *Int. J. Rock Mech. Min. Sci.* **2011**, *48*, 187–198. [CrossRef]
73. Vazaios, I.; Vlachopoulos, N.; Diederichs, M. DFN generation for Mechanical Stability Analysis of Underground Works. In Proceedings of the ITA WTC 2015 Congress and 41st General Assembly, Dubrovnik, Croatia, 22–28 May 2015.
74. Iglhaut, J.; Cabo, C.; Puliti, S.; Piermattei, L.; O'Connor, J.; Rosette, J. Structure from Motion Photogrammetry in Forestry: A Review. *Curr. For. Rep.* **2019**, *5*, 155–168. [CrossRef]
75. Jaroslaw, W. Research on Surveying Technology Applied for DTM Modelling and Volume Computation in Open pit Mines. *Min. Sci.* **2015**, *22*, 75–83.
76. Bemis, S.P.; Micklethwaite, S.; Turner, D.; James, M.R.; Akciz, S.; Thiele, S.T.; Bangash, H.A. Ground-based and UAV-Based photogrammetry: A multi-scale, high-resolution mapping tool for structural geology and paleoseismology. *J. Struct. Geol.* **2014**, *69*, 163–178. [CrossRef]
77. Patikova, A. Digital Photogrammetry in the practice of Open Pit Mining. The International Archives of the Photogrammetry. *Remote Sens. Spat. Inf. Sci.* **2004**, *34*, 1–4.
78. Rogers, S.; Bewick, R.; Brzovic, A.; Gaudreau, D. Integrating photogrammetry and discrete fracture network modelling for improved conditional simulation of underground wedge stability. In Proceedings of the Deep Mining 2017: Eighth International Conference on Deep and High Stress Mining, Perth, Australia, 3–6 March 2017.
79. Benton, D.J.; Seymour, J.B.; Boltz, M.S.; Raffaldi, M.J.; Finley, S.A. Photogrammetry in underground mining ground control-Lucky Friday mine case study. In Proceedings of the Deep Mining 2017: Eighth International Conference on Deep and High Stress Mining, Perth, Australia, 3–6 March 2017.
80. Li, X.; Liu, J.; Gong, W.; Xu, Y.; Bowa, V.M. A discrete fracture network based modeling scheme for analyzing the stability of highly fractured rock slope. *Comput. Geotech.* **2021**, *141*, 104558. [CrossRef]
81. Le, T.D.; Oh, J. Longwall face stability analysis from a discontinuum-Discrete Fracture Network modelling. *Tunn. Undergr. Space Technol.* **2022**, *124*, 104480. [CrossRef]
82. Singh, J.; Pradhan, S.P.; Singh, M.; Hruaikima, L. Control of structural damage on the rock mass characteristics and its influence on the rock slope stability along National Highway-07, Garhwal Himalaya, India: An ensemble of discrete fracture network (DFN) and distinct element method (DEM). *Bull. Eng. Geol. Environ.* **2022**, *81*, 1–23. [CrossRef]
83. Tóth, E.; Hrabovszki, E.; Schubert, F.; Tóth, T.M. Discrete fracture network (DFN) modelling of a high-level radio-active waste repository host rock and the effects on its hydrogeological behaviour. *J. Struct. Geol.* **2022**, *156*, 104556. [CrossRef]
84. Gao, X.; Zhang, Y.; Hu, J.; Huang, Y.; Liu, Q.; Zhou, J. Site-scale bedrock fracture modeling of a spent fuel reprocessing site based on borehole group in Northwest, China. *Eng. Geol.* **2022**, *304*, 106682. [CrossRef]
85. Zhu, W.; He, X.; Li, Y.; Lei, G.; Santoso, R.; Wang, M. Impacts of fracture properties on the formation and development of stimulated reservoir volume: A global sensitivity analysis. *J. Pet. Sci. Eng.* **2022**, *217*, 110852. [CrossRef]
86. Wu, N.; Liang, Z.; Zhang, Z.; Li, S.; Lang, Y. Development and verification of three-dimensional equivalent discrete fracture network modelling based on the finite element method. *Eng. Geol.* **2022**, *306*, 106759. [CrossRef]
87. Kuppusamy, V. Application of Discrete Fracture Networks (DFN's) to the design. In Proceedings of the Evolution of Geotech—25 Years of Innovation, Virtual Conference, 20–21 April 2021.
88. Fesomade, K.I.; Alewi, D.D.; Seidu, S.O.; Saka, S.O.; Osuide, B.I.; Ebidame, G.C.; Ugoh, M.C.; Animasaun, D.O. The Effect of Palm Kernel Shell Ash on the Mechanical and Wear Properties of White Cast Iron. *Adv. Technol. Mater.* **2020**, *45*, 20–27. [CrossRef]
89. Styles, T.; Coggan, J.; Pine, R. Stability Analysis of a Large Fractured Rock Slope Using a DFN-Based. In Proceedings of the Slope Stability 2011: International Symposium on Rock Slope Stability in Open Pit Mining and Civil Engineering, Vancouver, BC, Canada, 18–21 September 2011.
90. Foster, S.; West, L.; Bottrell, S.; Hildyard, M. A DFN Approach to Evaluating the Hydrogeological Significance of Lithostatic Unloading in Fractured Strata Around Open-Pit Workings. In Proceedings of the 2nd International Discrete Fracture Network Engineering Conference, Seattle, WA, USA, 20–22 June 2018.
91. Grenon, M.; Hadjigeorgiou, J. Implications of DFN Model Selection in Open Pit Bench Stability Analyses. *Boletín Geológicoy Minero* **2020**, *131*, 387–400. [CrossRef]
92. Kolapo, P.; Oniyide, G.O.; Said, K.O.; Lawal, A.I.; Onifade, M.; Munemo, P. An Overview of Slope Failure in Mining Operations. *Mining* **2022**, *2*, 19. [CrossRef]
93. Bonilla-Sierra, V.; Elmouttie, M.; Donzé, F.-V.; Scholtès, L. Composite wedge failure using photogrammetric measurements and DFN-DEM modelling. *J. Rock Mech. Geotech. Eng.* **2017**, *9*, 41–53. [CrossRef]
94. Rogers, S.F.; Elmo, D.; Webb, G.; Moreno, C. DFN Modelling of Major Structural Instabilities in a Large Open Pit for End of Life Planning Purposes. In Proceedings of the 50th U.S. Rock Mechanics/Geomechanics Symposium, Houston, TX, USA, 5–8 June 2016.
95. Kong, X.; Xia, Y.; Wu, X.; Wang, Z.; Yang, K.; Yan, M.; Li, C.; Tai, H. Discontinuity Recognition and Information Extraction of High and Steep Cliff Rock Mass Based on Multi-Source Data Fusion. *Appl. Sci.* **2022**, *12*, 11258. [CrossRef]
96. Elmo, D.; Rogers, S.; Stead, D.; Eberhardt, E. Discrete Fracture Network approach to characterise rock mass fragmentation and implications for geomechanical upscaling. *Min. Technol.* **2014**, *123*, 149–161. [CrossRef]

97. Junkin, W.R.; Ben-Awuah, E.; Fava, L. Incorporating DFN Analysis in Rock Engineering Systems Blast Fragmentation Models. In Proceedings of the 53rd U.S. Rock Mechanics/Geomechanics Symposium, New York City, NY, USA, 20–25 June 2019.
98. Guajardo, C. *Application of Discrete Fracture Network to Block Caving Fragmentation: A Hybrid Approach*. 2019. Available online: https://www.srk.com/en/publications/application-of-discrete-fracture-network-to-block-caving-fragmentation-a-hybrid-approach (accessed on 19 August 2022).
99. Xiao, Y.; Yang, C.; Li, J.; Zhou, K.; Lin, Y.; Sun, G. Estimation of the Mean Trace Length of Discontinuities in an Underground Drift Using Laser Scanning Point Cloud Data. *Sustainability* **2022**, *14*, 15650. [CrossRef]
100. Graaf, J.; Doumis, K.; Rogers, S. Towards the Implementation of Discrete Fracture Network Modelling as a Geotechnical Design Tool—Case Study of Callie Underground Mine. In Proceedings of the Third Australaisia Ground Control in Mining Conference, Sydney, Australia, 5–6 November 2014.
101. Starzec, P.; Tsang, C.-F. Use of fracture-intersection density for predicting the volume of unstable blocks in underground openings. *Int. J. Rock Mech. Min. Sci.* **2002**, *39*, 807–813. [CrossRef]
102. Elmo, D.; Moffitt, K.; Carvalho, J. Synthetic Rock Mass Modelling: Experience Gained and Lessons Learned. In Proceedings of the 50th U.S. Rock Mechanics/Geomechanics Symposium, Houston, TX, USA, 15–20 June 2016.
103. Esmaieli, K.; Hadjigeorgiou, J. Application of DFN–DEM modelling in addressing ground control issues at an underground mine. *Min. Technol.* **2015**, *124*, 138–149. [CrossRef]
104. Harthong, B.; Scholtes, L.; Donze´, F.-V. Strength characterization of rock masses, using a coupled DEM–DFN model. *Geophys. J. Int.* **2021**, *191*, 467–480. [CrossRef]
105. Javadi, M.; Sharifzadeh, M. Assessment of Inflow Possibility into Underground Excavations Using DFN and Percolation Concepts. In Proceedings of the 22nd World Mining Congress & Expo, Istanbul, Turkey, 11–16 September 2011.
106. Li, J.; Wang, Y.; Tan, Z.; Du, W.; Liu, Z. Study on Water Inflow Variation Law of No.1 Shaft Auxiliary Shaft in HighLiGongshan Based on Dual Medium Model. *Symmetry* **2021**, *13*, 930. [CrossRef]
107. Verga, F.; Giglio, G.; Masserano, F.; Ruvo, L. Validation of Near-Wellbore Fracture-Network Models With MDT. *SPE Reserv. Evaluation Eng.* **2002**, *5*, 116–125. [CrossRef]
108. Karatalov, N.; Stefaniak, A.; Vaughan, L. Advanced Modeling of Interwell-Fracturing Interference: An Eagle Ford Shale-Oil Study. In Proceedings of the Abu Dhabi International Petroleum Exhibition & Conference, Abu Dhabi, United Arab Emirates, 7–11 November 2017.
109. Mahmoodpour, S.; Singh, M.; Obaje, C.; Tangirala, S.K.; Reinecker, J.; Bär, K.; Sass, I. Hydrothermal Numerical Simulation of Injection Operations at United Downs, Cornwall, UK. *Geosciences* **2022**, *12*, 296. [CrossRef]
110. Gentier, S.; Rachez, X.; Ngoc, T.D.T.; Peter-Borie, M.; Souque, C. 3D flow modelling of the medium-term circulation test performed in the deep geothermal site of Soultz-sous-Forêts (France). In Proceedings of the World Geothermal Congress, Bali, Indonesia, 25–30 April 2010; p. 13.
111. Zhang, F. Effects of Hydraulic Fracture Crossing Natural Fractures Numerical study using hybrid discrete-continuum modelling. In Proceedings of the ISRM Congress 2015 Int'l Symposium on Rock Mechanics, San Francisco, CA, USA, 10–13 October 2015.
112. Viegas, G.; Urbancic, T.; Baig, A. Rupture Dynamics and Source Scaling Relations of Microseismic Hydraulic Fracture in Shale Reservouir. In Proceedings of the ISRM Congress 2015 Proceedings—Int'l Symposium on Rock Mechanics, San Francisco, CA, USA, 10–13 October 2015.
113. Wang, H. Discrete fracture networks modeling of shale gas production and revisit rate transient analysis in heterogeneous fractured reservoirs. *J. Pet. Sci. Eng.* **2018**, *169*, 796–812. [CrossRef]
114. Wang, B.; Fidelibus, C. An Open-Source Code for Fluid Flow Simulations in Unconventional Fractured Reservoirs. *Geosciences* **2021**, *11*, 106. [CrossRef]
115. Mahmoodpour, S.; Singh, M.; Turan, A.; Bär, K.; Sass, I. Hydro-Thermal Modeling for Geothermal Energy Extraction from Soultz-sous-Forêts, France. *Geosciences* **2021**, *11*, 464. [CrossRef]
116. Tester, J.W.; Brown, D.W.; Potter, R.M. *Hot Dry Rock Geothermal Energy: A New Energy Agenda for the 21st Century*; Technical Report, Los Alamos National Lab: Los Alamos, NM, USA, 1989.
117. Aydin, H.; Akin, S. Discrete Fracture Network Modeling of Alaşehir Geothermal Field. In Proceedings of the 44th Workshop on Geothermal Reservoir Engineering Stanford University, Stanford, CA, USA, 11–13 February 2019.
118. Jambayev, A.S. Discrete Fracture Network Modeling for a Carbonate Reservoir. Master's Dissertation, Colorado School of Mines, Denver, CO, USA, 2013.
119. Kohl, T.; Mégel, T. Predictive modeling of reservoir response to hydraulic stimulations at the European EGS site Soultz-sous-Forêts. *Int. J. Rock Mech. Min. Sci.* **2007**, *44*, 1118–1131. [CrossRef]
120. Cacciari, P.P.; Futai, M.M. Modeling a Shallow Rock Tunnel Using Terrestrial Laser Scanning and Discrete Fracture Networks. *Rock Mech. Rock Eng.* **2017**, *50*, 1217–1242. [CrossRef]
121. Wang, P.; Liu, C.; Qi, Z.; Liu, Z.; Cai, M. A Rough Discrete Fracture Network Model for Geometrical Modeling of Jointed Rock Masses and the Anisotropic Behaviour. *Appl. Sci.* **2022**, *12*, 1720. [CrossRef]
122. Ioannis, V.; Mark, D.; Nicholas, V. Factors affecting realism of DFNs for Mechanical Stability Analysis. In Proceedings of the International Discrete Fracture Network Engineering Conference, Vancouver, BC, Canada, 17–21 September 2014.
123. Saliu, M.A.; Komolafe, K. Investigating the Effect of Dolomite Exploitation on Groundwater Condition of Ikpeshi, Akoko—Edo, Edo State, Nigeria. *J. Earth Sci. Geotech. Eng.* **2014**, *4*, 137–147.

124. Pham, H.V.; Parashar, R.; Sund, N.; Pohlmann, K. Determination of fracture apertures via calibration of three-dimensional discrete-fracture-network models: Application to Pahute Mesa, Nevada National Security Site, USA. *Hydrogeol. J.* **2020**, *29*, 841–857. [CrossRef]
125. Bordas, J. Modeling Groundwater Flow and Contaminant Transporting Fractured Aquifers. Master's Thesis, Air Force Institute of Technology, Colombus, OH, USA, 2005.
126. Parker, B.L.; Cherry, J.A.; Chapman, S.W. Discrete Fracture Network Approach for Studying Contamination in Fractured Rock. *Aqua Mundi* **2012**, *12*, 101–116. [CrossRef]
127. Ogunsola, N.O.; Olaleye, B.M.; Saliu, M.A. Effects of weathering on some physical and mechanical properties of Ewekoro Limestone, South-western Nigeria. *Int. J. Eng. Appl. Sci.* **2017**, *4*, 257336.
128. Ogunsola, N.; Lawal, A.; Saliu, M. Federal University of Technology Akure Variations of physico-mechanical, mineralogical and geochemical properties of marble under the influence of weathering. *Min. Miner. Deposits* **2019**, *13*, 95–102. [CrossRef]
129. Ogunsola, N.O. The effects of weathering on some physical and mechanical properties of selected carbonate rocks in south-western Nigeria. Master's Dissertation, The Federal University of Technology Akure, Akure, Nigeria, 2016.
130. Goodman, R. *Introduction to Rock Mechanics*; Willey: New York, NY, USA, 1989.
131. Lisjak, A.; Grasselli, G. A review of discrete modeling techniques for fracturing processes in discontinuous rock masses. *J. Rock Mech. Geotech. Eng.* **2014**, *6*, 301–314. [CrossRef]
132. Martin, C.; Kaiser, P.; McCreath, D. Hoek-Brown parameters for predicting the depth of brittle failure around tunnels. *Can. Geotech. J.* **1999**, *1*, 136–151. [CrossRef]
133. Hoek, E.; Diederichs, M. Empirical estimates of rock mass modulus. *Int. J. Rock Mech. Min. Sci.* **2006**, *2*, 203–215. [CrossRef]
134. Hammah, R.E.; Yacoub, T.E.; Corkum, B.; Wibowo, F.; Curran, J.H. *Analysis of Blocky Rock Slopes with Finite Element Shear Strength Reduction Analysis*; Springer: Vancouver, BC, Canada, 2007; pp. 329–334.
135. Jarek, J.; Esterhuizen, S.G. *Use of the Mining Rock Mass Rating (MRMR) Classification: Industry Experience*; U.S. Department of Health and Human Services, Public Health Service, Centers for Disease Control and Prevention, National Institute for Occupational Safety and Health: Pittsburgh, PA, USA, 2007; pp. 73–78.
136. Sharif, L.K.; Elmo, D.; Stead, D. An Investigation of the Factors Controlling the Mechanical Behaviour of Slender Naturally Fractured Pillars. *Rock Mech. Rock Eng.* **2020**, *53*, 5005–5027. [CrossRef]
137. Shapka-Fels, T.; Elmo, D. Numerical Modelling Challenges in Rock Engineering with Special Consideration of Open Pit to Underground Mine Interaction. *Geosciences* **2022**, *12*, 199. [CrossRef]
138. Mayer, J. Applications of Uncertainty Theory to Rock Mechanics and Geotechnical Mine Design. Master's Thesis, Simon Fraser University, Burnaby, BC, Canada, 2015.
139. Elmo, D.; Stead, D. An Integrated Numerical Modelling–Discrete Fracture Network Approach Applied to the Characterisation of Rock Mass Strength of Naturally Fractured Pillars. *Rock Mech. Rock Eng.* **2009**, *43*, 3–19. [CrossRef]
140. Singh, J.; Pradhan, S.P.; Singh, M.; Yuan, B. Modified block shape characterization method for classification of fractured rock: A python-based GUI tool. *Comput. Geosci.* **2022**, *164*, 105125. [CrossRef]

eng

Review

A Review of the Metallogenic Mechanisms of Sandstone-Type Uranium Deposits in Hydrocarbon-Bearing Basins in China

Guihe Li [1,2], Jia Yao [2,*], Yiming Song [3], Jieyun Tang [3], Hongdou Han [3] and Xiangdong Cui [3]

1 School of Engineering and Applied Science, University of Pennsylvania, Philadelphia, PA 19104, USA; guiheli@seas.upenn.edu
2 College of Engineering and Physical Sciences, University of Wyoming, Laramie, WY 82071, USA
3 Research Institute of Exploration and Development, Liaohe Oilfield Company, CNPC, Panjin 124010, China; songyiming@petrochina.com.cn (Y.S.); tangjy1@petrochina.com.cn (J.T.); hanhd@petrochina.com.cn (H.H.); cuixd@petrochina.com.cn (X.C.)
* Correspondence: jyaouwyo@gmail.com

Abstract: As a valuable mineral resource, uranium is extensively utilized in nuclear power generation, radiation therapy, isotope labeling, and tracing. In order to achieve energy structure diversification, reduce dependence on traditional fossil fuels, and promote the sustainable development of energy production and consumption, research on the metallogenic mechanisms and related development technologies of uranium resources has been one of the focuses of China's energy development. Sandstone-type uranium deposits make up approximately 43% of all deposits in China, making them the most prevalent form of uranium deposit there. Sandstone-type uranium deposits and hydrocarbon resources frequently coexist in the same basin in China. Therefore, this study summarizes the spatial and chronological distribution, as well as the geological characteristics, of typical sandstone-type uranium deposits in China's hydrocarbon-bearing basins. From the perspectives of fluid action, geological structure, and sedimentary environment, the metallogenic mechanisms of sandstone-type uranium deposits in hydrocarbon-bearing basins are explored. According to the research, the rapid reduction effect of oil and gas in the same basin is a major factor in the generation of relatively large uranium deposits. Additionally, ions such as CO_3^{2-} and HCO_3^- in hydrothermal fluids of hydrocarbon-bearing basins, which typically originate from dispersed oil and gas, are more conducive to uranium enrichment and sedimentation. This study provides guidance for efficient sandstone-type uranium deposit exploration and production in hydrocarbon-bearing basins and helps to achieve significant improvements in uranium resource exploitation efficiency.

Keywords: hydrocarbon-bearing basins; sandstone-type uranium deposits; metallogenic mechanisms; hydrothermal fluids

check for updates

Citation: Li, G.; Yao, J.; Song, Y.; Tang, J.; Han, H.; Cui, X. A Review of the Metallogenic Mechanisms of Sandstone-Type Uranium Deposits in Hydrocarbon-Bearing Basins in China. *Eng* **2023**, *4*, 1723–1741. https://doi.org/10.3390/eng4020098

Academic Editors: Reza Rezaee and Yujie Yuan

Received: 28 April 2023
Revised: 13 June 2023
Accepted: 15 June 2023
Published: 19 June 2023

1. Introduction

Uranium is a rare mineral resource that is extensively distributed throughout the Earth's crust [1,2], but it typically exists at low concentrations, with an average abundance of about 2.7 ppm (parts per million) [3]. In addition, only a relatively small number of economically viable uranium deposits exist, and they are unevenly distributed [4,5]. Currently, global uranium production is about 54,224 tons [6] and is primarily from countries such as Kazakhstan (43%), Canada (15%), Namibia (11%), and Australia (8%) [7]. Uranium is an important unconventional energy resource primarily used for nuclear power generation [8–10]. Through uranium fission reactions in a nuclear reactor, enormous energy can be generated for electricity production. Nuclear power generation has the advantages of high efficiency and cleanliness [11,12], so it is particularly suitable for areas with high electricity demand, helping to reduce dependence on traditional fossil fuels, and thus reducing carbon emissions [13,14]. In addition, uranium also has other nuclear technology applications, such as radiation therapy [15,16] and isotope labeling and tracing [17,18]. The Organization for Economic Co-operation and Development (OECD) has provided information that the growing demand for clean energy transition in the future is anticipated to affect nuclear energy capacity, with the East Asia region expected to experience the largest increase in uranium demand [19,20]. To promote the growth of nuclear energy capacity and drive the increase in demand for uranium, it is crucial to acknowledge the benefits of nuclear energy in providing a secure, reliable, and predictable energy supply. Additionally, offering incentives to promote the diversification of low-carbon technologies can help to achieve these goals.

For countries with large populations and rapid economic development, such as China, ensuring energy security is crucial for sustainable development. In order to lower greenhouse gas emissions and combat climate change, the Chinese government has implemented a series of policies [21] to support the development of uranium resources, fully exploiting and utilizing domestic uranium resources to achieve energy diversification, reduce reliance on traditional fossil fuels, and promote sustainable energy production and consumption [22]. China previously aimed to adhere to a strategy referred to as the 'three-third rule' for its uranium supply, which involved one-third coming from domestic mining, one-third from direct international trades, and one-third from Chinese companies mining abroad. However, China's current approach has shifted towards a portfolio approach, where uranium is sourced from various locations based on feasibility and economic viability. The objective is to maximize uranium supply security by utilizing different sources, including those outlined in the 'three-third rule' [23]. Therefore, in recent years, China's uranium mining has also increased, with annual production reaching 1885 tons in 2018, contributing to roughly 4% of global uranium production and ranking among the top ten producers worldwide [24].

Sandstone-type uranium deposits, which are distinguished by their economical in-situ leaching, vast scale, and minimal environmental impact during mining, have become a key area of focus in China's ongoing exploration of uranium resources [25]. Consequently, research into the genesis, exploration, and mining of these deposits has gained significant traction in recent years. The presence of sandstone-type uranium deposits alongside oil and gas resources frequently occurs in China's basins [26–29], indicating a close relationship and coexistence between these resources. This can be attributed to the fact that sandstone, particularly high-porosity and permeability sandstone, functions as a storage space for oil and gas, as well as channels and storage sites for the transportation and enrichment of fluids carrying uranium [30–32]. In addition, the geological structures and sedimentary environments required by hydrocarbon generation and uranium mineralization share certain similarities [33]. The coexistence of hydrocarbon resources and sandstone-type uranium deposits within the same basin offers China a significant advantage in the efficient and economical exploitation of its energy resources. Developed oil and gas basins have relatively complete infrastructure and geological history data that can be directly utilized for uranium exploration and evaluation, leading to comprehensive resource utilization and a substantial improvement in economic efficiency.

However, it should be noted that the metallogenic mechanisms of sandstone-type uranium deposits within hydrocarbon-bearing basins may prove complex because of the presence of varying fluid compositions, geological structures, and sedimentary environments. These factors can have an impact on the enrichment and distribution of uranium. Given the unique features of basins containing both hydrocarbons and sandstone-type uranium deposits, it is imperative to conduct a comprehensive analysis of uranium mineralization in such basins. Therefore, this study presents a comprehensive overview of the geological properties, geographic distribution, and chronological variation of typical sandstone-type uranium deposits found in China's hydrocarbon-bearing basins. By integrating analyses of fluid action, geological structure, sedimentary environment, and other relevant factors, this study elucidates the metallogenic mechanisms underlying sandstone-type uranium deposits within China's hydrocarbon-bearing basins. Findings from this study can enable more precise assessments of uranium reserves, as well as shed light on the distribution patterns and factors underlying the formation of uranium deposits. The insights gained can inform the exploration and development strategies for sandstone-type uranium deposits, optimizing mining methods and beneficiation processes, and leading to improved efficiency and recovery rates.

2. Characteristics of Sandstone-Type Uranium Deposits in Hydrocarbon-Bearing Basins in China

Figure 1 illustrates the typical forms and distribution percentages of China's uranium resources, along with their respective locations. The four primary categories of uranium deposits [34] in China are depicted in the figure as follows: sandstone-type, accounting for 43.0% and widely distributed; granite-type, accounting for 22.9% and concentrated in southern China; volcanic-type, accounting for 17.6% and concentrated in eastern China; and C-Si-pelitic type, accounting for 8.7% and concentrated in central and southern China. These four primary categories together contribute to approximately 92.2% of the total confirmed uranium resources in the country [35]. Other types, including alkaline rock type, coal rock type, shale type, and phosphorite type, represent about 7.8% of the total [35]. The geological history of China reveals uranium mineralization events spanning the Paleoarchean, Early Paleozoic, Late Paleozoic, Triassic, Jurassic, Cretaceous, and Cenozoic eras, as depicted in Figure 2. The Cretaceous and Cenozoic eras were the main period of uranium mineralization [36]. The distribution of these mineralization ages was primarily influenced by China's tectonic environment [37,38], geological structural evolution [38], and the geochemical properties of uranium elements [39]. The formation of various mountain structures, fault zones, volcanic belts, and sedimentary basins in China, under the influence of tectonic movements, magmatic activity, and sedimentation, has provided suitable environments and conditions for uranium mineralization.

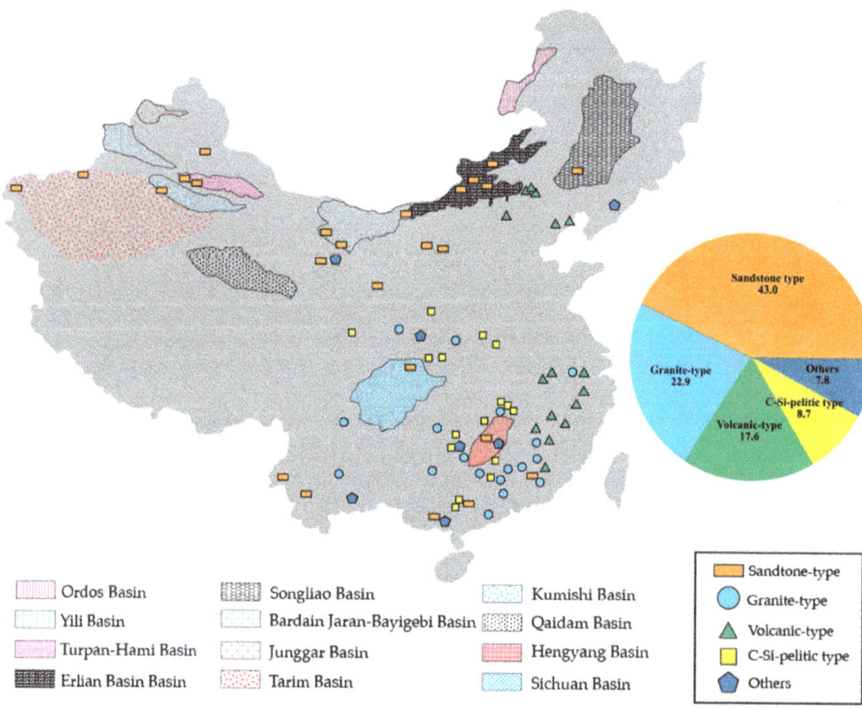

Figure 1. Typical types and distribution percentages and locations of uranium resources in China [40].

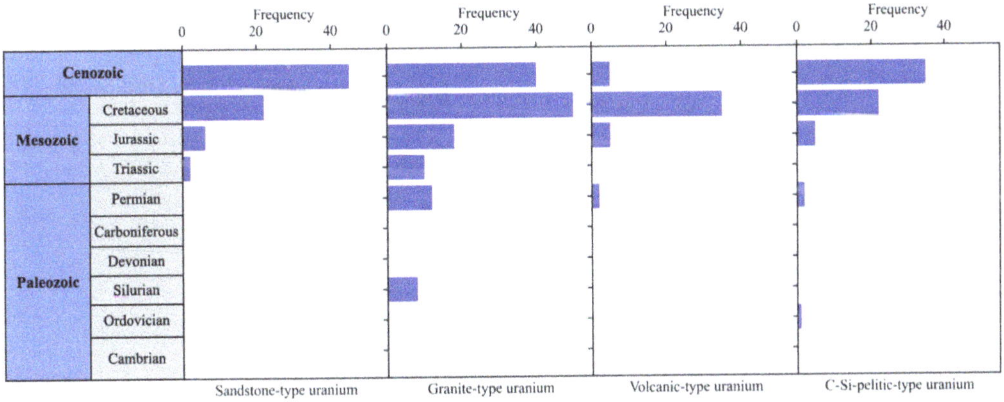

Figure 2. Statistical graph of ore-forming ages of four typical uranium resources in China [35].

Sandstone-type deposits, which make up the majority of China's confirmed uranium reserves, exhibit certain characteristics in terms of their overall spatial and temporal distribution. Firstly, they are characterized as "small but numerous". Although individual deposits are small, they are numerous, and have various mineralization geological ages, mineralization periods, and distribution areas [35]. Secondly, they are considered "lean but usable". Despite China's uranium resources being mainly of medium to low grades, the ore types are generally good, and the hydrometallurgical processing performance is favorable, making them technically and economically viable [35]. Lastly, they are "widely scattered but relatively concentrated". Although individual deposits have small scales, they are primarily concentrated in several mining areas and mineralized concentration areas [35].

2.1. Spatial Distribution Characteristics of Sandstone-Type Uranium Deposits in Hydrocarbon-Bearing Basins

Sandstone-type uranium deposits within hydrocarbon-bearing basins are primarily concentrated in China's large- and medium-sized basins, particularly in the northern region. Among them, six basins, including the Ordos Basin, Yili Basin, Turpan-Hami Basin, Erlian Basin, Songliao Basin, and Badain Jaran-Bayingebi Basin, are projected to hold 75% of China's entire sandstone-type uranium deposit resources [35]. All these basins are characterized by oil and gas production, with sandstone-type reservoirs being predominant, as well as some shale, mudstone, carbonate hydrocarbon reservoirs, and coalbed methane reservoirs. Table 1 presents the location distribution of the major hydrocarbon-bearing basins in China that are abundant in sandstone-type uranium deposits, along with the burial depth of the hydrocarbon reservoirs and sandstone-type uranium deposits.

Table 1. Distribution of major hydrocarbon-bearing basins and burial depth of their internal hydrocarbon reservoirs and sandstone-type uranium deposits [7,37,41–64].

Hydrocarbon-Bearing Basin	Basin Location	Burial Depth of Hydrocarbon Reservoirs (m)	Burial Depth of Sandstone-Type Uranium Deposits (m)	Resources (tU)
Ordos Basin	In northern China, covering parts of Shaanxi, Inner Mongolia, and Ningxia provinces	2000–4000	300–1500	80,100

Table 1. *Cont.*

Hydrocarbon-Bearing Basin	Basin Location	Burial Depth of Hydrocarbon Reservoirs (m)	Burial Depth of Sandstone-Type Uranium Deposits (m)	Resources (tU)
Yili Basin	In northern section of Xinjiang Uygur Autonomous Region	1000–4000	200–800	42,700
Turpan-Hami Basin	In northwest section of Xinjiang Uyghur Autonomous Region	3000–6000	100–800	10,100
Erlian Basin	In Inner Mongolia	1500–3500	500–1000	52,100
Songliao Basin	In northeastern China, spanning across parts of Liaoning, Jilin, and Heilongjiang provinces	2000–4000	500–1000	16,500
Badain Jaran-Bayingebi Basin	In northwestern China, spanning the provinces of Gansu, Ningxia, and Inner Mongolia	2000–5000	500–1000	7500
Junggar Basin	In northwestern China, covering parts of Xinjiang and Inner Mongolia	2500–15,000	150–1300	N/A
Tarim Basin	In Xinjiang Uygur Autonomous Region	3000–8000	500–3000	N/A
Kumishi Basin	In Xinjiang Uygur Autonomous Region	2000–5000	1000–2000	N/A
Qaidam Basin	In northwestern part of Qinghai province	3000–5000	<3000	N/A
Hengyang Basin	In Hunan province	1500–3000	<1000	N/A
Sichuan Basin	In southwestern China, covering parts of Sichuan, Chongqing, and Guizhou provinces	2000–7000	500–1000	5100

2.2. Temporal Distribution Characteristics of Sandstone-Type Uranium Deposits in Hydrocarbon-Bearing Basins

In China's hydrocarbon-bearing basins, uranium mineralization in sandstone-type deposits spans an extensive temporal range, from the early Mesozoic to the late Cenozoic periods [65]. Figure 2 indicates that the peak periods of uranium mineralization were during the Cretaceous and Cenozoic periods. Table 2 provides information on the formation periods of major hydrocarbon-bearing basins in China, as well as their internal hydrocarbon-bearing reservoirs and sandstone-type uranium deposits. According to Table 2, uranium mineralization primarily occurred in the western basins during the Neogene period, while in the central and eastern basins, it mainly occurred during the Cretaceous period [35].

Table 2. Formation periods of major hydrocarbon-bearing basins, as well as their internal hydrocarbon-bearing reservoirs and sandstone-type uranium deposits [34,42,48,59,66–87].

Hydrocarbon-Bearing Basin	Formation Period of Basin	Main Formation Period of Hydrocarbon-Bearing Layers	Main Mineralization Period of Sandstone-Type Uranium Deposits
Ordos Basin	Paleozoic to Cenozoic	Silurian to Triassic	Jurassic to early Cretaceous
Yili Basin	Late Paleozoic to early Mesozoic	Devonian to Triassic	Triassic to Jurassic
Turpan-Hami Basin	Late Paleozoic to Cenozoic	Permian to early Triassic	Cretaceous

Table 2. *Cont.*

Hydrocarbon-Bearing Basin	Formation Period of Basin	Main Formation Period of Hydrocarbon-Bearing Layers	Main Mineralization Period of Sandstone-Type Uranium Deposits
Erlian Basin	Mesozoic to Cenozoic	Cretaceous to Paleogene	Late Jurassic to early Cretaceous
Songliao Basin	Paleozoic to early Mesozoic	Carboniferous to Jurassic	Carboniferous to Permian
Badain Jaran-Bayingebi Basin	Cenozoic	Neogene	Early to middle Quaternary
Junggar Basin	Late Paleozoic to Mesozoic-Cenozoic	Cretaceous to Neogene	Cretaceous
Tarim Basin	Early Paleozoic to early Cenozoic	Jurassic and Cretaceous	Jurassic to Paleogene
Kumishi Basin	Cenozoic	Neogene	Neogene
Qaidam Basin	Mesozoic to Cenozoic	Neogene	Neogene to Quaternary
Hengyang Basin	Paleozoic	Devonian to Permian	Neogene
Sichuan Basin	Mesozoic to Cenozoic	Cretaceous to Paleogene	Neogene

2.3. Geological Characteristics of Sandstone-Type Uranium Deposits in Hydrocarbon-Bearing Basins

Table 3 summarizes the geological structural characteristics, as well as the sedimentary conditions, of hydrocarbon-bearing basins in China that are rich in sandstone-type uranium resources. Most of these basins have a high prevalence of fault structures, which is a significant factor in their formation and evolution. Some of these basins were formed through marine sedimentation, while others were formed through continental sedimentation that mainly involved lacustrine and fluvial sedimentation. Consequently, the hydrodynamic environment was extremely important during the development process of these basins.

Table 3. Geological structural characteristics and sedimentary conditions of major hydrocarbon-bearing basins [88–108].

Hydrocarbon-Bearing Basin	Geological Structure	Sedimentary Rock	Sedimentary Environment
Ordos Basin	Fault zone and multiple fault structures as primary structural units; Uplifts, and depressions as secondary structural units.	Sandstone, mudstone, coal seam	Continental sedimentation, including coal-bearing strata, ancient lake facies, river facies.
Yili Basin	Fault structures as primary structural units; Reverse faults, slopes, fractures, folds, and uplifts as secondary structural units.	Sandstone and mudstone	Marine sedimentation, including shallow marine facies, continental shelf facies, and marine facies.
Turpan-Hami Basin	Fault zone as primary structural unit; Reverse faults, blind thrust faults, folds, and uplifts as secondary structural units	Sandstone and mudstone	Continental sedimentation, including coal-bearing strata, river facies, and lake facies.
Erlian Basin	Stable structure with only slight structural deformations.	Sandstone, shale, coal seam	Continental sedimentation, including river facies, lake facies, and wind erosion.
Songliao Basin	Fault-rift zone as primary structural unit; Fault blocks, uplifts, depressions, and ancient buried hills as secondary structural units.	Sandstone and mudstone	Marine sedimentation, including shallow marine facies, continental shelf facies, and marine facies.

Table 3. *Cont.*

Hydrocarbon-Bearing Basin	Geological Structure	Sedimentary Rock	Sedimentary Environment
Badain Jaran-Bayingebi Basin	Fault zone as primary structural unit; Reverse faults, blind thrust faults, slopes, and uplifts as secondary structural units	Sandstone and mudstone	Continental sedimentation, including river facies and lake facies.
Junggar Basin	Fault zone as primary structural unit; Reverse faults, uplifts and depressions as secondary structural units.	Sandstone, mudstone, and carbonate rock	Continental sedimentation, including river facies, lake facies, and wind erosion.
Tarim Basin	Fault zone as primary structural unit; Reverse faults, blind thrust faults, slopes, and uplifts as secondary structural units.	Sandstone and mudstone	Marine sedimentation, including shallow marine facies, continental shelf facies, and marine facies.
Kumishi Basin	Fault zone as primary structural unit; Reverse faults, slopes, and uplifts as secondary structural units.	Sandstone and shale	Continental sedimentation, including river facies, lake facies, and wind erosion.
Qaidam Basin	Fault zone as primary structural unit; Reverse faults, uplifts and depressions as secondary structural units.	Sandstone, mudstone, carbonate rock, and shale.	Marine sedimentation, including shallow marine facies, continental shelf facies, and marine facies.
Hengyang Basin	Fault zone as primary structural unit; Reverse faults, normal faults, and strike-slip faults as secondary structural units.	Sandstone and mudstone	Continental sedimentation, including river facies and lake facies.
Sichuan Basin	Fault zone as primary structural unit; thrust faults, strike-slip faults, normal faults, and uplifts as secondary structural units.	Sandstone, mudstone, shale, and coal seam.	Marine to continental sedimentation.

3. Metallogenic Mechanisms of Sandstone-Type Uranium Deposits in Hydrocarbon-Bearing Basins in China

Sandstone-type uranium deposits are commonly a result of a specific stage in crustal evolution. The main sources of uranium minerals that are present in China's sandstone-type uranium deposits are sedimentary layers within the basin and nearby geological structures that contain significant concentrations of uranium minerals. The redox behavior of uranium ($U^{6+} \rightarrow U^{4+}$) is a fundamental principle followed by uranium mineralization. Uranium minerals are transported and enriched in the form of U^{6+} and ultimately deposited in the sandstone deposits in the form of U^{4+} compounds. Current research among Chinese scholars suggests a consensus regarding the infiltration of fluids in the vicinity of the basin's periphery, which leads to the formation of roll or plate uranium orebodies [109]. Some researchers have proposed that reducing fluids in deep basins are as important during the formation process of these deposits in light of recent advancements in prospecting and research [25]. Moreover, it is suggested that the basin's tectonic movement can cause uranium-bearing fluid to vertically migrate, facilitating uranium deposits to form. Figure 3 illustrates the metallogenic mechanisms involved in the formation process of sandstone-type uranium deposits within basins containing hydrocarbon resources. This study analyzes the mineralization process of sandstone-type uranium within hydrocarbon-bearing basins from the perspective of fluid action to better understand the metallogenic mechanisms involved in sandstone-type uranium deposit formation in such basins. The potential rational theory is clarified in that hydrocarbon-bearing basins are conducive to sandstone-type uranium mineralization by combining their structural characteristics and sedimentary environment.

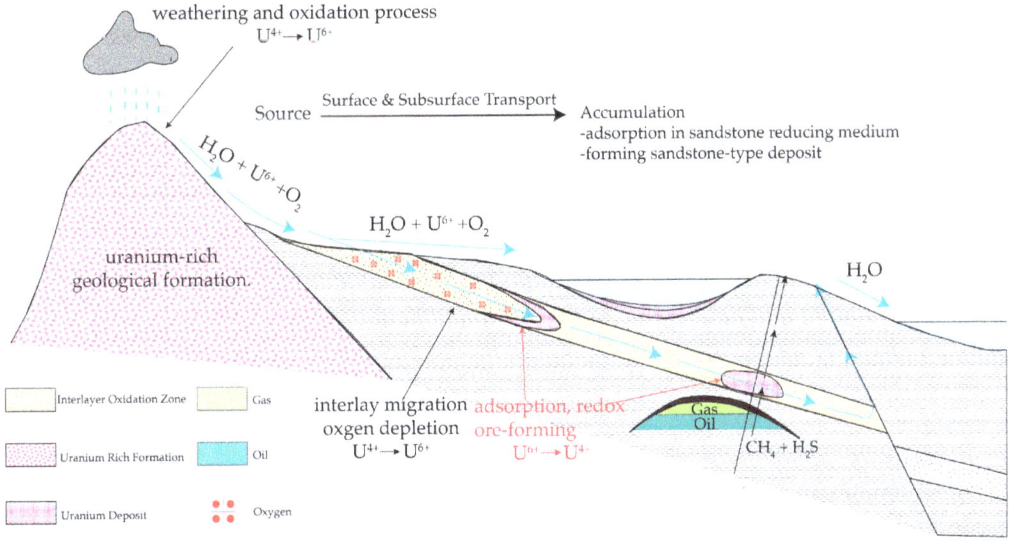

Figure 3. Metallogenic mechanisms of sandstone-type uranium deposits in hydrocarbon-bearing basins [110].

3.1. Fluid Action

The crucial role of fluids in uranium mineralization has been widely recognized in current research. Sandstone-type uranium deposits are formed with the significant involvement of surface water and groundwater. It has been discovered that uranium ore, oil, and gas, as well as low-temperature hydrothermal minerals, can be found in hydrocarbon-bearing basins with extensive sandstone-type uranium resources. This suggests the involvement of low-temperature hydrothermal fluids during uranium mineralization [111,112]. Therefore, this study offers a metallogenic analysis of sandstone-type uranium deposits within hydrocarbon-bearing basins, examining the effects of three fluid types: surface water and groundwater, oil and gas, and hydrothermal fluids.

3.1.1. Action of Surface Water and Groundwater

In uranium-rich hydrocarbon-bearing basins, sandstone-type uranium mineralization primarily occurs through the action of surface water and groundwater, which play a crucial role in providing the hydrodynamic force necessary for uranium mineralization. Under oxidizing conditions, surface water dissolves uranium minerals that are present within the rocks of the surrounding orogenic zone, while groundwater dissolves a significant quantity of uranium-bearing rock debris that is brought into the basin by weathering and denudation. These uranium minerals dissolved by surface water and groundwater can migrate with water flow to the tectonic slope zone, which is favorable for mineralization and enrichment. Under the influence of the siphon effect [113,114] and pulsation cycle mechanism [25], uranium-bearing fluids can promote uranium deposition in sandstone through the form of minerals or adsorptive precipitation. This occurs due to a slowing down of water flow, a decrease in oxygen content in water, and the occurrence of reduction reactions.

3.1.2. Action of Oil and Gas

Uranium deposits and hydrocarbon fields within the same basin are closely intertwined. The spatial location of uranium deposits and the origin of hydrocarbon fields are strongly related. In these hydrocarbon-bearing basins, when sandstones are rich in organic materials, such as fossil plant remains, these organic materials can undergo pyrolysis reactions under the influence of pressure and temperature over an extended length of time, leading to the formation of hydrocarbons. Additionally, these organic materials can also react with uranium-bearing oxidized groundwater as a reducing agent to form uranium deposits. This is because fossil plant remains are abundant in organic matter, including soluble organic matter, fixed organic matter, and structural organic matter, with the majority of it being humus. The organic matter can react with uranium-bearing oxidized groundwater either directly through reduction with bacterial as a catalyst [115], or indirectly through the production of biogenic hydrogen sulfide [116], a reducing agent that causes uranium to precipitate for the deposit formation. Therefore, sandstones that are rich in organic materials within hydrocarbon-bearing basins often become important sites for hydrocarbon generation and uranium deposit mineralization.

However, regarding production stratigraphy and mineralization location, sandstone-type uranium deposits are substantially distinct from hydrocarbon resources. Studies have shown that oil and gas reservoirs are typically located at lower or deeper sections of the sandstone-type uranium-bearing layer within the same basin [117]. In contrast, sandstone-type uranium is primarily mineralized at the top of hydrocarbon reservoirs, especially around the edges of hydrocarbon cap rocks [115]. This up-and-down superposition location association between sandstone-type uranium deposits and hydrocarbon is a key factor in the modification effect of hydrocarbons on uranium deposits.

The spatial distribution of uranium deposits above hydrocarbon reservoirs is determined by their source rock locations and formation processes. As shown in Figure 3, hydrocarbons are mainly derived from organic-rich source rocks, usually located underground or within sedimentary basins, while uranium is derived from uranium-rich minerals in surrounding rocks or debris. Hydrocarbon formation results from the accumulation and maturation of organic matter, while sandstone-type uranium deposition involves the uranium dissolution by surface water or groundwater under oxidizing conditions, the movement of uranium-bearing fluid, and the uranium precipitation under reducing conditions. In basins containing both hydrocarbon resources and sandstone-type uranium deposits, hydrocarbons are usually transported through fractures or unconformities to the overlying or shallow sandstone-type uranium-bearing layers and laterally migrate along the intra-stratigraphic sand body toward the decompression zone [117]. When hydrocarbons encounter oxygenated uranium-bearing fluids or uranium ore bodies, post-generation modification occurs [117] for uranium deposition. Therefore, sandstone-type uranium deposits are usually located above hydrocarbon reservoirs. Chemical analysis of rock samples from drill holes [118] and fluid inclusion tests [119,120] have revealed the presence of hydrocarbon gases, including hydrogen sulfide, carbon dioxide, and methane. These results suggest that hydrocarbon gases, with the function of rapid and local reduction, which leak from oil and gas reservoirs can be utilized as an essential favorable factor to promote larger deposits to form [115]. In addition, sandstone-type uranium deposits are more prone to form within areas with poor sealing at the margins of hydrocarbon reservoirs, as these conditions are more conducive to hydrocarbon infiltration.

Hydrocarbons possess various forms of reducing effects on uranium mineralization depending on their different formation timeframes [117]. Figure 4 presents the initial formation periods of these hydrocarbon-bearing basins, together with those of hydrocarbon reservoirs and uranium deposits contained therein. Within a given basin, the hydrocarbon generation and the sandstone-type uranium deposition may vary, with the possibility of occurring before, concurrently with, and after the latter. When oil and gas are generated earlier than uranium mineralization, they infiltrate the sandstone-type uranium-bearing layer to create a large-scale reducing environment. This increases the reducing capacity within the mineralized layer, favoring the precipitation and enrichment of uranium. When oil and gas are generated almost simultaneously with uranium mineralization, their infiltration not only increases the reducing capacity but also obstructs the upward migration of oxygenated and uranium-bearing fluids. This leads to the development of a uranium mineralization equilibrium interface, which is a favorable condition for enriched uranium deposits with large scales to form. When oil and gas are generated later than uranium mineralization, their infiltration and reducing modification occur outside the already-formed uranium deposit. This reduces the possible oxidation zone surrounding the deposit and protects the formed uranium mineral body. Additionally, carbonate minerals can form both inside and outside the mining area when hydrocarbons reduce uranium-bearing fluids. Therefore, the presence of carbonate minerals can serve as a valuable indicator or clue for exploring sandstone-type uranium resources.

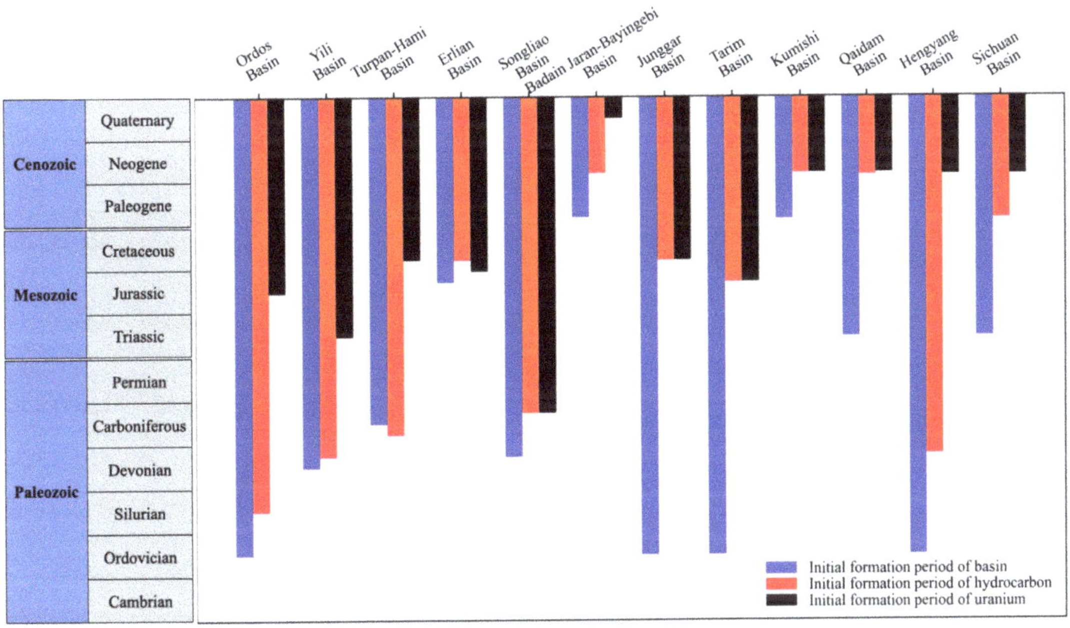

Figure 4. Initial formation periods of hydrocarbon-bearing basins and their internal hydrocarbon reservoirs and sandstone-type uranium deposits.

3.1.3. Action of Hydrothermal Fluids

Hydrothermal fluids form when groundwater at great depths undergoes dissolution of minerals and ions under high temperature and pressure conditions [121]. These fluids are primarily composed of water, dissolved minerals, and ions, which commonly include sodium, potassium, and iron ions.

Hydrothermal fluids found in hydrocarbon-bearing basins not only contain common ions but also CO_3^{2-} and HCO_3^- ions, which are typically sourced from dispersed oil and gas. In a slightly acidic environment, these ions in the hydrothermal fluids can more effectively dissolve U^{6+}. Soluble compounds such as uranyl carbonate complex ions ($[UO_2(CO_3)_3]^{4-}$) [122,123] easily form and dissolve in oxygen-containing groundwater, which contributes to the enrichment of uranium ores. These complex anions then react with reducers, such as organic carbon from dispersed oil and gas. As a result, U^{4+} can precipitate under reducing circumstances, which, in turn, further promotes uranium mineralization. Hence, these distinctive compositions of hydrothermal fluids within hydrocarbon-bearing basins play a pivotal role in the enrichment of uranium during its transportation, thus facilitating the development of larger-scale uranium deposits.

3.2. Geological Structure Effect

Tectonic events in geology refer to the deformation and movement of the Earth's plates, which can cause crustal movements and the formation and evolution of basins. These processes can result in phenomena such as basin uplift, subsidence, and deformation, which significantly affect hydrocarbon generation, transportation, and accumulation, as well as uranium deposition. Table 3 indicates that fault structures are predominant throughout the majority of China's hydrocarbon-bearing basins characterized by abundant sandstone-type uranium resources. This demonstrates that fault movements have a pivotal involvement during the formation process of uranium deposits. Secondary structural units commonly found within these basins are primarily composed of depressions and uplifts.

Faults, as the primary geological structural units, are of vital importance during the developmental and evolutionary processes of hydrocarbon-bearing basins, facilitating both vertical and horizontal migration and the movement of materials. They are able to control the distribution status of sand bodies within these basins and serve as a fundamental area during sandstone-type uranium accumulation [26]. Along with the primary structures, secondary structural units such as depressions and uplifts have the capability to further enhance the abundance of sandstone-type uranium deposits. This is because these units dominate the structural slope zones within the basins, which are the most favorable areas for the accumulation of sand, thereby contributing to the subsequent uranium accumulation for enrichment [124]. Geological processes such as folding and dissolution create favorable storage spaces within the sand bodies, characterized by pores, fractures, and solution cavities, which make it easier for uranium minerals to accumulate. Moreover, basin subsidence causes a significant inflow of groundwater, which promotes the dissolution of uranium in uranium-bearing rocks by groundwater, leading to uranium element enrichment. On the other hand, basin uplift creates a reducing environment for uranium-bearing groundwater, resulting in the conversion of U^{6+} to U^{4+} and precipitation in the form of a compound. Overall, tectonic events and the formation of secondary structures lead to regional differences in movement within the basins, resulting in the large-scale migration and seepage of groundwater and hydrocarbon-related fluids. These processes enhance the effects of groundwater migration and hydrocarbon reduction on uranium mineralization, further promoting the enrichment and mineralization of uranium minerals.

In China, the formation of certain uranium-rich hydrocarbon-bearing basins, such as Tarim Basin, Ordos Basin, and Sichuan Basin, is also influenced by magmatic activities [125]. The uranium-rich geological bodies within the magmatic zones surrounding these basins also significantly contribute to uranium mineralization at vast scales. These geological bodies are primarily composed of uranium-rich granites [126], which provide substantial material for uranium deposition in the uranium-bearing rock series of basins. When these granites intrude into sandstone or other uranium-bearing rocks, the uranium element in the granite is released through chemical reactions [127,128], and then reacts with the surrounding sandstone or other rocks to form uranium minerals. These minerals are further transported and enriched, ultimately forming uranium deposits. Therefore, these uranium-rich granites are crucial uranium sources and are essential catalysts during sandstone-type uranium deposition [129].

3.3. Sedimentary Environment Effect

Figure 4 illustrates that the formation of these hydrocarbon-bearing basins (or their host rocks) generally occurred earlier than their internal uranium mineralization. For non-hydrocarbon-bearing basins, there is no established correlation between basin formation and uranium mineralization. This is due to the fact that the formation time of a basin is determined by its geological background and structure. Typically, crustal movements and tectonic activities such as fault movements, uplift, and subsidence can induce basin formation. However, hydrocarbon-bearing basins usually have abundant sediment sources and suitable sedimentary environments, in addition to experiencing crustal movements and tectonic activities, which provide sufficient reserve space for the generation and enrichment of hydrocarbons. This abundance of sediment sources and suitable sedimentary environments also create favorable circumstances for uranium deposition. Thus, these factors contribute to the suitability of China's major hydrocarbon-bearing basins for sandstone-type uranium deposition.

Table 1 illustrates that the dominant sedimentary rock types in China's hydrocarbon-bearing basins are sandstone and mudstone. Sandstone serves as the primary medium for the transportation of uranium mineralizing fluids, including uranium-bearing fluids, oil and gas, and hydrothermal fluids. Additionally, it facilitates the accumulation and storage of uranium minerals. The quality of sandstone is governed by its internal heterogeneity and properties, as well as material composition. Sandstones with higher porosity and permeability are better suited for the transport and enrichment of uranium mineralizing fluids. Moreover, the presence of quartz within sandstones promotes a stronger adsorption capacity, thus making quartz-rich sandstones more conducive to the process of uranium mineralization. Mudstone is also a significant accompanying rock [130], and its clay minerals promote uranium deposition through adsorption. Additionally, some hydrocarbon-bearing basins contain carbonate rock, shale, and coal sedimentary rock types. The components of calcite and dolomite in carbonate rocks also employ adsorption [131] to facilitate uranium mineralization, whereas shale and coal seams primarily rely on the reduction effect of their associated organic matter and sulfides to promote uranium enrichment and mineralization.

Furthermore, sandstone-type uranium deposition within hydrocarbon-bearing basins involves hydrodynamics during marine or continental sedimentation. These basins are predominantly situated in system domains with facies characterized by river-dominated deltas and lake margins that are near the source material erosion. The near-source material erosion system [95,132] domains contain plentiful rock debris, organic matter, and other sedimentary materials, which facilitate the formation of thick sedimentary layers and provide ample raw materials, space, and favorable conditions for reduction and adsorption that promote the enrichment and generation of hydrocarbons and uranium. Furthermore, during the sedimentation process, hydrodynamic activities also facilitate the transportation and enrichment of hydrocarbons and uranium to a certain extent.

3.4. Effects of Other Factors

In addition to the primary factors discussed above that influence sandstone-type uranium deposition within hydrocarbon-bearing basins, there are other unique factors that can affect their formation, including paleo-environmental conditions, extreme climates, and biological activity. During the Early to Middle Proterozoic period, the sudden rise in atmospheric oxygen content [133] caused a large amount of previously accumulated uranium in a reduced state to oxidize, dissolve, and migrate, resulting in enrichment and mineralization. In extremely arid climates, strong evaporation can cause uranium to precipitate from sediment pore waters in ancient river valleys, forming uranium deposits [3]. Alternatively, fine sandstones can efficiently adsorb U^{6+} from lake aqueous media to form uranium deposits in environments with strong oxidizing conditions [3]. Low-level animals and plants that are rich in organic matter and propagate in shallow coastal seas can also become uranium deposit hosts and carriers, leading to uranium enrichment [134]. Therefore, these specific mineralization factors must also be considered when analyzing the potential for sandstone-type uranium deposition within hydrocarbon-bearing basins.

4. Conclusions

This study presents a comprehensive analysis and summary of the metallogenic mechanisms and influencing factors associated with sandstone-type uranium deposits within China's hydrocarbon-bearing basins based on their distinct features.

Sandstone-type uranium deposits, which make up the largest proportion of confirmed uranium reserves in China, are mainly distributed in the hydrocarbon-bearing basins in the northern part of the country. The metallogenic mechanisms underlying sandstone-type uranium deposits within these hydrocarbon-bearing basins are mainly influenced by fluid activity, geological structure, and sedimentary environment.

Fluids are of significance during sandstone-type uranium deposition within hydrocarbon-bearing basins. Along with the fundamental mechanism of uranium mineralization through oxidation and reduction by surface water and groundwater, hydrocarbon-bearing basins provide favorable conditions for uranium reduction sedimentation through the availability of organic matter, hydrocarbon gases, and hydrogen sulfide gases. Furthermore, the hydrothermal fluids in these basins can facilitate uranium enrichment and mineralization due to their ability to dissolve CO_3^{2-} and HCO_3^- ions, which are commonly sourced from dispersed oil and gas.

These hydrocarbon-bearing basins commonly have faults as the primary geological structures, and depressions and uplifts as the dominant secondary structures. These geological features provide excellent environments for uranium to accumulate for enrichment. Because of their high porosity and permeability, the sandstones in these basins can facilitate the enrichment and migration of uranium-bearing fluids. The mudstones in these basins promote the sedimentation of uranium ions in fluids through the adsorption capacity of their clay minerals. The formation of these basins involves marine or continental sedimentary processes with the involvement of hydrodynamics, providing sufficient materials, spaces, and favorable reduction and adsorption conditions to enable the enrichment and generation of hydrocarbons and uranium minerals.

The outcomes hold substantial significance in accurately evaluating potential uranium resources, as well as in identifying the distribution patterns and controlling factors of uranium mineral resources. These findings can be employed to guide the exploration and development of sandstone-type uranium resources, thereby enhancing the efficiency of uranium resource utilization.

Author Contributions: All the authors (G.L., J.Y., Y.S., J.T., H.H. and X.C.) have made substantial contributions to this work. G.L.: original draft writing; J.Y.: conceptualization, original draft writing, review and editing; Y.S., J.T., H.H. and X.C.: resources and visualization. All authors have read and agreed to the published version of the manuscript.

Funding: This research is supported by the Prospective Fundamental Technological Research Project of China National Petroleum Corporation, titled "Study on the mineralization mode and enrichment regularity of sandstone-type uranium deposits in hydrocarbon-bearing basins" (2021DJ5301).

Institutional Review Board Statement: Not applicable.

Informed Consent Statement: Not applicable.

Data Availability Statement: Not applicable.

Acknowledgments: The authors are grateful to the Research Institute of Exploration and Development, Liaohe Oilfield Company, CNPC, for the support.

Conflicts of Interest: The authors declare no conflict of interest.

References

1. Charalampides, G.; Vatalis, K.; Karayannis, V.; Baklavaridis, A. Environmental Defects And Economic Impact On Global Market Of Rare Earth Metals. *IOP Conf. Ser. Mater. Sci. Eng.* **2016**, *161*, 012069. [CrossRef]
2. Herring, J.S. Uranium and Thorium Resources. In *Nuclear Energy*; Springer: New York, NY, USA, 2013; pp. 463–490.
3. Leach, D.L.; Puchlik, K.P.; Glanzman, R.K. Geochemical Exploration for Uranium in Playas. *J. Geochem. Explor.* **1980**, *13*, 251–283. [CrossRef]
4. Zhang, F.; Jiao, Y.; Wu, L.; Rong, H. Relations between Pyrite Morphologies and Uranium Mineralization in the Shuanglong Region, Northern China. *Ore Geol. Rev.* **2022**, *141*, 104637. [CrossRef]
5. Zhang, F.; Wang, S.; Jiao, Y.; Wu, L.; Rong, H. Trapping of Uranium by Organic Matter within Sandstones during Mineralization Process: A Case Study from the Shuanglong Uranium Deposit, China. *Ore Geol. Rev.* **2021**, *138*, 104296. [CrossRef]
6. Peel, R. Nuclear Fuel Reserves. In *Kirk-Othmer Encyclopedia of Chemical Technology*; Wiley: New York, NY, USA, 2021; pp. 1–29.
7. NEA. *Uranium 2022: Resources, Production and Demand*; OECD Publishing: Pairs, France, 2023.
8. Hore-Lacy, I. Production of Byproduct Uranium and Uranium from Unconventional Resources. In *Uranium for Nuclear Power*; Elsevier: Amsterdam, The Netherlands, 2016; pp. 239–251.
9. Schaffer, M.B. Abundant Thorium as an Alternative Nuclear Fuel. *Energy Policy* **2013**, *60*, 4–12. [CrossRef]
10. Pearce, J.M. Limitations of Nuclear Power as a Sustainable Energy Source. *Sustainability* **2012**, *4*, 1173–1187. [CrossRef]
11. Warner, E.S.; Heath, G.A. Life Cycle Greenhouse Gas Emissions of Nuclear Electricity Generation. *J. Ind. Ecol.* **2012**, *16*, S73–S92. [CrossRef]
12. Sun, X.; Luo, H.; Dai, S. Ionic Liquids-Based Extraction: A Promising Strategy for the Advanced Nuclear Fuel Cycle. *Chem. Rev.* **2012**, *112*, 2100–2128. [CrossRef]
13. Brook, B.W.; Alonso, A.; Meneley, D.A.; Misak, J.; Blees, T.; van Erp, J.B. Why Nuclear Energy Is Sustainable and Has to Be Part of the Energy Mix. *Sustain. Mater. Technol.* **2014**, *1–2*, 8–16. [CrossRef]
14. Ryu, H.; Dorjragchaa, S.; Kim, Y.; Kim, K. Electricity-Generation Mix Considering Energy Security and Carbon Emission Mitigation: Case of Korea and Mongolia. *Energy* **2014**, *64*, 1071–1079. [CrossRef]
15. Nahar, S.N.; Pradhan, A.K.; Lim, S. K_α Transition Probabilities for Platinum and Uranium Ions for Possible X-ray Biomedical Applications. *Can. J. Phys.* **2011**, *89*, 483–494. [CrossRef]
16. Rump, A.; Eder, S.; Lamkowski, A.; Hermann, C.; Abend, M.; Port, M. A Quantitative Comparison of the Chemo- and Radiotoxicity of Uranium at Different Enrichment Grades. *Toxicol. Lett.* **2019**, *313*, 159–168. [CrossRef]
17. Malmon, A.G. High Resolution Isotope Tracing in Electron Microscopy Using Induced Nuclear Reactions. *J. Theor. Biol.* **1965**, *9*, 77–92. [CrossRef] [PubMed]
18. Brüske, A.; Martin, A.N.; Rammensee, P.; Eroglu, S.; Lazarov, M.; Albut, G.; Schuth, S.; Aulbach, S.; Schoenberg, R.; Beukes, N.; et al. The Onset of Oxidative Weathering Traced by Uranium Isotopes. *Precambrian Res.* **2020**, *338*, 105583. [CrossRef]
19. Von Hippel, D.F.; Hayes, P. Regional Cooperation for Nuclear Spent Fuel Management in East Asia: Costs, Benefits and Challenges—Part I. *J. Peace Nucl. Disarm.* **2018**, *1*, 305–343. [CrossRef]
20. Oxford Analytica. *Geopolitical Tensions Cloud Uranium Market*; Oxford Analytica: Oxford, UK, 2022.
21. Yao, J.; Han, H.; Yang, Y.; Song, Y.; Li, G. A Review of Recent Progress of Carbon Capture, Utilization, and Storage (CCUS) in China. *Appl. Sci.* **2023**, *13*, 1169. [CrossRef]
22. Liu, E.; Lu, X.; Wang, D. A Systematic Review of Carbon Capture, Utilization and Storage: Status, Progress and Challenges. *Energies* **2023**, *16*, 2865. [CrossRef]
23. Shang, D.; Geissler, B.; Mew, M.; Satalkina, L.; Zenk, L.; Tulsidas, H.; Barker, L.; El-Yahyaoui, A.; Hussein, A.; Taha, M.; et al. Unconventional Uranium in China's Phosphate Rock: Review and Outlook. *Renew. Sustain. Energy Rev.* **2021**, *140*, 110740. [CrossRef]
24. Guo, X.; Zhang, X.; Ren, D.; Lin, K. Research on Risk Management and Control Strategy of Uranium Resource Procurement in China. *Energy Sources Part A Recovery Util. Environ. Eff.* **2023**, *45*, 4178–4194. [CrossRef]
25. Jin, R.; Teng, X.; Li, X.; Si, Q.; Wang, W. Genesis of Sandstone-Type Uranium Deposits along the Northern Margin of the Ordos Basin, China. *Geosci. Front.* **2020**, *11*, 215–227. [CrossRef]
26. Peng, W.; Liu, Q.; Zhang, Y.; Jia, H.; Zhu, D.; Meng, Q.; Wu, X.; Deng, S.; Ma, Y. The First Extra-Large Helium-Rich Gas Field Identified in a Tight Sandstone of the Dongsheng Gas Field, Ordos Basin, China. *Sci. China Earth Sci.* **2022**, *65*, 874–881. [CrossRef]
27. Bai, H.; Wang, W.; Lu, Q.; Wang, W.; Feng, S.; Zhang, B. Geological Characteristics and Control Mechanism of Uranium Enrichment in Coal-Bearing Strata in the Yili Basin, Northwest China–Implications for Resource Development and Environmental Protection. *ACS Omega* **2022**, *7*, 5453–5470. [CrossRef] [PubMed]
28. Hu, S.; Hu, H.; Shi, E.; Tang, C.; Zhang, R.; Hao, Y. Seismic Interpretation of Sandstone-Type Uranium Deposits in the Songliao Basin, Northeast China. *Interpretation* **2022**, *10*, T665–T679. [CrossRef]
29. Zhu, Q.; Li, J.; Li, G.; Wen, S.; Yu, R.; Tang, C.; Feng, X.; Liu, X. Characteristics of Sandstone-Type Uranium Mineralization in the Hangjinqi Region of the Northeastern Ordos Basin: Clues from Clay Mineral Studies. *Ore Geol. Rev.* **2022**, *141*, 104642. [CrossRef]
30. Richards, M.C.; Issen, K.A.; Ingraham, M.D. A Coupled Elastic Constitutive Model for High Porosity Sandstone. *Int. J. Rock Mech. Min. Sci.* **2022**, *150*, 104989. [CrossRef]

31. Zhang, P.; Yu, C.; Zeng, X.; Tan, S.; Lu, C. Ore-Controlling Structures of Sandstone-Hosted Uranium Deposit in the Southwestern Ordos Basin: Revealed from Seismic and Gravity Data. *Ore Geol. Rev.* **2022**, *140*, 104590. [CrossRef]
32. Ren, Y.; Yang, X.; Hu, X.; Wei, J.; Tang, C. Mineralogical and Geochemical Evidence for Biogenic Uranium Mineralization in Northern Songliao Basin, NE China. *Ore Geol. Rev.* **2022**, *141*, 104556. [CrossRef]
33. Mukherjee, S.; Goswami, S.; Zakaulla, S. Geological Relationship between Hydrocarbon and Uranium: Review on Two Different Sources of Energy and the Indian Scenario. *Geoenergy Sci. Eng.* **2023**, *221*, 111255. [CrossRef]
34. Zhang, C.; Cai, Y.-Q.; Dong, Q.; Xu, H. Cretaceous–Neogene Basin Control on the Formation of Uranium Deposits in South China: Evidence from Geology, Mineralization Ages, and H–O Isotopes. *Int. Geol. Rev.* **2020**, *62*, 263–310. [CrossRef]
35. Cai, Y.; Zhang, J.; Li, Z.; Guo, Q.; Song, J.; Fan, H.; Liu, W.; Qi, F. Outline of Uranium Resources Characteristics and Metallogenetic Regularity in China. *Acta Geol. Sin.* **2015**, *89*, 1051–1069.
36. Hou, B.; Keeling, J.; Li, Z. Paleovalley-Related Uranium Deposits in Australia and China: A Review of Geological and Exploration Models and Methods. *Ore Geol. Rev.* **2017**, *88*, 201–234. [CrossRef]
37. Cheng, Y.; Wang, S.; Zhang, T.; Teng, X.; Ao, C.; Jin, R.; Li, H. Regional Sandstone-Type Uranium Mineralization Rooted in Oligo–Miocene Tectonic Inversion in the Songliao Basin, NE China. *Gondwana Res.* **2020**, *88*, 88–105. [CrossRef]
38. Han, X.; Wu, Z.; Ji, H.; Jiang, Z.; Guo, Y.; Lin, Z.; Hu, H.; Yin, D. Constraints of Tectonic Uplift and Denudation on Sandstone-Type Uranium Mineralization in Meso-Cenozoic Basins in Northern China: A Review. *Ore Geol. Rev.* **2021**, *139*, 104528. [CrossRef]
39. Ren, Y.; Yang, X.; Miao, P.; Hu, X.; Chen, Y.; Chen, L.; Zhao, H. Mineralogical and Geochemical Research on Pengyang Deposit: A Peculiar Eolian Sandstone-Hosted Uranium Deposit in the Southwest of Ordos Basin. *Ore Geol. Rev.* **2022**, *141*, 104571. [CrossRef]
40. Xu, D.; Chi, G.; Nie, F.; Fayek, M.; Hu, R. Diversity of Uranium Deposits in China—An Introduction to the Special Issue. *Ore Geol. Rev.* **2021**, *129*, 103944. [CrossRef]
41. Luo, J.L.; Morad, S.; Salem, A.; Ketzer, J.M.; Lei, X.L.; Guo, D.Y.; Hlal, O. Impact of Diagenesis on Reservoir-Quality Evolution in Fluvial and Lacustrine-Deltaic Sandstones: Evidence from Jurassic and Triassic Sandstones from the Ordos Basin, China. *J. Pet. Geol.* **2009**, *32*, 79–102. [CrossRef]
42. Jiao, Y.; Wu, L.; Rong, H.; Peng, Y.; Miao, A.; Wang, X. The Relationship between Jurassic Coal Measures and Sandstone-Type Uranium Deposits in the Northeastern Ordos Basin, China. *Acta Geol. Sin.-Engl. Ed.* **2016**, *90*, 2117–2132. [CrossRef]
43. Jin, R.; Yu, R.; Miao, P. *Basin Uranium Mineralization Law*; Springer: Singapore, 2023; pp. 325–356.
44. Liu, Z.Y.; Peng, S.P.; Qin, M.K.; Liu, H.X.; Geng, Y.Y.; Zhang, X.; Ding, B.; Xiu, X.Q. Origin and Role of Kaolinization in Roll-Front Uranium Deposits and Its Response to Ore-Forming Fluids in the Yili Basin, China. *Geofluids* **2018**, *2018*, 7847419. [CrossRef]
45. Chen, Y.; Li, J.; Wang, X.; Wang, Z.; Wei, Y.; Ren, J. Occurrence Characteristics and Influencing Factors of Uranium and Radon in Deep-Buried Thermal Storage Aquifers. *J. Radioanal. Nucl. Chem.* **2022**, *331*, 755–767. [CrossRef]
46. Qian, C.; Yang, S.; Wang, Y.; Wu, C.; Zhang, Y. Prediction and Modeling of Petrophysical Parameters of Deep-Buried, Low Permeability Glutenite Reservoirs in Yubei Area, Turpan-Hami Basin, China. *J. Pet. Sci. Eng.* **2021**, *207*, 109154. [CrossRef]
47. Liu, B.; Shi, Z.; Peng, Y.; Zhang, P.; Li, P. Sequence Stratigraphy of the Lower Cretaceous Uraniferous Measures and Mineralization of the Sandstone-hosted Tamusu Large Uranium Deposit, North China. *Acta Geol. Sin.-Engl. Ed.* **2022**, *96*, 167–192. [CrossRef]
48. Zuo, Y.; Wang, C.; Tang, S.; Hao, Q. Mesozoic and Cenozoic Thermal History and Source Rock Thermal Evolution of the Baiyinchagan Sag, Erlian Basin, Northern China. *J. Pet. Sci. Eng.* **2016**, *139*, 171–184. [CrossRef]
49. Wu, Q.; Wang, Y.; Li, Z.; Qiao, B.; Yu, X.; Huang, W.; Cao, C.; Li, Z.; Pan, Z.; Huang, Y. 2D and 3D Seismic Survey for Sandstone-Type Uranium Deposit and Its Prediction Patterns, Erlian Basin, China. *Minerals* **2022**, *12*, 559. [CrossRef]
50. Xi, K.; Cao, Y.; Jahren, J.; Zhu, R.; Bjørlykke, K.; Haile, B.G.; Zheng, L.; Hellevang, H. Diagenesis and Reservoir Quality of the Lower Cretaceous Quantou Formation Tight Sandstones in the Southern Songliao Basin, China. *Sediment. Geol.* **2015**, *330*, 90–107. [CrossRef]
51. Yao, J.; Li, G.; Wu, J. Application of In-Situ Combustion for Heavy Oil Production in China: A Review. *J. Oil Gas Petrochem. Sci.* **2018**, *1*, 69–72. [CrossRef]
52. Yao, J.; Song, Y. Dynamic Analysis Approach to Evaluate In-Situ Combustion Performance for Heavy Oil Production. *J. Oil Gas Petrochem. Sci.* **2019**, *2*, 42–47. [CrossRef]
53. Cao, B.; Luo, X.; Zhang, L.; Sui, F.; Lin, H.; Lei, Y. Diagenetic Evolution of Deep Sandstones and Multiple-Stage Oil Entrapment: A Case Study from the Lower Jurassic Sangonghe Formation in the Fukang Sag, Central Junggar Basin (NW China). *J. Pet. Sci. Eng.* **2017**, *152*, 136–155. [CrossRef]
54. Clayton, J.L.; Yang, J.; King, J.D.; Lillis, P.G.; Warden, A. Geochemistry of Oils from the Junggar Basin, Northwest China. *Am. Assoc. Pet. Geol. Bull.* **1997**, *81*, 1926–1944. [CrossRef]
55. Huang, S.; Qin, M.; Liu, Z.; He, Z.; Geng, Y. Litho-mineralogy, Geochemistry, and Chronology for the Genesis of the Kamust Sandstone-hosted Uranium Deposit, Junggar Basin, NW China. *Geol. J.* **2022**, *57*, 1530–1551. [CrossRef]
56. Liu, S.; Lei, X.; Feng, C.; Hao, C. Estimation of Subsurface Formation Temperature in the Tarim Basin, Northwest China: Implications for Hydrocarbon Generation and Preservation. *Int. J. Earth Sci.* **2016**, *105*, 1329–1351. [CrossRef]
57. Cai, C.; Qiu, N.; Soares, C.J.; Chen, H. Study on the Detrital Zircon Fission-track Ages from Natural Borehole Samples in the Bohai Bay and Tarim Basins with Different Thermal Backgrounds. *Geol. J.* **2021**, *56*, 4189–4200. [CrossRef]

58. Zhu, G.; Li, J.; Zhang, Z.; Wang, M.; Xue, N.; He, T.; Zhao, K. Stability and Cracking Threshold Depth of Crude Oil in 8000 m Ultra-Deep Reservoir in the Tarim Basin. *Fuel* **2020**, *282*, 118777. [CrossRef]
59. Liu, C.; Huang, L.; Zhao, H.; Wang, J.; Zhang, L.; Deng, Y.; Zhao, J.; Zhang, D.; Fan, C. Small-Scale Petroliferous Basins in China: Characteristics and Hydrocarbon Occurrence. *Am. Assoc. Pet. Geol. Bull.* **2019**, *103*, 2139–2175. [CrossRef]
60. Huang, C.; Yuan, X.; Song, C.; Yuan, J.; Ni, X.; Ma, X.; Zhang, S. Characteristics, Origin, and Role of Salt Minerals in the Process of Hydrocarbon Accumulation in the Saline Lacustrine Basin of the Yingxi Area, Qaidam, China. *Carbonates Evaporites* **2018**, *33*, 431–446. [CrossRef]
61. Ma, X.; Huang, C.; Shi, Y. Oil and Gas Enrichment Patterns and Major Controlling Factors for Stable and High Production of Tight Lacustrine Carbonate Rocks, a Case Study of Yingxi Area in Qaidam Basin, West China. *Carbonates Evaporites* **2019**, *34*, 1815–1831. [CrossRef]
62. He, Z.; Feng, J.; Luo, J.; Zeng, Y. Distribution, Exploitation, and Utilization of Intermediate-to-Deep Geothermal Resources in Eastern China. *Energy Geosci.* **2023**, *4*, 100187. [CrossRef]
63. Zhou, Q.; Xiao, X.; Tian, H.; Pan, L. Modeling Free Gas Content of the Lower Paleozoic Shales in the Weiyuan Area of the Sichuan Basin, China. *Mar. Pet. Geol.* **2014**, *56*, 87–96. [CrossRef]
64. Chen, S.; Zhu, Y.; Qin, Y.; Wang, H.; Liu, H.; Fang, J. Reservoir Evaluation of the Lower Silurian Longmaxi Formation Shale Gas in the Southern Sichuan Basin of China. *Mar. Pet. Geol.* **2014**, *57*, 619–630. [CrossRef]
65. Liu, C.; Qiu, X.; Wu, B.; Zhao, H. Subdivisions of the Central-East Asia Multi-Energy Minerals Metallogenetic Domain and Types of Those Basins. *Energy Explor. Exploit.* **2009**, *27*, 153–166. [CrossRef]
66. Tang, X.; Zhang, J.; Shan, Y.; Xiong, J. Upper Paleozoic Coal Measures and Unconventional Natural Gas Systems of the Ordos Basin, China. *Geosci. Front.* **2012**, *3*, 863–873. [CrossRef]
67. Wang, P.; Zhang, C.; Li, X.; Zhang, K.; Yuan, Y.; Zang, X.; Cui, W.; Liu, S.; Jiang, Z. Organic Matter Pores Structure and Evolution in Shales Based on the He Ion Microscopy (HIM): A Case Study from the Triassic Yanchang, Lower Silurian Longmaxi and Lower Cambrian Niutitang Shales in China. *J. Nat. Gas Sci. Eng.* **2020**, *84*, 103682. [CrossRef]
68. Jolivet, M.; Heilbronn, G.; Robin, C.; Barrier, L.; Bourquin, S.; Guo, Z.; Jia, Y.; Guerit, L.; Yang, W.; Fu, B. Reconstructing the Late Palaeozoic—Mesozoic Topographic Evolution of the Chinese Tian Shan: Available Data and Remaining Uncertainties. *Adv. Geosci.* **2013**, *37*, 7–18. [CrossRef]
69. Liu, D.; Jolivet, M.; Yang, W.; Zhang, Z.; Cheng, F.; Zhu, B.; Guo, Z. Latest Paleozoic–Early Mesozoic Basin–Range Interactions in South Tian Shan (Northwest China) and Their Tectonic Significance: Constraints from Detrital Zircon U–Pb Ages. *Tectonophysics* **2013**, *599*, 197–213. [CrossRef]
70. Chen, C.; Lu, H.; Jia, D.; Cai, D.; Wu, S. Closing History of the Southern Tianshan Oceanic Basin, Western China: An Oblique Collisional Orogeny. *Tectonophysics* **1999**, *302*, 23–40. [CrossRef]
71. Han, B.-F.; He, G.-Q.; Wang, X.-C.; Guo, Z.-J. Late Carboniferous Collision between the Tarim and Kazakhstan–Yili Terranes in the Western Segment of the South Tian Shan Orogen, Central Asia, and Implications for the Northern Xinjiang, Western China. *Earth Sci. Rev.* **2011**, *109*, 74–93. [CrossRef]
72. Novikov, I.S. Reconstructing the Stages of Orogeny around the Junggar Basin from the Lithostratigraphy of Late Paleozoic, Mesozoic, and Cenozoic Sediments. *Russ. Geol. Geophys.* **2013**, *54*, 138–152. [CrossRef]
73. Allen, M.B.; ªengör, A.M.C.; Natal'In, B.A. Junggar, Turfan and Alakol Basins as Late Permian to? Early Triassic Extensional Structures in a Sinistral Shear Zone in the Altaid Orogenic Collage, Central Asia. *J. Geol. Soc. Lond.* **1995**, *152*, 327–338. [CrossRef]
74. Xin, Z.; Fengjun, N.; Xuebin, S.; Fei, X.; Mangen, L.; Zhaobin, Y.; Chengyong, Z.; Zhibing, F. Relationships between Meso-Cenozoic Denudation in the Eastern Tian Shan and Uranium Mineralization in the Turpan-Hami Basin, NW China: Constraints from Apatite Fission Track Study. *Ore Geol. Rev.* **2020**, *127*, 103820. [CrossRef]
75. Zhao, X.; Jin, F.; Wang, Q.; Han, C.; Kang, R. Theory of Hydrocarbon Accumulation in Troughs within Continental Faulted Basins and Its Application: A Case Study in Jizhong Depression and Erlian Basin. *Acta Pet. Sin.* **2011**, *32*, 18–24.
76. Wang, Y.; Zhang, F.; Zhang, D.; Miao, L.; Li, T.; Xie, H.; Meng, Q.; Liu, D. Zircon SHRIMP U-Pb Dating of Meta-Diorite from the Basement of the Songliao Basin and Its Geological Significance. *Chin. Sci. Bull.* **2006**, *51*, 1877–1883. [CrossRef]
77. Lu, J.; Luo, Z.; Zou, H.; Li, Y.; Hu, Z.; Zhou, Z.; Zhu, J.; Han, M.; Zhao, L.; Lin, Z. Geochemical Characteristics, Origin, and Mechanism of Differential Accumulation of Natural Gas in the Carboniferous Kelameili Gas Field in Junggar Basin, China. *J. Pet. Sci. Eng.* **2021**, *203*, 108658. [CrossRef]
78. Chen, Y.; Miao, P.; Li, J.; Jin, R.; Zhao, H.; Chen, L.; Wang, C.; Yu, H.; Zhang, X. Association of Sandstone-Type Uranium Mineralization in the Northern China with Tectonic Movements and Hydrocarbons. *J. Earth Sci.* **2022**, *33*, 289–307. [CrossRef]
79. Allen, M.B.; Vincent, S.J.; Wheeler, P.J. Late Cenozoic Tectonics of the Kepingtage Thrust Zone: Interactions of the Tien Shan and Tarim Basin, Northwest China. *Tectonics* **1999**, *18*, 639–654. [CrossRef]
80. Sobel, E.R. Basin Analysis of the Jurassic–Lower Cretaceous Southwest Tarim Basin, Northwest China. *Geol. Soc. Am. Bull.* **1999**, *111*, 709–724. [CrossRef]
81. Shi, P.; Fu, B.; Ma, Y.; Guo, Q.; Xu, H. Remote Sensing Detection for Surface Anomalies Related to Hydracarbon in Bashibulake Uranium Ore, Sourthern Tianshan. In Proceedings of the 2016 IEEE International Geoscience and Remote Sensing Symposium (IGARSS), Beijing, China, 10–15 July 2016; IEEE: New York, NY, USA, 2016; pp. 6374–6377.

82. Ren, G.; Li, C.; Wu, C.; Zhang, H.; Wang, S.; Ren, Z.; Lei, Q.; Li, X. Late Quaternary Slip Rate and Kinematics of the Baoertu Fault, Constrained by 10Be Exposure Ages of Displaced Surfaces within Eastern Tian Shan. *Lithosphere* **2021**, *2021*, 7866920. [CrossRef]
83. Cheng, F.; Guo, Z.; Jenkins, H.S.; Fu, S.; Cheng, X. Initial Rupture and Displacement on the Altyn Tagh Fault, Northern Tibetan Plateau: Constraints Based on Residual Mesozoic to Cenozoic Strata in the Western Qaidam Basin. *Geosphere* **2015**, *11*, 921–942. [CrossRef]
84. Sun, P.; Guo, Z.; He, W.; Liu, W. Restoration of Eroded Thickness of the Neogene Strata in the Western Qaidam Basin and Its Significance for Oil and Gas Occurrence. *Acta Geol. Sin.-Engl. Ed.* **2017**, *91*, 1352–1362. [CrossRef]
85. Abudukeyumu, A.; Song, H.; Chi, G.; Li, Q.; Zhang, C. Quaternary Uranium Mineralization in the Qaidam Basin, Northern Tibetan Plateau: Insights from Petrographic and C-O Isotopic Evidences. *Ore Geol. Rev.* **2022**, *140*, 104628. [CrossRef]
86. Yan, Y.; Hu, X.; Lin, G.; Santosh, M.; Chan, L.-S. Sedimentary Provenance of the Hengyang and Mayang Basins, SE China, and Implications for the Mesozoic Topographic Change in South China Craton: Evidence from Detrital Zircon Geochronology. *J. Asian Earth Sci.* **2011**, *41*, 494–503. [CrossRef]
87. Liu, S.; Deng, B.; Zhong, Y. Unique Geologic Features of Burial and Superimposition of the Lower Paleozoic Shale Gas across the Sichuan Basin and Its Periphery. *Earth Sci. Front.* **2016**, *1*, 11–28.
88. Yang, M.; Li, L.; Zhou, J.; Qu, X.; Zhou, D. Segmentation and Inversion of the Hangjinqi Fault Zone, the Northern Ordos Basin (North China). *J. Asian Earth Sci.* **2013**, *70–71*, 64–78. [CrossRef]
89. Jin, R.; Yu, R.-A.; Yang, J.; Zhou, X.; Teng, X.; Wang, S.; Si, Q.; Zhu, Q.; Zhang, T. Paleo-Environmental Constraints on Uranium Mineralization in the Ordos Basin: Evidence from the Color Zoning of U-Bearing Rock Series. *Ore Geol. Rev.* **2019**, *104*, 175–189. [CrossRef]
90. Mei, Y.; Ren, S.; Zhang, P.; Ma, Y.; Ma, Y.; Fu, Y. Research on High-precision Gravity Scratch Analysis and Structure Features of Yili Basin, North-west China. *Geol. J.* **2019**, *54*, 1081–1089. [CrossRef]
91. Alexeiev, D.V.; Ryazantsev, A.V.; Kröner, A.; Tretyakov, A.A.; Xia, X.; Liu, D.Y. Geochemical Data and Zircon Ages for Rocks in a High-Pressure Belt of Chu-Yili Mountains, Southern Kazakhstan: Implications for the Earliest Stages of Accretion in Kazakhstan and the Tianshan. *J. Asian Earth Sci.* **2011**, *42*, 805–820. [CrossRef]
92. Jiang, S.; Li, S.; Somerville, I.D.; Lei, J.; Yang, H. Carboniferous–Permian Tectonic Evolution and Sedimentation of the Turpan-Hami Basin, NW China: Implications for the Closure of the Paleo-Asian Ocean. *J. Asian Earth Sci.* **2015**, *113*, 644–655. [CrossRef]
93. Zhu, W.; Ma, R.; Guo, J.; Sun, Y.; Guo, L.; Xu, M.; Hu, D. The Coupling of Sedimentary Characteristics and Tectonic Development of Turpan-Hami Basin and Adjacent Areas in Early Permian. *Geol. J. China Univ.* **2002**, *8*, 160–168.
94. Sun, T.; Wang, L.; Zhang, L.; Yang, H.; Zhang, Y. The Controlling Effect of Chair-Shaped Slope Break Landforms on Sedimentation in Continental Faulted Basins: A Case Study in the Xilinhaolai Area, Baiyinchagan Depression, Erlian Basin, China. *Interpretation* **2022**, *10*, T567–T580. [CrossRef]
95. Zou, C.; Zhao, W.; Jia, C.; Zhu, R.; Zhang, G.; Zhao, X.; Yuan, X. Formation and Distribution of Volcanic Hydrocarbon Reservoirs in Sedimentary Basins of China. *Pet. Explor. Dev.* **2008**, *35*, 257–271. [CrossRef]
96. Wang, P.-J.; Mattern, F.; Didenko, N.A.; Zhu, D.-F.; Singer, B.; Sun, X.-M. Tectonics and Cycle System of the Cretaceous Songliao Basin: An Inverted Active Continental Margin Basin. *Earth Sci. Rev.* **2016**, *159*, 82–102. [CrossRef]
97. Li, Y.; Lu, J.; Liu, X.; Wang, J.; Ma, W.; He, X.; Mou, F.; Li, X. Geochemistry and Origins of Natural Gas in the Hong-Che Fault Zone of the Junggar Basin, NW China. *J. Pet. Sci. Eng.* **2022**, *214*, 110501. [CrossRef]
98. Bian, W.; Hornung, J.; Liu, Z.; Wang, P.; Hinderer, M. Sedimentary and Palaeoenvironmental Evolution of the Junggar Basin, Xinjiang, Northwest China. *Paleobiodivers Paleoenviron* **2010**, *90*, 175–186. [CrossRef]
99. Teng, C.; Cai, Z.; Hao, F.; Cao, Z. Structural Geometry and Evolution of an Intracratonic Strike-Slip Fault Zone: A Case Study from the North SB5 Fault Zone in the Tarim Basin, China. *J. Struct. Geol.* **2020**, *140*, 104159. [CrossRef]
100. Shuichang, Z.; Baomin, Z.; Benliang, L.; Guangyou, Z.; Jin, S.; Xiaomei, W. History of Hydrocarbon Accumulations Spanning Important Tectonic Phases in Marine Sedimentary Basins of China: Taking the Tarim Basin as an Example. *Pet. Explor. Dev.* **2011**, *38*, 1–15. [CrossRef]
101. Fu, B.; Lin, A.; Kano, K.; Maruyama, T.; Guo, J. Quaternary Folding of the Eastern Tian Shan, Northwest China. *Tectonophysics* **2003**, *369*, 79–101. [CrossRef]
102. Yin, T.; Li, S. Application of Sulfur Isotopes for Analysing the Sedimentary Environment of Evaporite in Low-Altitude Intermountain Basins: A Case Study on the Kumishi Basin, Northwest China. *Carbonates Evaporites* **2022**, *37*, 11. [CrossRef]
103. He, W.; Barzgar, E.; Feng, W.; Huang, L. Reservoirs Patterns and Key Controlling Factors of the Lenghu Oil & Gas Field in the Qaidam Basin, Northwestern China. *J. Earth Sci.* **2021**, *32*, 1011–1021. [CrossRef]
104. Hu, S.; Cao, Y.; Huang, J.; Mou, Z. Discussion on Formation and Evolution of Jurassic Basin-Prototype for Qaidam Basin. *Exp. Pet. Geol.* **1999**, *21*, 189–195.
105. Chen, P. *Paleoenvironmental Changes during the Cretaceous in Eastern China*; Elsevier: Amsterdam, The Netherlands, 2000; Volume 17, pp. 81–90.

106. Sun, S.; Li, J.; Chen, H.; Peng, H.; Kenneth, J.H.; Shelton, J. Mesozoic and Cenozoic Sedimentary History of South China. *Am. Assoc. Pet. Geol. Bull.* **1989**, *73*, 1247–1269. [CrossRef]
107. Jiao, F.; Yang, Y.; Ran, Q.; Wu, G.; Liang, H. Distribution and Gas Exploration of the Strike–Slip Faults in the Central Sichuan Basin. *Nat. Gas Ind. B* **2022**, *9*, 63–72. [CrossRef]
108. Lu, G.; Chen, X.; Zou, H.; Preto, N.; Huang, X.; Wang, C.; Shi, Z.; Jin, X. Provenance of the First Terrigenous Sediments in the Western Sichuan Basin during the Late Triassic: Implications for Basin Evolution from Marine to Continental. *Mar. Pet. Geol.* **2023**, *147*, 105992. [CrossRef]
109. Jin, R.; Liu, H.; Li, X. Theoretical System of Sandstone-Type Uranium Deposits in Northern China. *J. Earth Sci.* **2022**, *33*, 257–277. [CrossRef]
110. Jiao, Y.; Wu, L.; Rong, H.; Zhang, F. Review of Basin Uranium Resources in China. *Earth Sci.-J. China Univ. Geosci.* **2021**, *46*, 2675. [CrossRef]
111. Min, M.-Z.; Luo, X.-Z.; Du, G.-S.; He, B.-A.; Campbell, A.R. Mineralogical and Geochemical Constraints on the Genesis of the Granite-Hosted Huangao Uranium Deposit, SE China. *Ore Geol. Rev.* **1999**, *14*, 105–127. [CrossRef]
112. Min, M.; Fang, C.; Fayek, M. Petrography and Genetic History of Coffinite and Uraninite from the Liueryiqi Granite-Hosted Uranium Deposit, SE China. *Ore Geol. Rev.* **2005**, *26*, 187–197. [CrossRef]
113. Hankins, D.E. Effect of Reactivity Addition Rate and of Weak Neutron Source on the Fission Yield of Uranium Solutions. *Nucl. Sci. Eng.* **1966**, *26*, 110–116. [CrossRef]
114. Tran, E.L.; Teutsch, N.; Klein-BenDavid, O.; Weisbrod, N. Uranium and Cesium Sorption to Bentonite Colloids under Carbonate-Rich Environments: Implications for Radionuclide Transport. *Sci. Total Environ.* **2018**, *643*, 260–269. [CrossRef]
115. Jaireth, S.; Mckay, A.; Lambert, I. Association of Large Sandstone Uranium Deposits with Hydrocarbons. *AusGeo News* **2009**, *89*, 1–6.
116. Spirakis, C.S. The Roles of Organic Matter in the Formation of Uranium Deposits in Sedimentary Rocks. *Ore Geol. Rev.* **1996**, *11*, 53–69. [CrossRef]
117. Liu, W.; Zhao, X.; Shi, Q.; Zhang, Z. Research on Relationship of Oil-Gas and Sandstone-Type Uranium Mineralization of Northern China. *Geol. China* **2017**, *44*, 279–287.
118. Cunningham, C.G.; Rasmussen, J.D.; Steven, T.A.; Rye, R.O.; Rowley, P.D.; Romberger, S.B.; Selverstone, J. Hydrothermal Uranium Deposits Containing Molybdenum and Fluorite in the Marysvale Volcanic Field, West-Central Utah. *Min. Depos.* **1998**, *33*, 477–494. [CrossRef]
119. Wilde, A.R.; Mernagh, T.P.; Bloom, M.S.; Hoffmann, C.F. Fluid Inclusion Evidence on the Origin of Some Australian Unconformity-Related Uranium Deposits. *Econ. Geol.* **1989**, *84*, 1627–1642. [CrossRef]
120. Ding, B.; Liu, H.; Zhang, C.; Liu, H.; Li, P.; Zhang, B. Mineralogy, Fluid Inclusion and H-O-C-S Stable Isotopes of Mengqiguer Uranium Deposit in the Southern Yili Basin, Xinjiang: Implication for Ore Formation. *Acta Geol. Sin.-Engl. Ed.* **2020**, *94*, 1488–1503. [CrossRef]
121. Heinrich, C.A. The Physical and Chemical Evolution of Low-Salinity Magmatic Fluids at the Porphyry to Epithermal Transition: A Thermodynamic Study. *Min. Depos* **2005**, *39*, 864–889. [CrossRef]
122. Szabó, Z.; Moll, H.; Grenthe, I. Structure and Dynamics in the Complex Ion $(UO_2)_2(CO_3)(OH)_3^-$. *J. Chem. Soc. Dalton Trans.* **2000**, 3158–3161. [CrossRef]
123. Bernhard, G.; Geipel, G.; Reich, T.; Brendler, V.; Amayri, S.; Nitsche, H. Uranyl(VI) Carbonate Complex Formation: Validation of the $Ca_2UO_2(CO_3)_3$(aq.) Species. *Radiochim. Acta* **2001**, *89*, 511–518. [CrossRef]
124. Yue, S.; Wang, G. Relationship between the Hydrogeochemical Environment and Sandstone-Type Uranium Mineralization in the Ili Basin, China. *Appl. Geochem.* **2011**, *26*, 133–139. [CrossRef]
125. Pei, S.; Zhao, J.; Sun, Y.; Xu, Z.; Wang, S.; Liu, H.; Rowe, C.A.; Toksöz, M.N.; Gao, X. Upper Mantle Seismic Velocities and Anisotropy in China Determined through Pn and Sn Tomography. *J. Geophys. Res.* **2007**, *112*, B05312. [CrossRef]
126. Wilson, M.R.; Åkerblom, G.V. Geological Setting and Geochemistry of Uranium-Rich Granites in the Proterozoic of Sweden. *Mineral. Mag.* **1982**, *46*, 233–245. [CrossRef]
127. Guthrie, V.A.; Kleeman, J.D. Changing Uranium Distributions during Weathering of Granite. *Chem. Geol.* **1986**, *54*, 113–126. [CrossRef]
128. Skierszkan, E.K.; Dockrey, J.W.; Mayer, K.U.; Beckie, R.D. Release of Geogenic Uranium and Arsenic Results in Water-Quality Impacts in a Subarctic Permafrost Region of Granitic and Metamorphic Geology. *J. Geochem. Explor.* **2020**, *217*, 106607. [CrossRef]
129. Parsons, I.; Lee, M.R.; Smith, J.V. Biochemical Evolution II: Origin of Life in Tubular Microstructures on Weathered Feldspar Surfaces. *Proc. Natl. Acad. Sci. USA* **1998**, *95*, 15173–15176. [CrossRef] [PubMed]
130. Zhou, Q.; Liu, S.; Xu, L.; Zhang, H.; Xiao, D.; Deng, J.; Pan, Z. Estimation of Radon Release Rate for an Underground Uranium Mine Ventilation Shaft in China and Radon Distribution Characteristics. *J. Environ. Radioact.* **2019**, *198*, 18–26. [CrossRef]
131. Ma, K.; Cui, L.; Dong, Y.; Wang, T.; Da, C.; Hirasaki, G.J.; Biswal, S.L. Adsorption of Cationic and Anionic Surfactants on Natural and Synthetic Carbonate Materials. *J. Colloid Interface Sci.* **2013**, *408*, 164–172. [CrossRef] [PubMed]
132. Guo, T. Key Controls on Accumulation and High Production of Large Non-Marine Gas Fields in Northern Sichuan Basin. *Pet. Explor. Dev.* **2013**, *40*, 150–160. [CrossRef]

133. Toens, P.D.; Andrews-Speed, C.P. The Time-Bound Character of Uranium Mineralising Processes, with Special Reference to the Proterozoic of Gondwana. *Precambrian Res.* **1984**, *25*, 13–36. [CrossRef]
134. Adams, S.; Smith, R. *Geology and Recognition Criteria for Sandstone Uranium Deposits in Mixed Fluvial-Shallow Marine Sedimentary Sequences, South Texas. Final Report*; OSTI.GOV: Oak Ridge, TN, USA, 1981.

eng MDPI

Article

The Development and Validation of Correlation Charts to Predict the Undisturbed Ground Temperature of Pakistan: A Step towards Potential Geothermal Energy Exploration

Tabish Ali [1,2,*], Waseem Haider [3], Muhammad Haziq [2,4], Muhammad Omar Khan [2,5] and Arif Hussain [2,3,*]

1 Department of Civil & Environmental Engineering, Hanyang University, Seoul 04763, Republic of Korea
2 School of Civil and Environmental Engineering, National University of Sciences and Technology, Islamabad 44000, Pakistan
3 Department of Electrical and Computer Engineering, Sungkyunkwan University, Suwon 16419, Republic of Korea; haider@skku.edu
4 ANS Associates Pvt. Limited, Lahore 54000, Pakistan
5 EMJMD, MSc Geospatial Technologies, Westfälische Wilhelms Universität, 48149 Münster, Germany
* Correspondence: taoab386@gmail.com (T.A.); engrarif14@skku.edu (A.H.)

Abstract: As a country, Pakistan is mostly dependent on fossil fuels for fulfilling its energy demand, which is expensive, as well as being environmentally unfriendly. It is high time that the country decides to shift from fossil fuels to renewable energy resources like geothermal, wind, solar, etc., to cater for global warming issues. Pakistan has a lot of potential geothermal sites, as the location of Pakistan lies on several fault lines and hot springs, thus making it very easy to extract the temperature from deep inside the earth and harness it for Geothermal Energy. Also, a sound knowledge of ground temperature is essential to use geothermal energy, which is obtained by drilling boreholes and putting in sensors. However it becomes a very expensive and labor intensive procedure. Therefore, to avoid the huge cost for drilling boreholes, particularly for ground temperature analysis, a numerical approach has been considered for determining ground temperature. Furthermore, correlation charts between air and ground temperatures have been developed, as there were no proper studies on the ground temperature of Pakistan. Then, with the help of a boreholes drilled in the National University of Sciences and Technology, Islamabad, Pakistan, the actual ground and numerically calculated temperatures have been compared. The results show a temperature error margin in the range between 0.27% for higher depths of about 5.6 m and 7.3% near the surface of about 2.7 m. Thus, it is shown that the proposed method is easy to implement and better than large scale testing methods for the depths at which geothermal energy is extracted.

Keywords: geothermal energy; ground temperature; correlation chart; soil types; temperature equation

Citation: Ali, T.; Haider, W.; Haziq, M.; Khan, M.O.; Hussain, A. The Development and Validation of Correlation Charts to Predict the Undisturbed Ground Temperature of Pakistan: A Step towards Potential Geothermal Energy Exploration. *Eng* **2023**, *4*, 1837–1850. https://doi.org/10.3390/eng4030104

Academic Editor: Reza Rezaee

Received: 20 April 2023
Revised: 27 June 2023
Accepted: 28 June 2023
Published: 30 June 2023

1. Introduction

Sustainability, considering the carbon neutral energy production, is the primary goal of the global energy market. Moreover, responsible consumption of natural resources and renewable energy are the dire need of this era, according to the United Nation's Sustainable development goals [1]. Pakistan is rich in renewable energy resources, and still, it is facing energy crises like the shortage of electricity and load shedding [2,3]. Also, to generate electricity it imports fuel, which is a huge hindrance to its economic development. The conventional energy sources like coal, gas, diesel, and fossil fuels are on the verge of ending. Pakistan needs to consider alternative renewable energy sources [4]. The conventional methods are hazardous to the environment due to the release of ozone depletion and greenhouse gases (GHG). These two phenomena are mostly responsible for global warming and climate change, thus causing extreme weather conditions, glacier melting and flooding [5,6]. On the other hand, renewable sources like wind, hydro and geothermal energy are clean energy technologies which produce less GHG and are environmentally friendly. Furthermore, these are a counter measure for global warming to fulfill energy demands [7–9]. Pakistan has a lot of geothermal potential due to its geology, which can be harvested for energy purposes. This research discusses the potential of geothermal energy in Pakistan, where all the energy resources are being used for energy generation except the geothermal energy to cater global warming. The statistics of the available energy sources previously studied illustrate that the energy sources are mainly composed of conventional sources and fossil fuels [10]. Thus, it is becoming very difficult to meet the energy requirements of the country, as the prices of fossil fuels are increasing day by day. The average annual energy demand of Pakistan is also increasing at the rate of 8% to 10% annually [11]. The board of investment Pakistan provides the figure of 22,797 MW as the installed power capacity. But the generation stands approximately between 12,000 MW and 13,000 MW per day [10]. The China–Pakistan Economic Corridor (CPEC) is opening ways for investors in Pakistan, increasing the energy demand abruptly. These commercial enterprise ventures can be powered by the energy generated by themselves using geo-thermal energy.

Pakistan is located within the latitude and longitude of 30.37° N, 69.34° E with a land of about 800,000 km^2. The northeast to southwest extent of the country is about 1700 km, and its East–West width is approximately 1000 km. The geomorphology of Pakistan varies from lofty mountains of the Himalayas, Karakorum, Hindukush, and Pamirs in the north to the fascinating coastline of the Arabian Sea in the south. In between the northern and southern extreme ends of the country, notable and unique bended north–south-oriented mountain ranges exist centrally bounded by the fertile plains of the 3000-long River Indus and western part of the famous Thar Desert on the eastern side, and by the Chagai volcanic arc, vast tectonic depression of Kharan, and the westward swinging mountain ranges of the Makran flysch basin. Pakistan is a huge museum of geological formations with a variety of rocks and formations. Pakistan's geothermal energy resources are located in the following four regions: northern Pakistan, consisting of Gilgit Baltistan; northwestern and South Baluchistan; Indus Basin and West Sindh; and southwestern and northern Punjab [12]. The sources, location and outlet reservoir temperature of some of the geothermal springs in Pakistan are Murtazaabad Springs in Hunza (237 °C), Budelas Valey Springs in Hunza (159 °C), Dasu Springs in Kohistan (200 °C), Tatta Pani in Astore (200 °C), Sassi in Skardu (200 °C), Karsaz Spring in Karachi (70–145 °C), Manghopir Spring in Karachi (70–145 °C), Kharan and Kohe Sultan Spring in Kharan and Chaghi (200–300 °C) [12,13]. Besides the regions mentioned, there are other spots as well, where single or multiple geothermal springs exist, and a geological survey of Pakistan has issued a list of 12 regions to be investigated [14]. These regions are blessed with un-tapped geothermal energy resources. The only need is to exploit and utilize them efficiently and responsibly [15]. But the concept of geothermal energy is rare in Pakistan. Furthermore, to determine the geothermal potential, ground temperature data is required, but the ground temperature of Pakistan is not studied in detail. Thus, a numerical approach is needed to find out the temperature at higher depths, in order to avoid the cost of drilling boreholes and using heavy instrumentations to measure the ground temperature. By now, it is evident that Pakistan has enough resources to generate and use the geothermal energy, but many geothermal potential areas are still unexplored and those which are known are not being used to fulfill the energy needs. Thus, there is a dire need to study the geothermal potential and ground temperature of Pakistan in detail.

The literature shows that ground temperature is influenced significantly by the air temperature at shallower depths. As the depth increases, a lag between air and ground temperature is observed. However, this lag becomes constant after a particular depth. At this depth, the temperature remains almost the same for the whole year. Thus, the difference between the air and ground temperature is utilized for heating purposes during the winter season and for cooling during the summer season. The greater the difference between air and ground temperature, the greater the efficiency of the geothermal system installed [16]. It is evident that the heat energy of the earth, stored in the ground, is a renewable source and can be extracted easily through environmentally friendly techniques [17]. The subsurface ground temperature profile is a prominent determinant of various thermo-active and latent heat applications. These include geothermal energy extraction and storage through contact mechanisms with structural elements, heating of airport runways, and the closed systems, using tunnel lining and support as energy absorbers [18,19]. Geothermal energy is now widely regarded as a global green energy source [20]. Moreover, the Ground Temperature (GT) profile is also being utilized to effectively predict the biological, zoological and pathological life cycle processes that occur underground. These complex activities have their major roots in the earth, acting as a fluctuating heat source with variable vertical temperature distribution and thermal diffusivity, at various times of the year. Geothermal resources are, at present, mainly used for generating electricity and air conditioning by means of District Heating (DH) and geothermal heat pump systems [21]. These resources are normally classified as renewable energy sources, because they are maintained by a continuous energy current [22]. Geothermal technologies use renewable energy resources to generate electricity or heating and cooling while producing very low levels of greenhouse-gas emissions [23–25]. Researchers have also explored novel and innovative approaches to invest in geothermal energy. Furthermore, the same study states that intelligently schemes for the modeling of interfacial tension in CO_2 brine systems has implications for CO_2 trapping. This research emphasizes the importance of understanding CO_2 brine system dynamics to develop effective methods for carbon dioxide storage. The global interest in studying geothermal energy is evident, as it can contribute to broader carbon capture and storage initiatives [26]. Thus, as a solution of global warming, this research focuses on the potential of geothermal energy in Pakistan by mapping its ground temperature at different depths with the help of newly generated temperature correlation charts between air and ground that have never performed before for this region. Section 2 discusses the numerical and empirical approach, while Section 3 analyzes the temperature chart of respective regions and concludes with discussion at end.

2. Methodology

2.1. Determination of Ground Temperature by Numerical Approach

Kusuda formulated a correlation, in which it was described that ground temperature is a function of the depth below the surface and time of year [16,27]. The thermo-physical properties of soils and rocks make them an excellent propagator of latent heat in the ground. The equation developed by Kusuda is as follows:

$$T_{soil}(D, t_{year}) = T_{mean} - T_{amp} \times exp\left(-D\sqrt{\frac{\pi}{365 \times \alpha}}\right) \times cos\left[\frac{2\pi}{365}\left(t_{year} - t_{shift} - \frac{D}{2}\sqrt{\frac{365}{\pi \times \alpha}}\right)\right] \quad (1)$$

where:

T_{soil} (D, t_{year}) = Ground Temperature
T_{mean} = average surface, i.e., air temperature
T_{amp} = Amplitude of surface temperature [(max. air temperature − min. air temperature)/2]
D = Depth below the surface (surface = 0)
α = Thermal diffusivity of the soil
t_{year} = current time (day)
t_{shift} = day of the year of minimum surface temperature

This correlation has been widely used to effectively predict the ground temperature profile, over the term of many decades. The correlation assumes that thermal diffusivity varies with different depths, irrespective of the time of year. The time period can be as short as one day (86,400 s) to as long as a year (3.1536×10^7 s). Applying this equation for Islamabad Pakistan, the profile of ground temperature Figure 1 is obtained. The ground temperatures were calculated by considering the soil thermal diffusivity equivalent to clay (0.0267 m^2/day) because the predominant soil in the Islamabad region is made up of clay, which is also confirmed by the bore logs. The air temperature data was obtained from the Pakistan Meteorological Department (PMD).

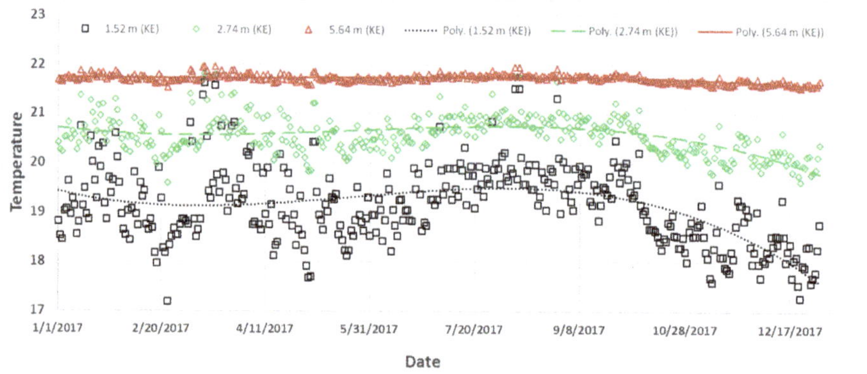

Figure 1. Ground temperature at different depths using Kusuda Equation.

2.2. Determination of Ground Temperature by Experimental Approach

Previous studies show that the Kusuda formula might not provide accurate results everywhere because of different spatial temperature variation and the equatorial location [28]. Thus, it is necessary to verify and validate this equation for Pakistan before its application. The validation of this equation was carried out by comparing its results obtained with the borehole drilled inside the university, especially for ground temperature measurement purposes. Figure 2 shows the actual ground temperatures. The comparison between numerical results and experimental results shows an error margin in the range between 0.27% for higher depths to 7.3% near the surface. Figure 3 shows that the approach works perfectly for depths greater than about 2 m and the error margin decreases as the depth increases. At lower depths, the results from field and equation vary significantly because shallow depths are affected greatly by surface weather conditions [16].

Figure 2. Ground Temperature at different depths from thermometers in the boreholes.

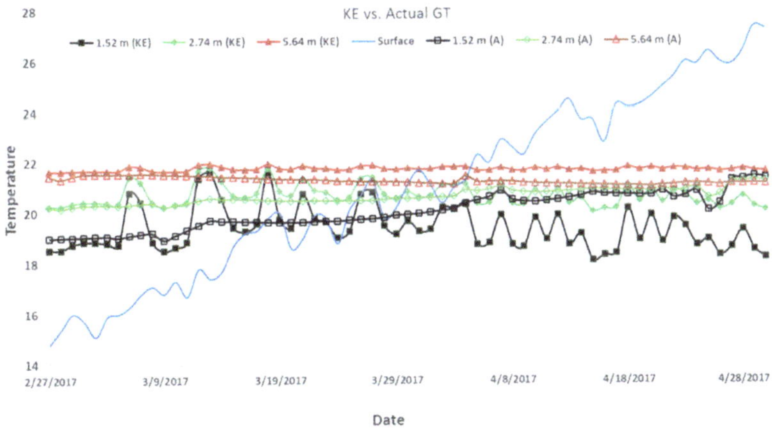

Figure 3. Comparison of Numerical approach (Kusuda Equation) and Experimental Results.

3. Results

Temperature Correlation

Air vs. Ground Temperature (A/G) Charts:

The Kusuda formula has been widely used to effectively predict the ground temperature profile, over the term of many decades. However, a very important contemplation of Kusuda's equation is the hypothesis of defining the variation of the ratio of air and ground temperatures with depth, as shown in Figure 4a–d, which is the main contribution of the authors. After the formulation of the ground temperature graphs, as shown in Figure 5, A/G (Ratio between air and ground temperature) charts were developed to make this numerical approach practical. Thus, four charts of both fine (clay) and coarse (sand) classifications of soil and rock types (gravel and shale) have been shown in Figure 4. In order to make these correlation charts more diversified to account for different geographical locations of Pakistan, further major soil and rock classification were considered. This allowed us to make the zoning of the country into different ground temperature potential areas. Other charts of rock and soil with different moisture content, etc., are shown in Figure 6. The A/G graphs have been plotted by taking into account the average monthly air and ground temperatures on the y-axis and uniformly increasing ground depth on the x-axis. Time interval taken is one complete month, over the course of a complete year. Different types of soils and rocks have been investigated in the postulation of this hypothesis; these include: sand (unsaturated and saturated), clay (unsaturated and saturated), silt, peat, gravel, sandy gravel, loam, shale, granite, dolomite, quartz, sandstone and limestone, which correspond to meteorologically recorded average monthly temperatures of eleven separate districts, from all across Pakistan. These areas include Rawalpindi, Bahawalpur, Lahore, Multan, Faisalabad, Quetta, Peshawar, Skardu, Jhelum, Zhob and Karachi. It can be seen that the slope of monthly air/ground temperatures vs. depth is steeper for the first 2 m below the ground. However, it is observed that the A/G ratio becomes relatively constant, exceeding the depths of 6 m. It is due to the high thermal inertia of the soil, the fluctuations in ground temperature are diminished with higher depths and the ground temperature remains relatively constant [29]. The increase in fluctuation of A/G (Air temperature/Ground temperature) values shows that there is some dominant contribution of surface temperatures at shallow depths. At higher depths, the ground temperature becomes constant, thus making minimizing the effect of aerial temperatures and making the A/G value consistent. This trend is also confirming the error margins encountered during the verification of Kusuda's Equation (percentage of error decreases with increase in depth underground). The A/G graphs also indicate some abnormal behavior of rock lithology. This is due to the reason that rocks of different categories have higher thermal diffusivity values than different soil types. The ability of rocks to conduct more heat energy as compared to fine soils has made us believe that they show major fluctuations in their A/G values at shallow depths because of larger influence of surface temperatures, thus contributing in errors to some extent.

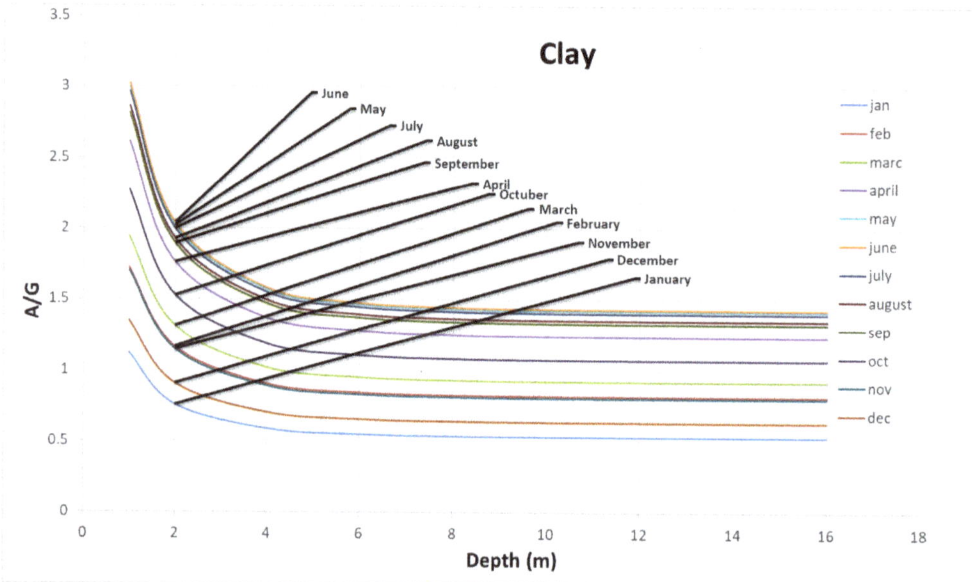

(**a**) A/G chart for clay

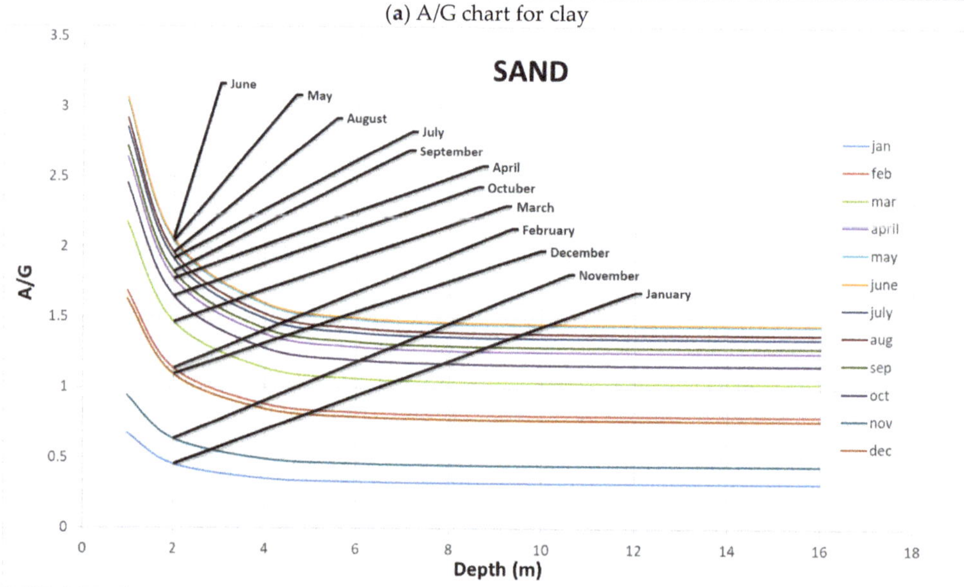

(**b**) A/G chart for sand

Figure 4. *Cont.*

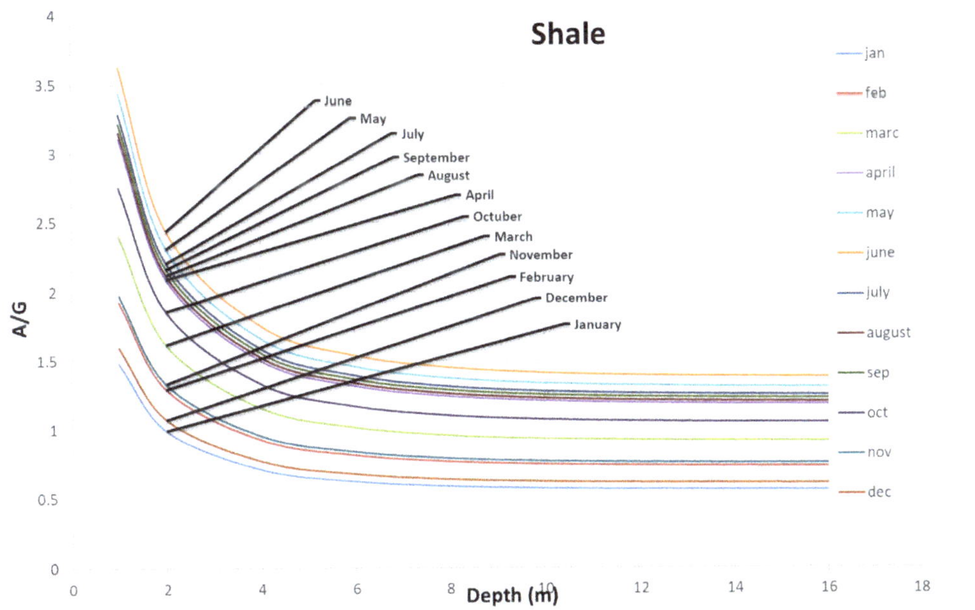

(c) A/G chart for shale

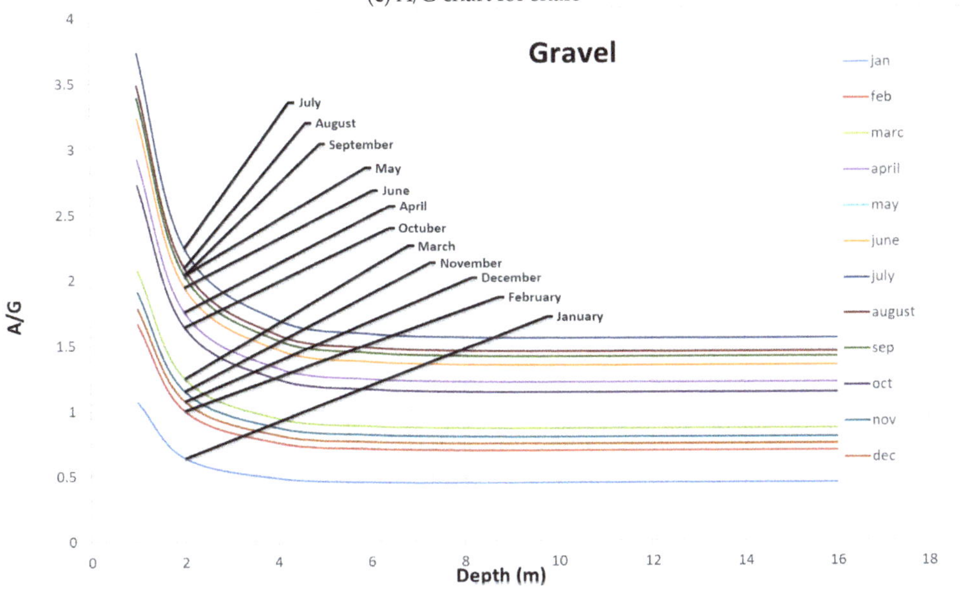

(d) A/G chart for gravel

Figure 4. Ground temperature correlation charts. (**a**) Clay (**b**) Sand (**c**) Shale (**d**) Gravel.

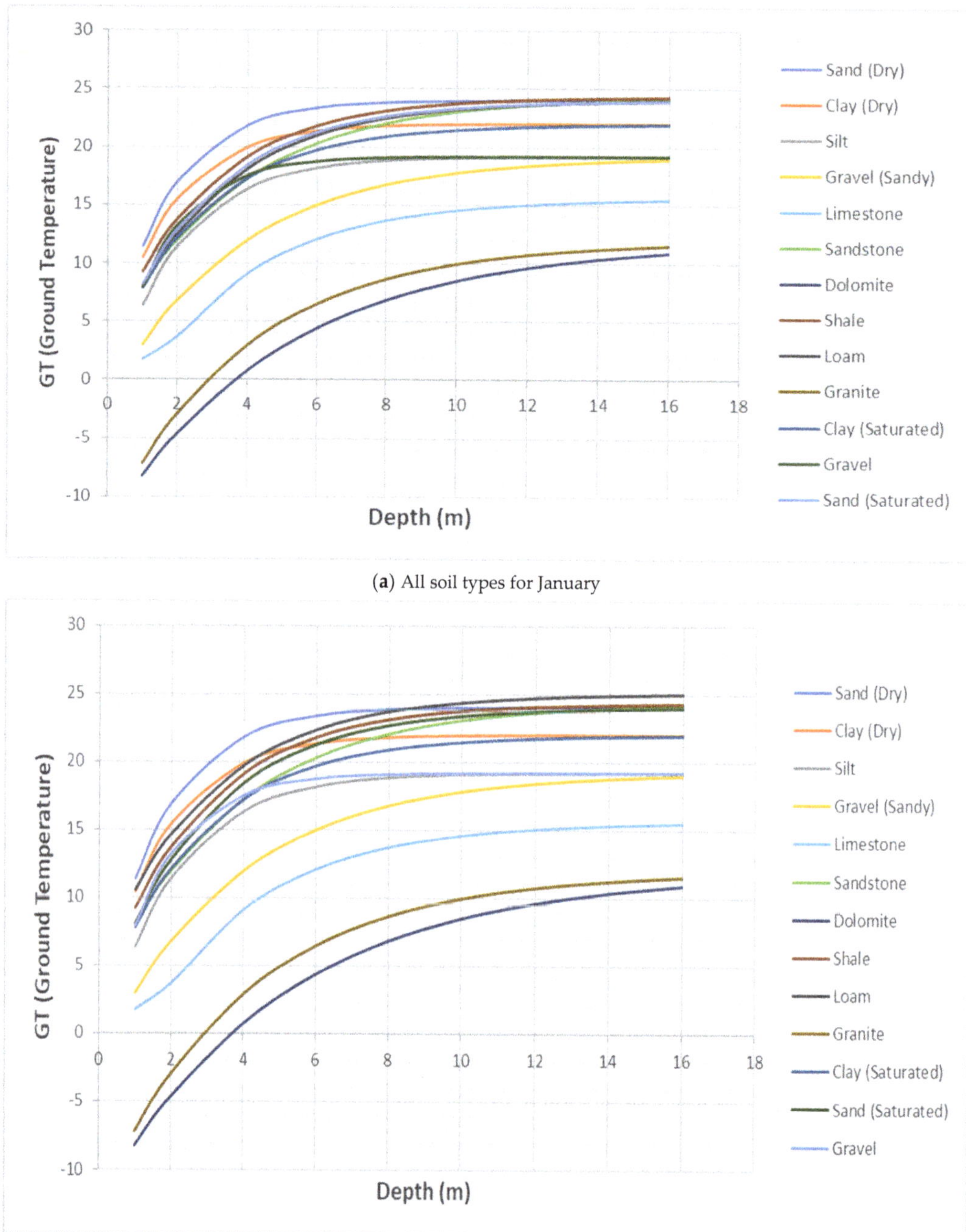

(**a**) All soil types for January

(**b**) All soil types for June

Figure 5. Initial Temperature Graphs obtained for different soil types for (**a**) January and (**b**) June.

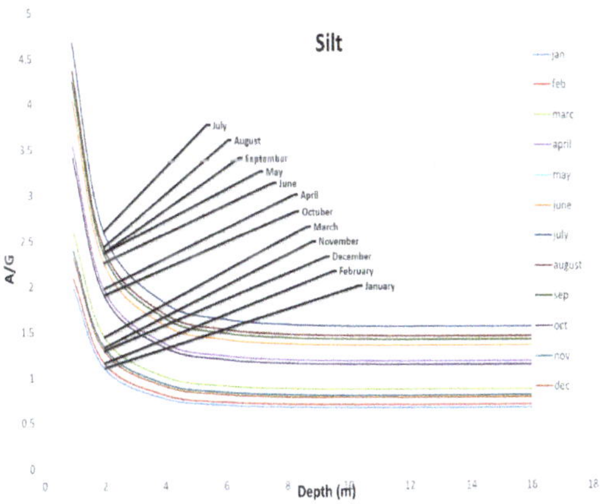

(**a**) A/G chart for silt

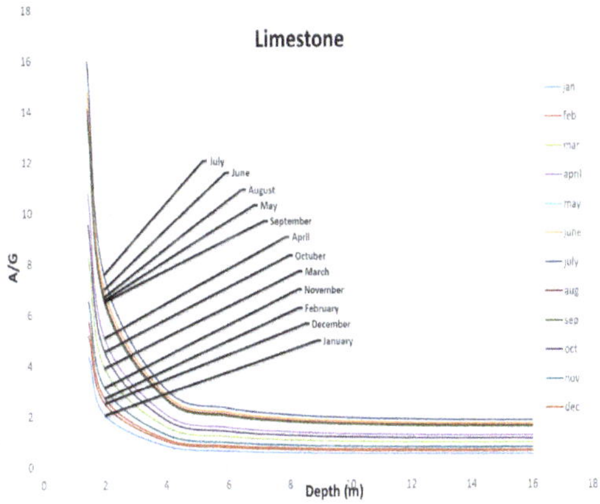

(**b**) A/G chart for limestone

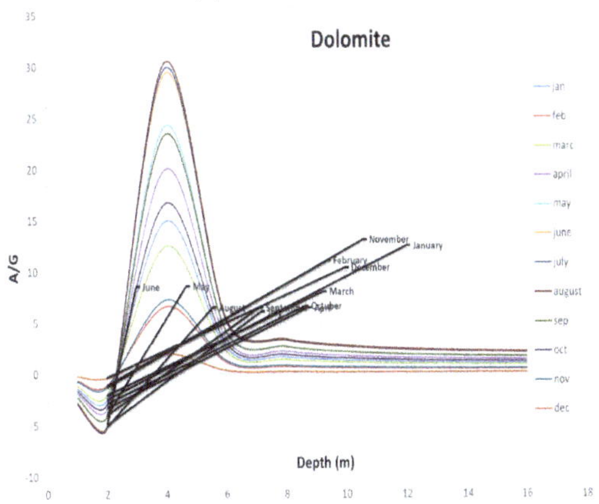

(**c**) A/G chart for dolomite

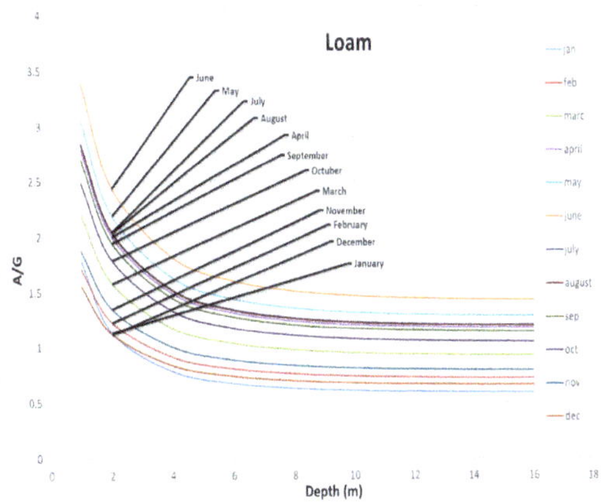

(**d**) A/G chart for loam

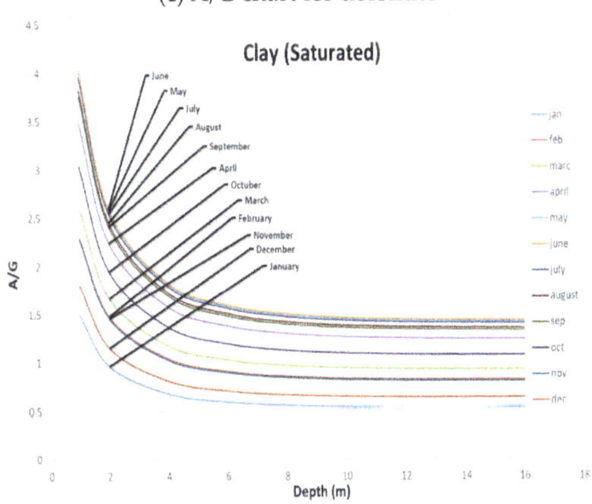

(**e**) A/G chart for saturated clay

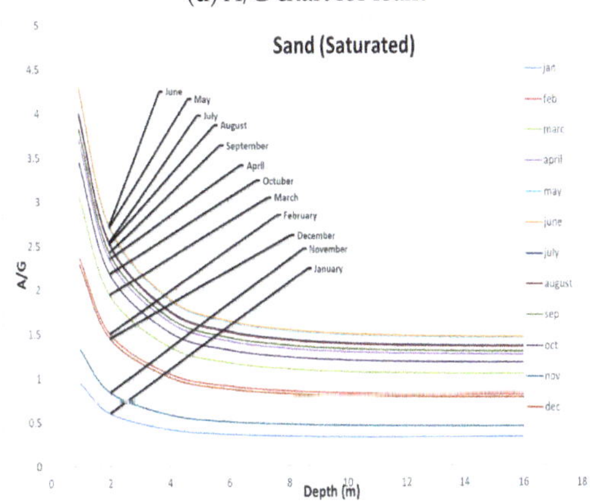

(**f**) A/G chart for satured sand

Figure 6. *Cont.*

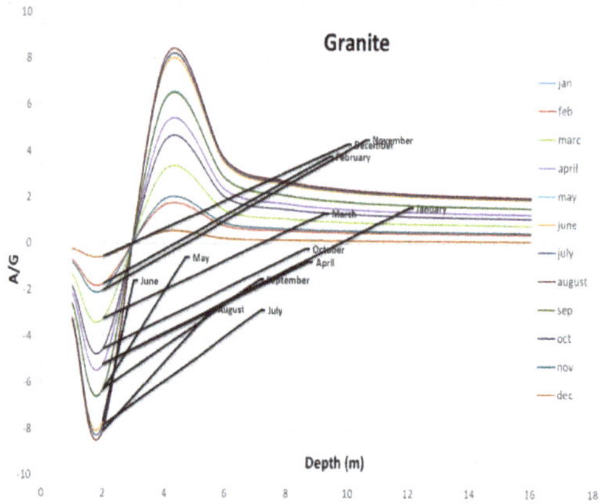

(**g**) A/G chart for granite

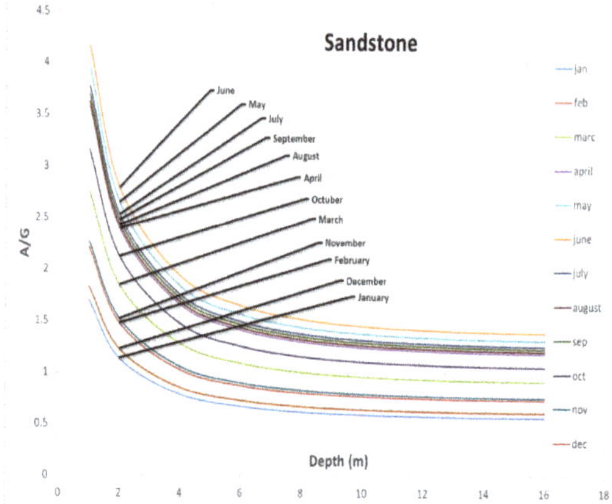

(**h**) A/G chart for sandstone

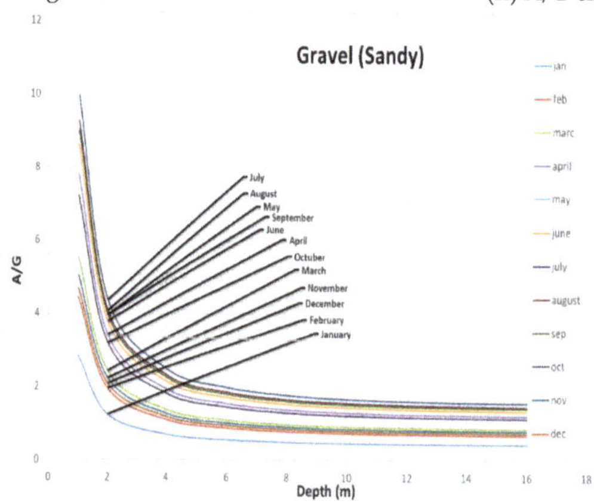

(**i**) A/G chart for sandy gravel

Figure 6. A/G charts for various types of soils, rocks and moisture conditions.

Temperature Mapping of Pakistan:

Resulting soil and rock temperature envelopes, formulated from the annual A/G ratio vs. Depth graphs, have been used to develop a comparison between each soil envelope. Figure 7 shows the resulting annual graphs of each soil and rock, modelled over a period of one year. It can be perceived from the resultant graphs that the A/G vs. depth is highest for clay (dry), sand (dry), loam, shale and sandstone for the month of June; silt, gravel, sandy gravel and limestone for the month of July; and dolomite and granite for the month of August. However, the lower most temperature fluctuates between the winter months of January and December for all of the different soil and rock types under investigation. It can be inferred from the hypothesis that the ambient temperature of the month under study and resulting ground temperature greatly influences soil and rock temperature envelopes, when thermal diffusivities of the said geological materials are kept constant. Conversely, the soil can have a better heat-exchanging capacity through engineered soil mixing/replacement techniques, in order to obtain the desired geothermal energy storage/transfer potential of soil. This will greatly influence in achieving the energy capacity for any kind of energy requirements of infrastructures.

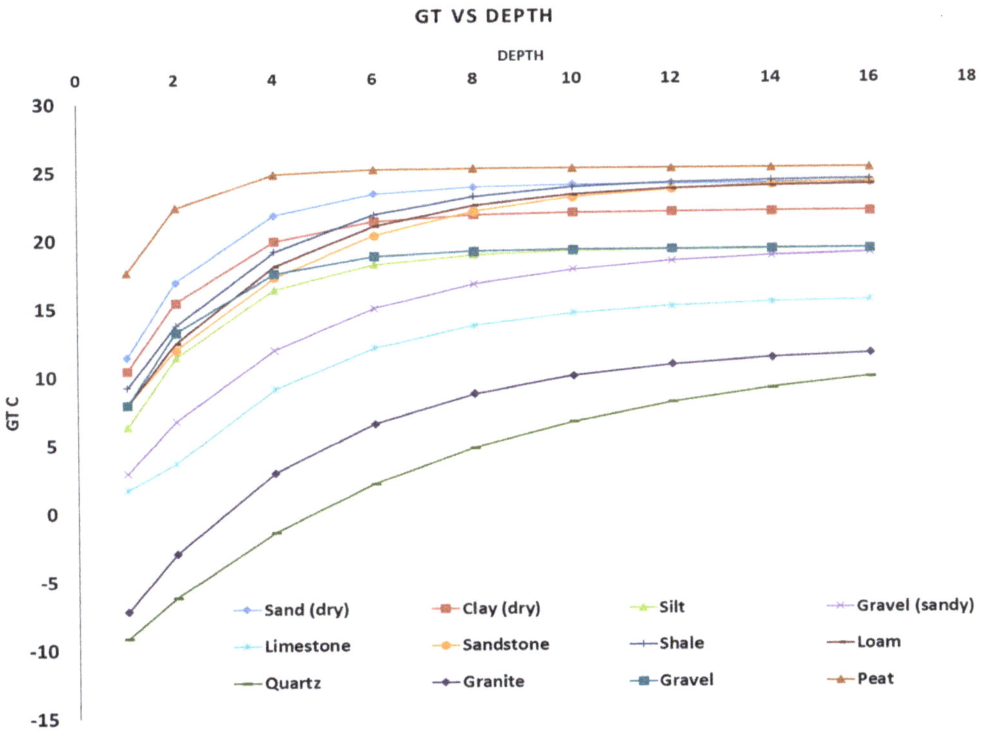

Figure 7. Ground Temperature envelope for all soils and rocks considered.

Figure 8 depicts the work of the authors who used A/G charts for the temperature zoning of Pakistan, based on the annual ground temperatures obtained considering the predominant soil type in the mentioned region. The soils-type map was obtained from the soil survey of Pakistan. The ground depth considered for temperature mapping is 16 m. Since most of the A/G curves become flat at this depth, this was considered as a benchmark for zoning. Also, this depth is usually considered for most of the commercial buildings in Pakistan for foundations, hence allowing the designers to have access to these temperatures charts for their foundation depth and to use them in geothermal heating and cooling of the buildings over the year. The zoning map will allow the designers to work out different ways to improve the energy requirements and efficiency of the buildings and other infra works. Form the charts, it can also be observed that increasing the moisture content of a particular soil can also influence the heat exchanging capability of that particular soil. This phenomenon will open new dimensions for developing both heat transfer and storage capabilities of soils and rock types, respectively. Moreover, the charts are also provide ease to the user, so that they can determine ground temperatures without doing any extensive calculations.

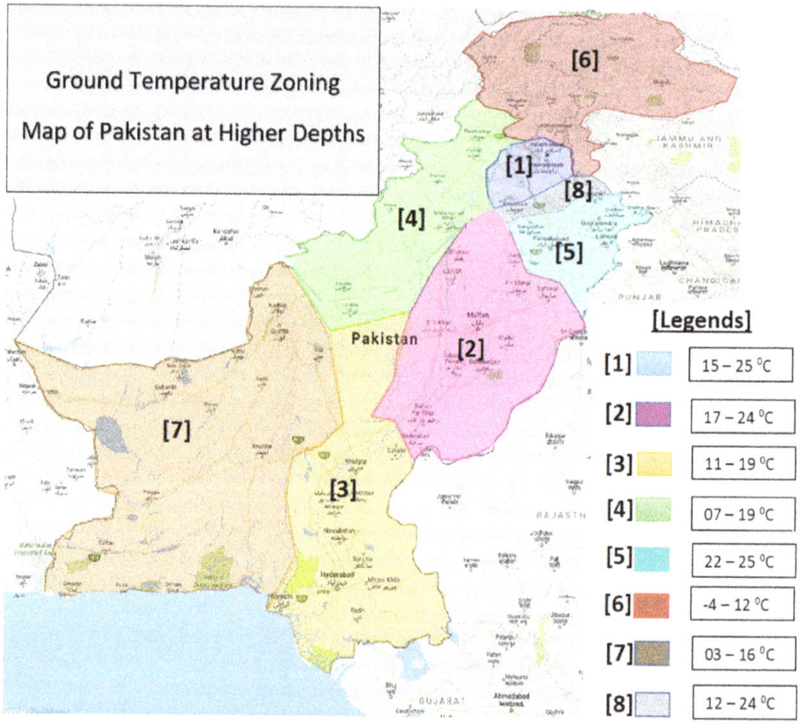

Figure 8. Zoning of Pakistan based on Ground Temperature for Geothermal potential.

4. Discussion and Conclusions

Geothermal energy is a clean energy technology which plays a key role in reducing greenhouse gases, thus providing a solution for global warming while fulfilling the energy needs. The authors discussed the geothermal energy potential of Pakistan and provided a new correlation chart to understand the ground energy in terms of temperature. It is also shown that the type of soil and its characteristics governs the transmission of ground temperature between soil and the structural foundation element. Moreover, the morphological, rheological and geological characteristics of the soil and rock type, greatly affects A/G vs. depth charts. The thermal properties of the material under study are significantly affected by the ground temperature profile. Overall, using the A/G charts is both environmentally friendly and inexpensive, considering both time and labor cost. These charts also depict that there is substantial geothermal energy potential, considering the ground temperatures of Pakistan, that is waiting to be explored. Thus, the geothermal potential of Pakistan is easily conceptualized by using the A/G charts. The study is concluded as:

1. Rock types at a shallow depth have a highly fluctuating air/ground temperature profile compared to the same rock types at higher depths. Significant fluctuations of the A/G pattern is observed in the case of dolomite and other sedimentary rocks.
2. Air/ground temperature ratio for igneous and sedimentary rocks (granite and dolomite) is higher than soil and soft rocks.
3. A/G trend, with regard to limestone, sandstone and shale, is similar to that of soils even though they are sedimentary rocks.
4. Kusuda formula might not provide accurate results everywhere because of different spatial temperature variation and equatorial location. Kusuda equation has limita-

tions at shallower depths. Aerial temperature greatly influences the A/G (Aerial Temperature/Ground Temperature) values at shallower depths.

5. The zoning map will allow the designers to workout different ways to improve the energy requirements and efficiency of the buildings and other infra works. Form the charts, it can also be observed that increasing the moisture content of a particular soil can also influence the heat-exchanging capability of that particular soil.
6. The thermal characteristics of soil and rock types greatly influence their heat exchanging and storage capacity. Heat-exchanging capabilities of the same soil can be changed by modifying its physical characteristics or mixing it with other soils.

For future studies, it is recommended that other cities' data must also be validated for any inconsistencies, as the boreholes were drilled only in one city. In this way, air/ground temperature charts made for other cities of Pakistan and the zoning conducted for geothermal potential by considering the ground temperature, can become more reliable.

Author Contributions: Conceptualization, T.A. and W.H.; methodology, T.A.; software, T.A.; validation, T.A., W.H. and A.H.; formal analysis, A.H.; investigation, T.A., M.H. and M.O.K.; resources, A.H.; data curation, T.A.; writing—original draft preparation, T.A.; writing—review and editing, W.H., M.H. and M.O.K.; visualization, W.H.; supervision, A.H.; project administration, T.A.; funding acquisition, A.H. All authors have read and agreed to the published version of the manuscript.

Funding: This research received no external funding.

Informed Consent Statement: Not applicable.

Data Availability Statement: The data will be available on request.

Conflicts of Interest: The authors declare no conflict of interest.

References

1. Alcaraz, M.; García-Gil, A.; Vázquez-Suñé, E.; Velasco, V. Use rights markets for shallow geothermal energy management. *Appl. Energy* **2016**, *172*, 34–46. [CrossRef]
2. Arshad, F.; Shamshad, M. Energy Crisis in Pakistan: Socio-Economic Implications and the Way Forward. *Ann. Soc. Sci. Perspect.* **2022**, *3*, 105–115. [CrossRef]
3. Ali, M.; Wazir, R.; Imran, K.; Ullah, K.; Janjua, A.K.; Ulasyar, A.; Khattak, A.; Guerrero, J.M. Techno-economic assessment and sustainability impact of hybrid energy systems in Gilgit-Baltistan, Pakistan. *Energy Rep.* **2021**, *7*, 2546–2562. [CrossRef]
4. Kalair, A.; Abas, N.; Saleem, M.S.; Kalair, A.R.; Khan, N. Role of energy storage systems in energy transition from fossil fuels to renewables. *Energy Storage* **2021**, *3*, e135. [CrossRef]
5. Kumari, S. Chapter-5 Global Warming: Causes, Effects and Solutions. In *Emerging Trends in Agriculture Sciences*; Integrated Publications: New Delhi, India, 2021; p. 67.
6. Wang, L.K.; Wang, M.-H.S.; Wang, N.-Y.; Wong, J.O. Effect of Global Warming and Climate Change on Glaciers and Salmons. In *Integrated Natural Resources Management*; Springer: Cham, Switzerland, 2021; pp. 1–36.
7. Nautiyal, A.; Ramlal, A. Clean Energy Sources for a Better and Sustainable Environment of Future Generations. In *Energy: Crises, Challenges and Solutions*; John Wiley & Sons Ltd.: Hoboken, NJ, USA, 2021; pp. 151–168.
8. Dincer, I.; Acar, C. A review on clean energy solutions for better sustainability. *Int. J. Energy Res.* **2015**, *39*, 585–606. [CrossRef]
9. Evans, A.; Strezov, V.; Evans, T.J. Assessment of sustainability indicators for renewable energy technologies. *Renew. Sustain. Energy Rev.* **2009**, *13*, 1082–1088. [CrossRef]
10. Farooqui, S.Z. Prospects of renewables penetration in the energy mix of Pakistan. *Renew. Sustain. Energy Rev.* **2014**, *29*, 693–700. [CrossRef]
11. Khan, H.A.; Pervaiz, S. Technological review on solar PV in Pakistan: Scope, practices and recommendations for optimized system design. *Renew. Sustain. Energy Rev.* **2013**, *23*, 147–154. [CrossRef]
12. Hossein, S.; Bukhari, S. Country Update Paper on Pakistan. In Proceedings of the World Geothermal Congress, Melbourne, Australia, 25–29 April 2015; p. 7.
13. Ahmad, M.; Rafique, M.; Iqbal, N.; Fazil, M. *Investigation of Origin, Subsurface Processes and Reservoir Temperature of Geothermal Springs Around Koh-i-Sultan Volcano, Chagai, Pakistan*; Pakistan Institute of Nuclear Science and Technology: Nilore, Pakistan, 2009.
14. Bakht, M.S. An overview of geothermal resources of pakistan. In Proceedings of the World Geothermal Congress, Khyushu-Tohoku, Japan, 28 May–10 June 2000; pp. 77–83.
15. Ahmad, I.; Rashid, A. Study of geothermal energy resources of Pakistan for electric power generation. *Energy Sources Part A Recovery Util. Environ. Eff.* **2010**, *32*, 826–838. [CrossRef]

16. Florides, G.; Kalogirou, S. Measurements of Ground Temperature at Various Depths. In Proceedings of the SET 2004, 3rd International Conference on Sustainable Energy Technologies, Nottingham, UK, 28–30 June 2004; pp. 1–6.
17. Limberger, J.; Boxem, T.; Pluymaekers, M.; Bruhn, D.; Manzella, A.; Calcagno, P.; Beekman, F.; Cloetingh, S.; van Wees, J.-D. Geothermal energy in deep aquifers: A global assessment of the resource base for direct heat utilization. *Renew. Sustain. Energy Rev.* **2018**, *82*, 961–975. [CrossRef]
18. Brandl, H. Thermo-active ground-source structures for heating and cooling. *Procedia Eng.* **2013**, *57*, 9–18. [CrossRef]
19. Pascal, C.; Elvebakk, H.; Olesen, O. An assessment of deep geothermal resources in Norway. In Proceedings of the Abstracts and Proceedings, World Geothermal Congress, Bali, Indonesia, 25–29 April 2010.
20. Smith, D.P. Fitting geothermal energy into the energy transition. *Neth. J. Geosci.* **2019**, *98*, e6. [CrossRef]
21. Manzella, A.; Bonciani, R.; Allansdottir, A.; Botteghi, S.; Donato, A.; Giamberini, S.; Lenzi, A.; Paci, M.; Pellizzone, A.; Scrocca, D. Environmental and social aspects of geothermal energy in Italy. *Geothermics* **2018**, *72*, 232–248. [CrossRef]
22. Axelsson, G. Production capacity of geothermal systems. In Proceedings of the Workshop for Decision Makers on the Direct Heating Use of Geothermal Resources in Asia, Tanjin, China, 11–18 May 2008; p. 14.
23. Fridleifsson, I.B.; Bertani, R.; Huenges, E.; Lund, J.W.; Ragnarsson, A.; Rybach, L. The possible role and contribution of geothermal energy to the mitigation of climate change. In Proceedings of the IPCC Scoping Meeting on Renewable Energy Sources, Luebeck, Germany, 20–25 January 2008; Citeseer: Princeton, NJ, USA, 2008; Volume 20, pp. 59–80.
24. Sowizdzal, A. Geothermal energy resources in Poland–Overview of the current state of knowledge. *Renew. Sustain. Energy Rev.* **2017**, *82*, 4020–4027. [CrossRef]
25. Litjens, G.B.M.A.; Worrell, E.; Van Sark, W.G.J.H.M. Lowering greenhouse gas emissions in the built environment by combining ground source heat pumps, photovoltaics and battery storage. *Energy Build.* **2018**, *180*, 51–71. [CrossRef]
26. Adalı, Z.; Dinçer, H.; Eti, S.; Mikhaylov, A.; Yüksel, S. Identifying new perspectives on geothermal energy investments. In *Multidimensional Strategic Outlook on Global Competitive Energy Economics and Finance*; Emerald Publishing Limited: Bingley, UK, 2022.
27. Kusuda, T.; Archenbach, P. Earth temperature and thermal diffusivity at selected stations in the United States. *ASHRAE Trans.* **1965**, *71*, 61–75.
28. Muhammad, I.S.; Baharun, A.; Ibrahim, H.S. Investigation of ground temperature for heat sink application in kuching, sarawak, malaysia. *J. Civ. Eng.* **2016**, *7*, 20–29. [CrossRef]
29. Florides, G.; Kalogirou, S. Annual ground temperature measurements at various depths. In Proceedings of the 8th REHVA World Congress, Proceedings of CLIMA 2005, Lausanne, Switzerland, 9–12 October 2005; pp. 1–6.

MDPI
St. Alban-Anlage 66
4052 Basel
Switzerland
www.mdpi.com

Eng-Advances in Engineering Editorial Office
E-mail: eng@mdpi.com
www.mdpi.com/journal/eng

www.ingramcontent.com/pod-product-compliance
Lightning Source LLC
Chambersburg PA
CBHW040746250926
51850CB00007BB/1423
* 9 7 8 3 7 2 5 8 1 0 1 0 9 *